# Springer Optimization and Its Applications

VOLUME 87



*Aims and Scope*
Optimization has been expanding in all directions at an astonishing rate during the last few decades. New algorithmic and theoretical techniques have been developed, the diffusion into other disciplines has proceeded at a rapid pace, and our knowledge of all aspects of the field has grown even more profound. At the same time, one of the most striking trends in optimization is the constantly increasing emphasis on the interdisciplinary nature of the field. Optimization has been a basic tool in all areas of applied mathematics, engineering, medicine, economics, and other sciences.

The series *Springer Optimization and Its Applications* publishes undergraduate and graduate textbooks, monographs and state-of-the-art expository work that focus on algorithms for solving optimization problems and also study applications involving such problems. Some of the topics covered include nonlinear optimization (convex and nonconvex), network flow problems, stochastic optimization, optimal control, discrete optimization, multi-objective programming, description of software packages, approximation techniques and heuristic approaches.

For further volumes:
http://www.springer.com/series/7393

Vladimir F. Demyanov • Panos M. Pardalos
Mikhail Batsyn
Editors

# Constructive Nonsmooth Analysis and Related Topics

*Editors*
Vladimir F. Demyanov
Saint Petersburg State University
Saint Petersburg, Russia

Mikhail Batsyn
National Research University
Higher School of Economics
Nizhny Novgorod, Russia

Panos M. Pardalos
Department of Industrial
and Systems Engineering
University of Florida
Gainesville, FL, USA

ISSN 1931-6828
ISBN 978-1-4614-8614-5 ISBN 978-1-4614-8615-2 (eBook)
DOI 10.1007/978-1-4614-8615-2
Springer New York Heidelberg Dordrecht London

Library of Congress Control Number: 2013950901

Mathematics Subject Classification: 49J52, 58C20, 90C48



Printed on acid-free paper

Springer is part of Springer Science+Business Media (www.springer.com)

*Immortal and indestructible, surrounds all and directs all.*

*Anaximandros. Fragment. 610–546 BCE.*

# Preface

The necessity to solve optimization problems was one of the main incentives for the appearance and development of various methods of mathematics which, in turn, provided the tools and devices for solving such problems. Initially, only particular problems were studied, and for solving each of them, a specific method was to be invented. With the rise of mathematical analysis, the main tool for which is the notion of gradient (derivative in the one-dimensional case) introduced by I. Newton and G. Leibniz, it became possible to solve a wide class of optimization problems, described by differentiable (smooth) functions. Nonsmooth problems were persistently knocking at the door, and P.L. Chebyshev opened the way for them, formulating and solving (140 years ago) the problem of polynomials least deviating from zero (the famous Chebyshev polynomials). In doing this, he employed the existing tools of (smooth) mathematical analysis. P.L. Chebyshev is therefore considered the godfather of nonsmooth analysis (NSA).

However, it was quite clear that for solving a wider class of nonsmooth problems, specific tools were required. U. Dini introduced the notion of directional derivative. The family of directionally differentiable functions is much broader than that of smooth functions. For the study of this family of functions, NSA came into being. NSA is an extension and generalization of the classical (smooth) mathematical analysis. A breakthrough in this area was made in 1962 when J.-J. Moreau introduced the notion of subdifferentials for convex functions. Since then, NSA has become a well-established branch of modern mathematics, and nondifferentiable optimization (NDO) has developed into a rich source of methods and algorithms for solving a variety of optimization problems (both smooth and nonsmooth).

This volume contains a collection of papers based on lectures and reports delivered at the International Conference on Constructive Nonsmooth Analysis (CNSA) held in St. Petersburg (Russia) from June 18 to 23, 2012. The conference was organized to mark the 50th anniversary of the birth of NSA and NDO and was dedicated to J.-J. Moreau and the late B.N. Pshenichnyi, A.M. Rubinov, and N.Z. Shor whose contributions to NSA and NDO remain invaluable.

The book is organized as follows. The first four chapters are devoted to the theory of NSA. A novel application of the models of nonstandard set theory in NSA is

Jean Jacques Moreau

Boris Nikolaevich Pshenichny

Alexander Moiseevich Rubinov

Naum Zuselevich Shor

presented in one of these chapters. Another chapter introduces a generalization of the Demyanov difference for the case of real Hausdorff topological vector spaces. In the next chapter the author proves that the three equivalent properties of set-valued mappings between metric spaces, linear openness, metric regularity, and the Aubin property of the inverse, admit separable reduction or are separably determined.

The next four chapters contain new results in nonsmooth mechanics and calculus of variations. One of these results is a regularization method in NDO and its application to hemivariational inequalities. Another contribution provides optimal parameters and controls for multibody mechanical systems moving over a horizontal plane in the presence of dry friction. The dynamics of such systems is described by differential equations with nonsmooth right-hand sides. The next result is a new method of steepest descent for two-dimensional problems of calculus of variations. Finally, a new theoretical result is obtained for the Lipschitz lower semicontinuity as a quantitative stability property for set-valued maps. A solvability, stability, and sensitivity condition for perturbed optimization problems with quasidifferentiable data is provided.

The next five chapters are related to NDO. They include a new method for solving bi-level vector extremum problems; sufficient conditions for the well-posedness of lexicographic vector equilibrium problems in metric spaces; analysis of the problem of the best approximate separation of two finite sets in the linear case; the alternance form of the necessary optimality conditions in the general case of directionally differentiable functions.

This book concludes with four interesting and important historical chapters. One of these chapters provides an introduction of abstract subdifferentials in the field of analysis of second-order generalized derivatives. Two more chapters present tributes to the giants of NSA, convexity, and optimization: Alexandr Alexandrov, Leonid Kantorovich, and also Alex Rubinov, all three from St. Petersburg. The last chapter provides an overview and important snapshots of the 50-year history of convex analysis and optimization.

We would like to take this opportunity to thank all authors of the chapters, the anonymous referees, and Elizabeth Loew and Razia Amzad from Springer for their professional advice and help during the production of the book.

Saint Petersburg, Russia — Vladimir F. Demyanov
Gainesville, FL, USA — Panos M. Pardalos
Nizhny Novgorod, Russia — Mikhail Batsyn

# Contents

# Use Model Theory in Nonsmooth Analysis

**S.S. Kutateladze**

**Abstract** This is a short invitation to using the models of nonstandard set theory in nonsmooth analysis. The techniques of the infinitesimal and Boolean-valued versions of analysis are illustrated by the concept of infinitesimal optimality and the operator-valued Farkas lemma.

**Keywords** Nonstandard models • Nonstandard set theory • Infinitesimal optimality • Farkas lemma

## 1 Introduction

There are significant features distinguishing the modern mathematics whose epoch we count from the turn of the twentieth century.

Mathematics becomes logic. Logic is the calculus of truth and proof. The ideas of mathematical logic have penetrated into many sections of science and technology. Logic organizes and orders our ways of thinking, manumitting us from conservatism in choosing the objects and methods of research. Logic of today is a fine instrument and institution of mathematical freedom. Logic liberates mathematics by model theory. Model theory evaluates and counts truth and proof.

Another crucial circumstance is the universal mathematization of knowledge. Mathematical ideas have trespassed the frontiers of the exact sciences and imbued the humanitarian sphere, primarily, politics, sociology, and economics. Social events are principally volatile and possess a high degree of uncertainty. Economic processes utilize a wide range of the admissible ways of production, organization, and management. The nature of nonunicity in economics transpires: The genuine

S.S. Kutateladze
Sobolev Institute of Mathematics, 4 Koptuyg Avenue, Novosibirsk 630090, Russia
e-mail: sskut@math.nsc.ru

V.F. Demyanov et al. (eds.), *Constructive Nonsmooth Analysis and Related Topics*, Springer Optimization and Its Applications 87, DOI 10.1007/978-1-4614-8615-2_1,

interests of human beings cannot fail to be contradictory. The unique solution is an oxymoron in any nontrivial problem of economics which refers to the distribution of goods between a few agents. It is not by chance that the social sciences and instances of humanitarian mentality invoke the numerous hypotheses of the best organization of production and consumption, the most just and equitable social structure, the codices of rational behavior and moral conduct, etc. The art of decision making has become a science in the twentieth century. The presence of many contradictory conditions and conflicting interests is the main particularity of the social situations under control of today. Management by objectives is an exceptional instance of the stock of rather complicated humanitarian problems of goal agreement which has no candidates for a unique solution.

Nonstandard models of analysis form one of the areas of mathematics that appears in the twentieth century. It still resembles a youngster seeking ends, means, and opportunities. This article touches a few opportunities that are open up in nonsmooth analysis by model theory. Most attention is paid to the nonstandard concept of infinitesimal optimality and the Boolean-valued interpretation of the Farkas lemma.

## 2 Infinitesimal Optimality

Let us restrict the discussion of vector optimization to convex problems. Under vector optimization we mean the search of optima under conflicting goals or multiple criteria decision making. Technically this reduces to working in ordered vector space setting.

Assume that $X$ is a vector space, $E$ is an ordered vector space, $E^\bullet := E \cup \{+\infty\}$, $f\colon X \to E^\bullet$ is a convex operator, and $C \subset X$ is a convex set. We define a *vector (convex) program* to be a pair $(C, f)$ and write it as

$$x \in C, \quad f(x) \to \inf.$$

A vector program is also commonly called a *multiple objective* or *multiple criteria optimization problem*. An operator $f$ is the *objective of the program* and $C$, the *constraint*. The points $x \in C$ are referred to as *feasible elements*. The above notation of a vector program reflects the fact that we consider the following extremal problem: find a greatest lower bound of the values of $f$ on $C$. In the case $C = X$ we speak of an *unconstrained problem*.

Constraints in an extremal problem can be posed in different ways, for example, in the form of equation or inequality. Let $g\colon X \to F^\bullet$ be a convex operator, $\Lambda \in L(X, Y)$, and $y \in Y$, where $Y$ is a vector space and $F$ is an ordered vector space. If the constraints $C_1$ and $C_2$ have the form

$$C_1 := \{x \in C : g(x) \le 0\},$$

$$C_2 := \{x \in X : g(x) \le 0,\ \Lambda x = y\},$$

then instead of $(C_1, f)$ and $(C_2, f)$, we, respectively, write $(C, g, f)$ and $(\Lambda, g, f)$, or more expressively,

$$x \in C, \ \ g(x) \leq 0, \ \ f(x) \to \inf;$$
$$\Lambda x = y, \ \ g(x) \leq 0, \ \ f(x) \to \inf.$$

An element $e := \inf_{x \in C} f(x)$ (if exists) is the *value* of $(C, f)$. It is clear that $e = -f^*(0)$, where $f^* : L(X, E) \to \overline{E}$, $f^*(T) = \sup\{Tx - f(x) : x \in C\}$. A feasible element $x_0$ is an *ideal optimum* or a *solution* if $e = f(x_0)$. Thus, $x_0$ is an ideal optimum if and only if $f(x_0)$ is the least element of the image $f(C)$, i.e., $f(C) \subset f(x_0) + E^+$.

We can immediately see from the definitions that $x_0$ is a solution of the unconstrained problem $f(x) \to \inf$ if and only if the zero operator belongs to the *subdifferential* $\partial f(x_0)$, i.e.,

$$f(x_0) = \inf_{x \in X} f(x) \ \ \leftrightarrow \ \ 0 \in \partial f(x_0).$$

The difference between *local* and *global* optima is not essential for us, since we will consider only the problems of minimizing convex operators on convex sets. Indeed, let $x_0$ be an ideal local optimum for the program $(C, f)$ in the following (very weak) sense: there exists a set $U \subset X$ such that $0 \in \operatorname{core} U$ and

$$f(x_0) = \inf\{f(x) : x \in C \cap (x_0 + U)\}.$$

Given an arbitrary $h \in C$, choose $0 < \varepsilon < 1$ so as to have $\varepsilon(h - x_0) \in U$. Then $z \in C \cap (x_0 + U)$ for $z := x_0 + \varepsilon(h - x_0) = (1 - \varepsilon)x_0 + \varepsilon h$, whence $f(x_0) \leq f(z)$. Hence, $f(x_0) \leq (1 - \varepsilon) f(x_0) + \varepsilon f(h)$ or $f(x_0) \leq f(h)$.

Considering simple examples, we can check that ideal optimum is extremely rare. This circumstance impels us to introduce various concepts of optimality suitable for these or those classes. Among them is *approximate optimality* which is useful even in a scalar situation (i.e., in problems with a scalar objective function).

Fix a positive element $\varepsilon \in E$. A feasible point $x_0$ is an *$\varepsilon$-solution* of $(C, f)$ if $f(x_0) \leq e + \varepsilon$, where $e$ is the value of the program. Thus, $x_0$ is an $\varepsilon$-solution of the program $(C, f)$ if and only if $x_0 \in C$ and $f(x_0) - \varepsilon$ is a lower bound of the image $f(C)$, or which is the same, $f(C) + \varepsilon \subset f(x_0) + E^+$. It is obvious that a point $x_0$ is an $\varepsilon$-solution of the unconstrained problem $f(x) \to \inf$ if and only if zero belongs to $\partial_\varepsilon f(x_0)$, the *$\varepsilon$-subdifferential* of $f$ at $x_0$:

$$f(x_0) \leq \inf_{x \in X} f(x) + \varepsilon \ \ \leftrightarrow \ \ 0 \in \partial_\varepsilon f(x_0).$$

We call a set $\mathcal{A} \subset C$ a *generalized $\varepsilon$-solution* of the program $(C, f)$ whenever $\inf_{x \in \mathcal{A}} f(x) \leq e + \varepsilon$, where, as above, $e$ is the value of the program. If $\varepsilon = 0$, then we speak simply of a *generalized solution*. Of course, a generalized $\varepsilon$-solution always exists (for instance, $\mathcal{A} = C$); but we however try to choose it as least as possible. An inclusion-minimal generalized $\varepsilon$-solution $\mathcal{A} = \{x_0\}$ is an ideal $\varepsilon$-optimum.

**Theorem 1.** *Each generalized $\varepsilon$-solution is an $\varepsilon$-solution of some vector convex program.*

*Proof.* To demonstrate, we need recalling the concept of canonical operator.

Consider a Dedekind complete vector lattice $E$ and an arbitrary nonempty set $\mathcal{A}$. Denote by $l_\infty(\mathcal{A}, E)$ the set of all order bounded mappings from $\mathcal{A}$ into $E$; i.e., $f \in l_\infty(\mathcal{A}, E)$ if and only if $f\colon \mathcal{A} \to E$ and $\{f(\alpha) : \alpha \in \mathcal{A}\}$ is order bounded in $E$. It is easy to verify that $l_\infty(\mathcal{A}, E)$, endowed with the coordinate-wise algebraic operations and order, is a Dedekind complete vector lattice. The operator $\varepsilon_{\mathcal{A},E}$ acting from $l_\infty(\mathcal{A}, E)$ into $E$ by the rule

$$\varepsilon_{\mathcal{A},E} : f \mapsto \sup\{f(\alpha) : \alpha \in \mathcal{A}\} \quad (f \in l_\infty(\mathcal{A}, E))$$

is the *canonical sublinear operator* relative to $\mathcal{A}$ and $E$. We write $\varepsilon_{\mathcal{A}}$ instead of $\varepsilon_{\mathcal{A},E}$, when $E$ is clear from the context.

Let $\Delta_{\mathcal{A}} := \Delta_{\mathcal{A},E}$ be the embedding of $E$ into $l_\infty(\mathcal{A}, E)$ which assigns the constant mapping $\alpha \mapsto e$ ($\alpha \in \mathcal{A}$) to every element $e \in E$ so that $(\Delta_{\mathcal{A}} e)(\alpha) = e$ for all $\alpha \in \mathcal{A}$.

Turning back to demonstration, consider the operator $\mathscr{F}\colon X^{\mathcal{A}} \to E^{\mathcal{A}} \cup \{+\infty\}$ acting for $\chi \in X^{\mathcal{A}}$ by the rule $\mathscr{F}(\chi)\colon \alpha \in \mathcal{A} \mapsto f(\chi(\alpha))$ if $\operatorname{Im}\chi \subset \operatorname{dom} 1f$, and $\mathscr{F}(\chi) = +\infty$ otherwise. Let $\chi_0 \in X^{\mathcal{A}}$ and $\chi_0(\alpha) = \alpha$ ($\alpha \in \mathcal{A}$), and suppose without loss of generality that $\mathscr{F}(\chi_0) \in l_\infty(\mathcal{A}, E)$.

Now take $\mu \in \partial\varepsilon_{\mathcal{A}}(-\mathscr{F}(\chi_0))$, where $\varepsilon_{\mathcal{A}}\colon l_\infty(\mathcal{A}, E) \to E$ is the canonical operator. By [1, 2.1.4(3)], we have

$$\mu \geq 0,\ \mu \circ \Delta_{\mathcal{A},E} = I_E,$$
$$\mu \circ \mathscr{F}(\chi_0) = -\varepsilon_{\mathcal{A}}(-\mathscr{F}(\chi_0)) = \inf_{\alpha \in \mathcal{A}} f(\alpha).$$

If $\mathcal{A}$ is a generalized $\varepsilon$-solution, then

$$\mu \circ \mathscr{F}(\chi) \geq -\varepsilon_{\mathcal{A}}(-\mathscr{F}(\chi)) = \inf_{\alpha \in \mathcal{A}} f(\alpha) \geq \inf_{x \in C} f(x)$$
$$\geq \inf_{\alpha \in \mathcal{A}} f(\alpha) - \varepsilon = \mu(\mathscr{F}(\chi_0)) - \varepsilon$$

for $\chi \in C^{\mathcal{A}}$. Consequently, $\chi_0$ is an $\varepsilon$-solution of the program

$$\chi \in C^{\mathcal{A}},\ \ \mathscr{F}(\chi) \to \inf.$$

Conversely, if $\chi_0$ is an $\varepsilon$-solution of the last problem, then

$$\mu \circ \mathscr{F}(\chi_0) \leq \mu \circ \mathscr{F} \circ \Delta_{\mathcal{A},X}(x) + \varepsilon = \mu \circ \Delta_{\mathcal{A},E} \circ f(x) + \varepsilon = f(x) + \varepsilon$$

for every $x \in C$. Thus, the following relations hold:

$$\inf_{\alpha \in \mathcal{A}} f(\alpha) = \mu \circ \mathscr{F}(\chi_0) \leq \inf_{x \in C} f(x) + \varepsilon,$$

i.e., $\mathcal{A}$ is a generalized $\varepsilon$-solution of the program $(C, f)$. The proof of the theorem is complete. □

From what was said above we can conclude, in particular, that $\mathcal{A} \subset X$ is a generalized $\varepsilon$-solution of the unconstrained problem $f(x) \to \inf$ if and only if the following system of equations is compatible:

$$\mu \in L^+(l_\infty(\mathcal{A}, E), E), \quad \mu \circ \Delta_{\mathcal{A}, E} = I_E;$$
$$\mu \circ \mathscr{F}(\chi_0) = \inf_{\alpha \in \mathcal{A}} f(\alpha), \quad 0 \in \partial_\varepsilon(\mu \circ \mathscr{F})(\chi_0).$$

The above concepts of optimality are connected with the infimum of the objective function on the set of feasible elements, i.e., with the value of the program. The notion of minimal element leads to a principally different concept of optimality.

Here it is convenient to assume that $E$ is a *preordered vector space*, i.e., the cone of positive elements is not necessarily sharp. Thereby the subspace $E_0 := E^+ \cap (-E^+)$, generally speaking, does not reduce to the zero element alone. Given $u \in E_0$, we denote

$$[u] := \{v \in E : u \leq v, \ v \leq u\}.$$

The record $u \sim v$ means that $[u] = [v]$.

A feasible point $x_0$ is *Pareto $\varepsilon$-optimal* in the program $(C, f)$ if $f(x_0)$ is a minimal element of $f(C) + \varepsilon$, i.e., if $(f(x_0) - E^+) \cap (f(C) + \varepsilon) \subset [f(x_0)]$. In detail, the Pareto $\varepsilon$-optimality of a point $x_0$ means that $x_0 \in C$ and for every point $x \in C$ the inequality $f(x_0) \geq f(x) + \varepsilon$ implies $f(x_0) \sim f(x) + \varepsilon$. If $\varepsilon = 0$, then we simply speak of the Pareto optimality. Studying the Pareto optimality, we often use the *scalarization method*, i.e., the reduction of the program under consideration to a scalar extremal problem with a single objective. Scalarization proceeds in different ways. We will consider one of the routes.

Suppose that the preorder $\leq$ on $E$ is defined as follows:

$$u \leq v \ \leftrightarrow \ (\forall l \in \partial q) \ lu \leq lv,$$

where $q \colon E \to \mathbb{R}$ is a sublinear functional. This is equivalent to the fact that the cone $E^+$ has the form $E^+ := \{u \in E : (\forall l \in \partial q) \ lu \geq 0\}$. Then a feasible point $x_0$ is Pareto $\varepsilon$-optimal in the program $(C, f)$ if and only if for every $x \in C$ either $f(x_0) \sim f(x) + \varepsilon$ or there exists a functional $l \in \partial q$ for which $l(f(x_0)) < l(f(x) + \varepsilon)$. In particular, a Pareto $\varepsilon$-optimal point $x_0 \in C$ satisfies

$$\inf_{x \in C} q(f(x) - f(x_0) + \varepsilon) \geq 0.$$

The converse is not true, since the last inequality is equivalent to a weaker concept of optimality. Say that a point $x_0 \in C$ is *Pareto weakly $\varepsilon$-optimal* if for every $x \in C$ there exists a functional $l \in \partial q$ such that $l(f(x) - f(x_0) + \varepsilon) \geq 0$, i.e., if for any $x \in C$ the system of strict inequalities $l(f(x_0)) < l(f(x) + \varepsilon)$ $(l \in \partial q)$ is not compatible. As we can see, Pareto weak $\varepsilon$-optimality is equivalent to the fact that $q(f(x) - f(x_0) + \varepsilon) \geq 0$ for all $x \in C$ and this concept is not trivial only in the case $0 \notin \partial q$.

The role of $\varepsilon$-subdifferentials is revealed, in particular, by the fact that for a sufficiently small $\varepsilon$ an $\varepsilon$-solution can be considered as a competitor for a "practical optimum," "practically exact" solution to the initial problem.

The rules for calculating $\varepsilon$-subdifferentials yield a formal apparatus for calculating the limits of exactness for a solution to the extremal problem but do not agree completely with the practical methods of optimization in which simplified rules for "neglecting infinitesimals" are employed.

Let us illustrate this by example. Recall that some cones $K_1$ and $K_2$ in a topological vector space $X$ are *in general position* provided that

(1) the algebraic span of $K_1$ and $K_2$ is some subspace $X_0 \subset X$; i.e., $X_0 = K_1 - K_2 = K_2 - K_1$;
(2) the subspace $X_0$ is complemented; i.e., there exists a continuous projection $P\colon X \to X$ such that $P(X) = X_0$;
(3) $K_1$ and $K_2$ constitute a nonoblate pair in $X_0$.

Finally, observe that the two nonempty convex sets $C_1$ and $C_2$ are *in general position* if so are their Hörmander transforms $H(C_1)$ and $H(C_2)$.

**Theorem 2.** [1]*Let $f_1 : X \times Y \to E^\bullet$ and $f_2\colon Y \times Z \to E^\bullet$ be convex operators and $\delta, \varepsilon \in E^+$. Suppose that the convolution*

$$f_2 \vartriangle f_1 = \inf\{f_1(x,y) + f_2(y,z) \mid y \in Y\}$$

*is $\delta$-exact at some point $(x,y,z)$; i.e., $\delta + (f_2 \vartriangle f_1)(x,y) = f_1(x,y) + f_2(y,z)$. If, moreover, the convex sets* epi $(f_1,Z)$ *and* epi $(X, f_2)$ *are in general position, then*

$$\partial_\varepsilon (f_2 \vartriangle f_1)(x,y) = \bigcup_{\substack{\varepsilon_1 \geq 0,\ \varepsilon_2 \geq 0,\\ \varepsilon_1 + \varepsilon_2 = \varepsilon + \delta}} \partial_{\varepsilon_2} f_2(y,z) \circ \partial_{\varepsilon_1} f_1(x,y).$$

In practice $\varepsilon$ is viewed as an actual infinitesimal, which is happily formalized within infinitesimal analysis by A. Robinson and his followers. Model theory suggests the concept of infinitesimal solution within Nelson's theory of internal sets.

Distinguish some downward-filtered subset $\mathscr{E}$ of $E$ that is composed of positive elements. Assuming $E$ and $\mathscr{E}$ standard, define the *monad* $\mu(\mathscr{E})$ of $\mathscr{E}$ as $\mu(\mathscr{E}) := \bigcap\{[0,\varepsilon] \mid \varepsilon \in {}^\circ\mathscr{E}\}$. The members of $\mu(\mathscr{E})$ are *positive infinitesimals* with respect

[1]Cp. [1, Theorem 4.2.8].

to $\mathscr{E}$. As usual, ${}^{\circ}\mathscr{E}$ denotes the external set of all standard members of $\mathscr{E}$, the *standard part* of $\mathscr{E}$.

Assume that the monad $\mu(\mathscr{E})$ is an external cone over ${}^{\circ}\mathbb{R}$ and, moreover, $\mu(\mathscr{E}) \cap {}^{\circ}E = \{0\}$. In application, $\mathscr{E}$ is usually the filter of order units of $E$. The relation of *infinite proximity* or *infinite closeness* between the members of $E$ is introduced as follows:

$$e_1 \approx e_2 \;\leftrightarrow\; |e_1 - e_2| \in \mu(\mathscr{E}).$$

Now

$$Df(x_0) := \bigcap_{\varepsilon \in {}^{\circ}\mathscr{E}} \partial_{\varepsilon} f(x_0) = \bigcup_{\varepsilon \in \mu(\mathscr{E})} \partial_{\varepsilon} f(x_0),$$

which is the *infinitesimal subdifferential* of $f$ at $x_0$. The elements of $Df(x_0)$ are *infinitesimal subgradients* of $f$ at $x_0$.

Assume that there exists a limited value $e := \inf_{x \in C} f(x)$ of the program $(C, f)$. A feasible point $x_0$ is called an *infinitesimal solution* if $f(x_0) \approx e$, i.e., if $f(x_0) \leq f(x) + \varepsilon$ for every $x \in C$ and every standard $\varepsilon \in \mathscr{E}$. Using the definition of infinitesimal subdifferential, we see that a point $x_0 \in X$ is an infinitesimal solution of the unconstrained problem $f(x) \to \inf$ if and only if $0 \in Df(x_0)$.

Consider some *Slater regular program*

$$\Lambda x = \Lambda \bar{x}, \quad g(x) \leq 0, \quad f(x) \to \inf;$$

i.e., first, $\Lambda \in L(X, \mathfrak{X})$ is a linear operator with values in some vector space $\mathfrak{X}$ and the mappings $f\colon X \to E^{\bullet}$ and $g\colon X \to F^{\bullet}$ are convex operators (for the sake of convenience we assume that $\operatorname{dom}(f) = \operatorname{dom}(g) = X$); second, $F$ is an Archimedean ordered vector space and $E$ is a standard Dedekind complete vector lattice of bounded elements; and, at last, the element $-g(x)$ with some feasible point $x$ is a strong order unit in $F$.

**Theorem 3.** [2]*A feasible point $x_0$ is an infinitesimal solution of a Slater regular program if and only if the following system of conditions is compatible:*

$$\beta \in L^{+}(F, E), \quad \gamma \in L(\mathfrak{X}, E), \quad \beta g(x_0) \approx 0,$$
$$0 \in Df(x_0) + D(\beta \circ g)(x_0) + \gamma \circ \Lambda.$$

The models of infinitesimal analysis reside in many places (cp. [2] and the reference within). The complete details of use of infinitesimal analysis in vector optimization are collected in [3].

---

[2]Cp. [3, Sect. 5.7].

## 3 The Boolean-Valued Farkas Lemma

Boolean-valued models were invented for simplifying the Cohen method of forcing. We will demonstrate how the technique of these models may be applied to simultaneous linear inequalities with operators. Recall that the Farkas lemma plays a key role in linear programming and the relevant areas of optimization.

Assume that $X$ is a real vector space and $Y$ is a *Kantorovich space*, i.e., a Dedekind complete vector lattice. Let $\mathbb{B} := \mathbb{B}(Y)$ be the *base* of $Y$, i.e., the complete Boolean algebra of positive projections in $Y$; and let $m(Y)$ be the *universal completion* of $Y$. Let $L(X,Y)$ denote the space of linear operators from $X$ to $Y$. In case $X$ is furnished with some $Y$-seminorm on $X$, by $L^{(m)}(X,Y)$, we mean the *space of dominated operators* from $X$ to $Y$. As usual,

$$\{T \leq 0\} := \{x \in X \mid Tx \leq 0\}, \qquad \text{Ker}\,(T) = T^{-1}(0) \qquad \text{for } T : X \to Y.$$

**Kantorovich's Theorem.** *Find $\mathfrak{X}$ satisfying*

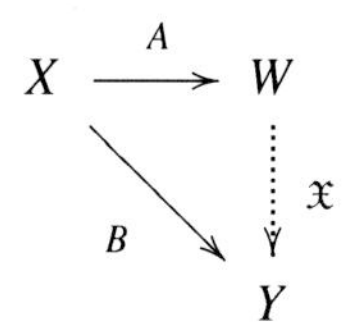


(1) $(\exists \mathfrak{X})\ \mathfrak{X}A = B \leftrightarrow \text{Ker}\,(A) \subset \text{Ker}\,(B)$.
(2) *If $W$ is ordered by $W_+$ and $A(X) - W_+ = W_+ - A(X) = W$, then*[3]

$$(\exists \mathfrak{X} \geq 0)\ \mathfrak{X}A = B \leftrightarrow \{A \leq 0\} \subset \{B \leq 0\}.$$

**Alternative Theorem.** *Let $X$ be a $Y$-seminormed real vector space, with $Y$ a Kantorovich space. Assume that $B$ and $A_1,\dots,A_N$ belong to $L^{(m)}(X,Y)$.*

*Then one and only one of the following holds:*

(1) *There are $x \in X$ and $b, b' \in \mathbb{B}$ such that $b' \leq b$ and*

$$b'Bx > 0,\ bA_1x \leq 0,\ \dots,\ bA_Nx \leq 0.$$

(2) *There are positive orthomorphisms $\alpha_1,\dots,\alpha_N \in \text{Orth}(m(Y))$ such that*

$$B = \sum_{k=1}^{N} \alpha_k A_k.$$

[3] Cp. [1, p. 51].

**Farkas Lemma.** *Let $X$ be a $Y$-seminormed real vector space, with $Y$ a Kantorovich space. Assume given some dominated operators $A_1, \dots, A_N, B \in L^{(m)}(X,Y)$ and elements $u_1, \dots, u_N, v \in Y$. Assume further that the simultaneous inhomogeneous operator inequalities $A_1x \le u_1, \dots, A_Nx \le u_N$ are consistent. Then the following are equivalent:*

(1) *For all $b \in \mathbb{B}$ the inhomogeneous operator inequality $bBx \le bv$ is a consequence of the simultaneous inhomogeneous operator inequalities $bA_1x \le bu_1, \dots, bA_Nx \le bu_N$, i.e.,*

$$\{bB \le bv\} \supset \{bA_1 \le bu_1\} \cap \dots \cap \{bA_N \le bu_N\}.$$

(2) *There are positive orthomorphisms $\alpha_1, \dots, \alpha_N \in \mathrm{Orth}(m(Y))$ satisfying*

$$B = \sum_{k=1}^{N} \alpha_k A_k; \quad v \ge \sum_{k=1}^{N} \alpha_k u_k.$$

These theorems are obtained by using Boolean-valued models of set theory. The latter were invented for simplifying Cohen's final solution of the problem of the cardinality of the continuum within ZFC. The honor of creation of these models belong to Scott, Solovay, and Vopěnka.[4]

Takeuti coined the term "Boolean-valued analysis" for applications of the models to analysis. Scott wrote in 1969[5]: "We must ask whether there is any interest in these nonstandard models aside from the independence proof; that is, do they have any mathematical interest? The answer must be yes, but we cannot yet give a really good argument."

In 2009 he added[6]: "At the time, I was disappointed that no one took up my suggestion. And then I was very surprised much later to see the work of Takeuti and his associates[7]. I think the point is that people have to be trained in Functional Analysis in order to understand these models. I think this is also obvious from your book and its references. Alas, I had no students or collaborators with this kind of background, and so I was not able to generate any progress."

Boolean-valued models reveal that each mathematical result has many interpretations that are invisible from the start. For instance, all $L_p$ spaces may be considered as subspaces of the reals in a suitable Boolean-valued model. Some details may clarify the matter.

Let $\mathbb{B}$ be a complete Boolean algebra. Given an ordinal $\alpha$, put

$$V_\alpha^{(\mathbb{B})} := \{x \mid (\exists \beta \in \alpha)\, x\colon \mathrm{dom}(x) \to \mathbb{B} \ \&\ \mathrm{dom}(x) \subset V_\beta^{(\mathbb{B})}\}.$$

---

[4]Cp. [7].

[5]Cp. [5].

[6]Letter of April 29, 2009 to S. S. Kutateladze.

[7]Cp. [4]

The *Boolean-valued universe* $\mathbb{V}^{(\mathbb{B})}$ is

$$\mathbb{V}^{(\mathbb{B})} := \bigcup_{\alpha \in \mathrm{On}} V_\alpha^{(\mathbb{B})},$$

with On the class of all ordinals.

The truth value $[\![\varphi]\!] \in \mathbb{B}$ is assigned to each formula $\varphi$ of ZFC relativized to $\mathbb{V}^{(\mathbb{B})}$. The phrase "$x$ satisfies $\varphi$ inside $\mathbb{V}^{(\mathbb{B})}$" means that $[\![\varphi(x)]\!] = \mathbb{1}$, with $\mathbb{1}$ the top of $\mathbb{B}$. Application of the so-called Frege–Russell–Scott trick makes the Boolean-valued universe *separated*: $x = y \leftrightarrow [\![x = y]\!] = \mathbb{1}$ for all $x, y \in \mathbb{V}^{(\mathbb{B})}$.

The *descent* $x{\downarrow}$ of $x \in \mathbb{V}^{(\mathbb{B})}$ is defined as

$$x{\downarrow} := \{t \mid t \in \mathbb{V}^{(\mathbb{B})} \ \& \ [\![t \in x]\!] = \mathbb{1}\}.$$

The class $x{\downarrow}$ is a set. If $x$ is a nonempty set inside $\mathbb{V}^{(\mathbb{B})}$, then

$$(\exists z \in x{\downarrow}) \ [\![(\exists t \in x) \ \varphi(t)]\!] = [\![\varphi(z)]\!].$$

The *ascent* functor acts in the opposite direction.

There is an object $\mathscr{R}$ inside $\mathbb{V}^{(\mathbb{B})}$ modeling $\mathbb{R}$, i.e.,

$$[\![\mathscr{R} \text{ is the reals}]\!] = \mathbb{1}.$$

Let $\mathscr{R}{\downarrow}$ be the descent of the carrier $|\mathscr{R}|$ of the algebraic system

$$\mathscr{R} := (|\mathscr{R}|, +, \cdot, 0, 1, \leq)$$

inside $\mathbb{V}^{(\mathbb{B})}$. Implement the descent of the structures on $|\mathscr{R}|$ to $\mathscr{R}{\downarrow}$ as follows:

$$\begin{aligned}
x + y = z &\ \leftrightarrow\ [\![x + y = z]\!] = \mathbb{1};\\
xy = z &\ \leftrightarrow\ [\![xy = z]\!] = \mathbb{1};\\
x \leq y &\ \leftrightarrow\ [\![x \leq y]\!] = \mathbb{1};\\
\lambda x = y &\ \leftrightarrow\ [\![\lambda^{\wedge} x = y]\!] = \mathbb{1} \quad (x, y, z \in \mathscr{R}{\downarrow},\ \lambda \in \mathbb{R}).
\end{aligned}$$

**Gordon Theorem.** [8] *$\mathscr{R}{\downarrow}$ with the descended structures is a universally complete vector lattice with base $\mathbb{B}(\mathscr{R}{\downarrow})$ isomorphic to $\mathbb{B}$.*

This beautiful result makes it possible to derive the Farkas lemma for operators by descending the classical version from a suitable Boolean-valued universe to vector space environment. The complete details on Boolean-valued analysis are collected in [6, 7]. About application to nonsmooth analysis, refer to [8, 9].

---

[8] Cp. [7, p. 349].

## References

[1] Kusraev, A.G., Kutateladze, S.S.: Subdifferential Calculus: Theory and Applications. Nauka, Moscow (2007)
[2] Kanovei, V., Reeken, M.: Nonstandard Analysis: Axiomatically. Springer, Berlin (2004)
[3] Gordon, E.I., Kusraev, A.G., Kutateladze S.S.: Infinitesimal Analysis: Selected Topics. Nauka, Moscow (2011)
[4] Takeuti, G.: Two Applications of Logic to Mathematics. Iwanami Publishers and Princeton University Press, Tokyo and Princeton (1978)
[5] Scott, D.: Boolean models and nonstandard analysis. In: Applications of Model Theory to Algebra, Analysis, and Probability, pp. 87–92. Holt, Rinehart, and Winston, New York (1969)
[6] Bell, J.L.: Set Theory: Boolean Valued Models and Independence Proofs. Clarendon Press, Oxford (2005)
[7] Kusraev, A.G., Kutateladze, S.S.: Introduction to Boolean Valued Analysis. Nauka, Moscow (2005)
[8] Kusraev, A.G., and Kutateladze, S.S.: Boolean methods in positivity. J. Appl. Indust. Math. **2**(1), 81–99 (2008)
[9] Kutateladze, S.S.: Boolean trends in linear inequalities. J. Appl. Indust. Math. **4**(3), 340–348 (2010)

# Demyanov Difference in Infinite-Dimensional Spaces

**Jerzy Grzybowski, Diethard Pallaschke, and Ryszard Urbański**

**Abstract** In this paper we generalize the Demyanov difference to the case of real Hausdorff topological vector spaces. We prove some classical properties of the Demyanov difference. In the proofs we use a new technique which is based on the properties given in Lemma 1. Due to its importance it will be called the *preparation lemma*. Moreover, we give connections between Minkowski subtraction and the union of upper differences. We show that in the case of normed spaces the Demyanov difference coincides with classical definitions of Demyanov subtraction.

**Keywords** Minkowski subtraction • Demyanov difference • Pairs of closed bounded convex sets

## 1 Introduction

For a Hausdorff topological vector space $(X,\tau)$ let us denote by $\mathcal{A}(X)$ the set of all nonempty subsets of $X$, by $\mathcal{B}^*(X)$ the set of all nonempty bounded subsets of $X$, by $\mathcal{C}(X)$ the set of all nonempty closed convex subsets of $X$, by $\mathcal{B}(X) = \mathcal{B}^*(X) \cap \mathcal{C}(X)$ the set of all bounded closed convex sets of $X$, and by $\mathcal{K}(X)$ the set of all nonempty compact convex subsets of $X$. (Note that we consider only vector spaces over the reals.) Recall that for $A, B \in \mathcal{A}(X)$ the *algebraic sum* is

J. Grzybowski • R. Urbański
Faculty of Mathematics and Computer Science, Adam Mickiewicz University, Umultowska 87, PL-61-614 Poznań, Poland
e-mail: jgrz@amu.edu.pl; rich@amu.edu.pl

D. Pallaschke (✉)
Institute of Operations Research, University of Karlsruhe (KIT), Kaiserstr. 12, D-76128 Karlsruhe, Germany
e-mail: diethard.pallaschke@kit.edu

V.F. Demyanov et al. (eds.), *Constructive Nonsmooth Analysis and Related Topics*, Springer Optimization and Its Applications 87, DOI 10.1007/978-1-4614-8615-2_2, 

defined by $A+B = \{x=a+b \mid a \in A \text{ and } b \in B\}$ and for $\lambda \in \mathbb{R}$ and $A \in \mathcal{A}(X)$ the *multiplication* is defined by $\lambda A = \{x=\lambda a \mid a \in A\}$.

The *Minkowski sum* for $A,B \in \mathcal{A}(X)$ is defined by

$$A \dot{+} B = \text{cl}(\{x=a+b \mid a \in A \text{ and } b \in B\}),$$

where $\text{cl}(A)=\bar{A}$ denotes the closure of $A \subset X$ with respect to $\tau$. For compact convex sets, the Minkowski sum coincides with the algebraic sum, i.e., for $A,B \in \mathcal{K}(X)$, we have $A \dot{+} B = A+B$. $A \in \mathcal{B}(X)$ is called a *summand* of $B \in \mathcal{B}(X)$ if there exists a set $C \in \mathcal{B}(X)$ such that $A \dot{+} C = B$. In quasidifferential calculus of Demyanov and Rubinov, [2] pairs of bounded closed convex sets are considered. More precisely, for a Hausdorff topological vector space $X$ two pairs $(A,B),(C,D) \in \mathcal{B}^2(X)=\mathcal{B}(X) \times \mathcal{B}(X)$ are called *equivalent* if $B \dot{+} C = A \dot{+} D$ holds and $[A,B]$ denotes the equivalence class represented by the pair $(A,B) \in \mathcal{B}^2(X)$.

For $A \in \mathcal{B}(X)$ we denote by $\text{ext}(A)$ the set of its extremal points and by $exp(A)$ the set of its exposed points (see [7]) . Next, for $A,B \in \mathcal{A}(X)$, we define $A \vee B = \text{cl conv}(A \cup B)$, where $\text{conv}(A \cup B)$ denotes the convex hull of $A \cup B$. We will use the abbreviation $A \dot{+} B \vee C$ for $A \dot{+} (B \vee C)$ and $C+d$ instead of $C+\{d\}$ for all bounded closed convex sets $A,B,C \in \mathcal{A}(X)$ and a point $d \in X$.

The elements of $\mathcal{A}(X)$ are ordered by inclusion, i.e. for $A,B \in \mathcal{A}(X)$ we define $A \leq B$ if and only if $A \subseteq B$. Note that for $A,B \in \mathcal{B}(X)$ the set $A \vee B$ is the maximum of $A$ and $B$ in this order. In general the minimum does not exist, but if $A \cap B \neq \emptyset$, then the intersection is the minimum in $\mathcal{B}(X)$.

A distributivity relation between the Minkowski sum and the maximum operation is expressed by the *Pinsker formula* (see [9]) which is stated in a more general for in [7] as

**Proposition 1.** *Let $(X,\tau)$ be a Hausdorff topological vector space, $A,B,C \in \mathcal{A}(X)$ and $C$ be a convex set. Then*

$$(A \dot{+} C) \vee (B \dot{+} C) = C \dot{+} (A \vee B).$$

The Minkowski–Rådström–Hörmander theorem on the cancellation property for bounded closed convex subsets in Hausdorff topological vector spaces states that for $A,B,C \in \mathcal{B}(X)$ the inclusion $A \dot{+} B \subseteq B \dot{+} C$ implies $A \subseteq C$. A generalization which is due to R. Urbański [12] states

**Theorem 1.** *Let $X$ be a Hausdorff topological vector space. Then, for any $A \in \mathcal{A}(X)$, $B \in \mathcal{B}^*(X)$ and $C \in \mathcal{C}(X)$, the inclusion*

$$A+B \subseteq C \dot{+} B \quad \textit{implies} \quad A \subseteq C. \qquad \textbf{(olc)}$$

This implies that $\mathcal{B}(X)$ endowed with the Minkowski sum "$\dot{+}$" and the ordering induced by inclusion is a commutative ordered *semigroup* (i.e. an ordered set endowed with a group operation, without having inverse elements), which satisfies the order cancellation law and contains $\mathcal{K}(X)$ as a sub-semigroup.

### *1.1 Separation*

Let $A, B$, and $S$ be nonempty subsets of a (real) vector space $X$. Recall that $S$ *separates* the sets $A$ and $B$ if $[a,b] \cap S \neq \emptyset$ for every $a \in A$ and $b \in B$. In [8] the following theorem is given:

**Theorem 2.** *Let $A, B$ be nonempty subsets of a Hausdorff topological vector space $X$ and assume that $A \vee B \in \mathcal{B}(X)$. Moreover, let $S$ be a closed convex subset of $X$. Then $S$ separates the sets $A$ and $B$ if and only if*

$$A + B \subset A \vee B \dot{+} S.$$

The assumption $A \vee B \in \mathcal{B}(X)$ is essential. Note that in general the convex hull of two bounded sets $A$ and $B$ is not bounded.

## 2 Differences of Convex Sets

The main object of this paper is to study the subtraction introduced by Demyanov [2].

Before focusing to the Demyanov difference, let us first discuss the *Minkowski (Pontryagin, Hukuhara) difference* $A, B \in \mathcal{A}(X)$. This is given by

$$A \dot{-} B = \{x \in X \mid B + x \subseteq A\} = \bigcap_{b \in B} (A - b) \qquad (*)$$

and commonly called *Pontryagin difference* of $A$ and $B$.

#### (a) The Minkowski–Pontryagin–Hukuhara Difference

The Minkowski–Pontryagin–Hukuhara ([3, 10]) difference can be naturally defined for ordered semigroups. This was done in [6], where the authors defined for F-semigroups $(S, +, \leq)$ (which are an abstraction of an ordered semigroup with cancellation property and Pinsker rule) the so-called abstract differences.

For completeness we repeat their results. Therefore, let $S$ be a commutative semigroup endowed with an ordering $\leq$. Then system $(S, +, \leq)$ is called an *ordered semigroup*. For $a, b \in S$ let us denote by $a \vee b$ the *smallest upper bound* (if exists) of elements $a$ and $b$ and, analogously, by $a \wedge b$ the *greatest lower bound* of $a$ and $b$. Now we assume that $S$ satisfies the following axioms:

(S1) For every $a, b, s \in S$, if $a + s \leq b + s$, then $a \leq b$.
(S2) For every $a, b, s \in S$, if $a \leq b$, then $a + s \leq b + s$.
(S3) If $a \leq s$ and $b \leq s$ for some $s \in S$, then there exists $a \vee b$.
(S4) If $s \leq a$ and $s \leq b$ for some $s \in S$, then there exists $a \wedge b$.
(S5) If for some $a, b, s \in S$ there exists $a \vee b$ and $(a+s) \vee (b+s)$, then $(a \vee b) + s \leq (a+s) \vee (b+s)$.

An *ordered semigroup* $S$ $(S,+,\leq)$ which satisfies (S1)–(S5) is called a *F-semigroup.*

**Definition 1.** Let S be a $F$-semigroup and let $a, b \in S$. The *abstract difference* of two elements $a$ and $b$ is the greatest element (if exists) of the set $\mathcal{D}(a,b) = \{x : x + b \leq a\}$ and is denoted by $a \overset{*}{-} b$.

The abstract difference [6] has the following properties:

(A1) If $a \overset{*}{-} b$ exists, then $b + (a \overset{*}{-} b) \leq a$.
(A2) If $a = b + c$, then $c = a \overset{*}{-} b$.
(A3) If there exists the neutral element $0 \in S$, then for every $a \in S$,
one has $a \overset{*}{-} a = 0$.
(A4) If $a \leq b$, and for some $c \in S$ , $a \overset{*}{-} c$, $b \overset{*}{-} c$ exists, then $a \overset{*}{-} c \leq b \overset{*}{-} c$.
(A5) If $b \leq c$, and for some $a \in S$ , $a \overset{*}{-} b$, $a \overset{*}{-} c$ exists, then $a \overset{*}{-} c \leq a \overset{*}{-} b$.
(A6) If $a \overset{*}{-} b$ exists, then for every $c \in S$, $(a + c) \overset{*}{-} (b + c) = a \overset{*}{-} b$.

In order to prove a formula similar to $(*)$ for the abstract difference, we need additionally the following axiom:

(S6) For any family $\{x_\alpha\}_{\alpha \in \Lambda}$ of elements of $S$ if there exists $\vee_\alpha x_\alpha$, then for every $c \in S$ there exists $\vee_\alpha (c + x_\alpha)$ and $c + \vee_\alpha x_\alpha \leq \vee_\alpha (c + x_\alpha)$.

Then (see [6], Theorem 3.2) holds:

**Theorem 3.** *Let $S$ be a $F$-semigroup which satisfies the axiom (S6) and let $b = \bigvee_\alpha x_\alpha$. Moreover assume that there exist $a \overset{*}{-} b$, $a \overset{*}{-} x_\alpha$, and $\bigwedge_\alpha (a \overset{*}{-} x_\alpha)$, then*

$$a \overset{*}{-} b = \bigwedge_\alpha (a \overset{*}{-} x_\alpha).$$

### (b) The Demyanov Difference

Demyanov original subtraction $A \dot{\dot{-}} B$ of compact convex subsets in finite-dimensional space is defined with the help of the Clarke subdifferential (see [1]) of the difference of support functions, i.e.

$$A \dot{\dot{-}} B = \partial_{\mathrm{cl}} (p_A \;-\; p_B)\Big|_0,$$

where $p_A$ and $p_B$ are the support functions of $A$ and $B$, i.e.$p_A(x) = \max_{a \in A} \langle a, x \rangle$

This can be equivalently formulated by

$$A \dot{\dot{-}} B = \overline{\mathrm{conv}} \{a - b | a \in A, b \in B, a + b \in \exp(A + B)\},$$

where $\exp(A+B)$ are the exposed points of $A+B$. For the proof see [11] Proposition 2 and note that every exposed point of $A+B$ is the unique sum of an exposed point of $A$ with an exposed point of $B$.

The difference $A \ddot{-} B$ is well defined by this equality for $A, B \in \mathcal{K}(X)$, where $X$ is a normed vector space, because by Klee's generalization of the Krein–Milman theorem (see [5,7]) every compact convex set of a normed vector space is the closed convex hull of its exposed points. To extend the definition of the difference $A \ddot{-} B$ to locally convex topological vector spaces, the set of exposed points will be replaced by the set of extremal points of $A+B$. This leads to the following generalization of the Demyanov difference:

**Definition 2.** Let $(X, \tau)$ be a locally convex vector space and $\mathcal{K}(X)$ the family of all nonempty compact convex subsets of $X$. Then, for $A, B \in \mathcal{K}(X)$, the set

$$A \ddot{-} B = \overline{\text{conv}}\{a - b | a \in A, b \in B, a + b \in \text{ext}(A+B)\} \in \mathcal{K}(X)$$

is called the *Demyanov difference* of $A$ and $B$.

This is a canonical generalization of the above definition, because for every $A, B \in \mathcal{K}(X)$ every extremal point $z \in \text{ext}(A+B)$ has a unique decomposition $z = x + y$ into the sum of two extreme points $x \in \text{ext}(A)$ and $y \in \text{ext}(B)$ (see [4], Proposition 1).

## 3 Upper Differences and a Preparation Lemma

For $A, B \subset X$ we define *upper difference* $\mathcal{E}_{A,B}$ as the family

$$\mathcal{E}_{A,B} = \{C \in \mathcal{C}(X) \mid A \subset B \dot{+} C\},$$

where $\mathcal{C}(X)$ is the family of all nonempty closed convex subsets of $X$ and $B \dot{+} C = \overline{B + C}$.

**Lemma 1.** *Let $X$ be a Hausdorff topological vector space, $A$ be closed convex, and $B$ be bounded subset of $X$. Then for every bounded subset $M$ we have*

$$A \dot{+} M = \bigcap_{C \in \mathcal{E}_{A,B}} (B \dot{+} C \dot{+} M).$$

*Proof.* By the definition of the family $\mathcal{E}_{A,B}$ we have

$$A \dot{+} M \subset \bigcap_{C \in \mathcal{E}_{A,B}} (B \dot{+} C \dot{+} M).$$

Let us fix any $m \in M$ and $b \in B$. Denote $C' = M - m \dot{+} A - b$. Notice that $C' \in \mathcal{E}_{A,B}$. Then

$$\bigcap_{C\in\mathcal{E}_{A,B}} (B\dot{+}C\dot{+}M) \subset B\dot{+}C'\dot{+}M = B\dot{+}M - m\dot{+}A - b\dot{+}M.$$

Hence

$$b+m+\bigcap_{C\in\mathcal{E}_{A,B}} (B\dot{+}C\dot{+}M) \subset B\dot{+}M\dot{+}M\dot{+}A.$$

Therefore,

$$B\dot{+}M\dot{+}\bigcap_{C\in\mathcal{E}_{A,B}} (B\dot{+}C\dot{+}M) \subset B\dot{+}M\dot{+}M\dot{+}A,$$

and by the law of cancellation we get

$$\bigcap_{C\in\mathcal{E}_{A,B}} (B\dot{+}C\dot{+}M) \subset A\dot{+}M.$$

□

In particular, from Lemma 1 follows that $A=\bigcap_{C\in\mathcal{E}_{A,B}}(B\dot{+}C)$.

**Proposition 2.** *Let $X$ be a Hausdorff topological vector space, $E$ and $A$ be closed convex, and $B$ be bounded subset of $X$. Then $E\dot{+}A\dot{-}B=\bigcap_{C\in\mathcal{E}_{A,B}}(E\dot{+}C)$.*

*Proof.* Let $x\in E\dot{+}A\dot{-}B$. Then $B+x\subset E\dot{+}A$ and for every $C\in\mathcal{E}_{A,B}$ we have $B+x\subset E\dot{+}A\subset E\dot{+}B\dot{+}C$. Now by the order law of cancellation we get $x\in E\dot{+}C$.

To prove the reverse inclusion note that, by Lemma 1, we have

$$B\dot{+}\bigcap_{C\in\mathcal{E}_{A,B}} (E\dot{+}C) \subset \bigcap_{C\in\mathcal{E}_{A,B}} (B\dot{+}E\dot{+}C) = E\dot{+}A.$$

Hence $\bigcap_{C\in\mathcal{E}_{A,B}}(E\dot{+}C)\subset E\dot{+}A\dot{-}B$. □

In particular, from Proposition 2 follows that $A\dot{-}B=\bigcap\mathcal{E}_{A,B}$.

**Lemma 2.** *If $A\dot{+}(-B)\in\mathcal{C}(X)$ and the set $B$ is compact, then $A\dot{+}(-B)\in\mathcal{E}_{A,B}$ and for all $C\in\mathcal{E}_{A,B}$ we have $(A\dot{+}(-B))\cap C\in\mathcal{E}_{A,B}$.*

*Proof.* At first we observe that $A\dot{+}(-B)\in\mathcal{E}_{A,B}$. Now let $C\in\mathcal{E}_{A,B}$ and $a\in A$. Then $a=b+c$ for some $b\in B$ and $c\in C$. Hence $a=b+(a-b), a-b\in C$ and we have $A\subset B+C\cap(A\dot{+}(-B))$. □

**Lemma 3.** *Let $X$ be a Hausdorff topological vector space and subsets $A$ and $B_i$, $i\in I$ of $X$ be closed. Moreover, let $A$ be compact or $B_i$ be compact for some $i\in I$ and the family $\{B_i\}_{i\in I}$ is directed with respect to inclusion "$\supset$" then*

$$A+\bigcap_{i\in I}B_i=\bigcap_{i\in I}(A+B_i).$$

*Proof.* We have $x \in \bigcap_{i\in I}(A+B_i)$ if and only if $0 \in A - x + B_i$ for all $i \in I$, but it is equivalent to $(x-A)\cap B_i \neq \emptyset$ for all $i \in I$. Since the family of compact sets $\{(x-A)\cap B_i\}_{i\in I}$ has the finite intersection property, we have $\bigcap_{i\in I}((x-A)\cap B_i) \neq \emptyset$. Hence $(x-A)\cap\bigcap_{i\in I} B_i \neq \emptyset$ and $x \in A + \bigcap_{i\in I} B_i$. □

**Theorem 4.** *Let $X$ be a Hausdorff topological vector space. If $B$ or some $C \in \mathcal{E}_{A,B}$ is compact, then $C$ contains a minimal $C_0 \in \mathcal{E}_{A,B}$.*

*Proof.* Let $\{C_i\}_{i\in I} \subset C$ be a chain in $\mathcal{E}_{A,B}$. Since the family $\{C_i\}_{i\in I}$ is a chain, $C_0 = \bigcap_{i\in I} C_i \neq \emptyset$. By Lemma 3 we have

$$A \subset \bigcap_{i\in I}(B+C_i) = B + \bigcap_{i\in I} C_i;$$

hence $C \subset C_0 = \bigcap_{i\in I} C_i \in \mathcal{E}_{A,B}$. Now by the Kuratowski–Zorn lemma $C_0$ is a minimal element of $\mathcal{E}_{A,B}$. □

Using Theorem 2 we obtain the following proposition:

**Proposition 3.** *Let $X$ be a Hausdorff topological vector space $C_1, C_2 \in \mathcal{E}_{A,B} \cap \mathcal{B}(X)$. Then*

$$\mathcal{E}_{C_1+C_2, C_1\vee C_2} \subset \mathcal{E}_{A,B}.$$

*Proof.* Let $C \in \mathcal{E}_{C_1+C_2, C_1\vee C_2}$. Then by Theorem 2 the set $C$ separates the set $C_1$ and $C_2$. Now for any fixed $a \in A$ we have $a = b_1 + c_1 = b_2 + c_2$ for some $b_1, b_2 \in B$ and $c_1 \in C_1, c_2 \in C_2$. Since $[c_1, c_2] \cap C \neq \emptyset$, $a \in B + C$. □

**Theorem 5.** *Let $X$ be a Hausdorff topological vector space and pairs $(A,B), (C,D) \in \mathcal{B}^2(X)$. Then $[A,B] = [C,D]$ if and only if $\mathcal{E}_{A,B} = \mathcal{E}_{C,D}$.*

*Proof.* Given any $C' \in \mathcal{E}_{A,B}$, then $A \subset B \dot{+} C'$. Since pairs $(A,B)$ and $(C,D)$ are equivalent, $B \dot{+} C = A \dot{+} D \subset B \dot{+} D \dot{+} C'$. Hence by the law of cancellation $C' \in \mathcal{E}_{C,D}$.

Now let $\mathcal{E}_{A,B} = \mathcal{E}_{C,D}$; then by Lemma 1 we have

$$A \dot{+} D = \bigcap_{C'\in\mathcal{E}_{A,B}} (D \dot{+} B \dot{+} C') = \bigcap_{C'\in\mathcal{E}_{C,D}} (D \dot{+} B \dot{+} C') = B \dot{+} C. \qquad \square$$

## 4 The Generalized Demyanov Difference

In this section we extend the definition of the Demyanov difference to arbitrary Hausdorff topological vector spaces.

Let us first denote the family of inclusion minimal elements of $\mathcal{E}_{A,B}$ by $m\mathcal{E}_{A,B}$.

Now we define a new subtraction by $A \overset{D}{-} B = \overline{\text{conv}} \bigcup m\mathcal{E}_{A,B}$. We will show that $A \overset{D}{-} B$ is a generalization of Demyanov difference. Notice that $A \overset{D}{-} B$ is well defined for $A, B \in \mathcal{K}(X)$, where $X$ is a Hausdorff topological vector space.

**Lemma 4.** *Let $A, B \in \mathcal{K}(X)$. The set $D$ belongs to $\mathcal{E}_{B,A}$ if and only if $0 \in \bigcap_{C \in \mathcal{E}_{A,B}}(C+D)$*

*Proof.* Let $C \in \mathcal{E}_{A,B}$, $D \in \mathcal{E}_{B,A}$. Then $B \subset A+D \subset B+C+D$. By order law of cancellation $0 \in C+D$. On the other hand, if $D \in \mathcal{K}(X)$ and $0 \in \bigcap_{C \in \mathcal{E}_{A,B}}(C+D)$, then by Lemma 1

$$B \subset B + \bigcap_{C \in \mathcal{E}_{A,B}} (C+D) \subset \bigcap_{C \in \mathcal{E}_{A,B}} (B+C+D) = A+D.$$

Hence $D \in \mathcal{E}_{B,A}$. Therefore, $D \in \mathcal{E}_{B,A}$ if and only if $0 \in \bigcap_{C \in \mathcal{E}_{A,B}}(C+D)$. □

**Proposition 4.** *Let $A, B \in \mathcal{K}(X)$. Then $A \overset{D}{-} B = -(B \overset{D}{-} A)$.*

*Proof.* Let $C \in \mathcal{E}_{A,B}, D \in \mathcal{E}_{B,A}$. Denote $D' = D \cap (-\overline{\text{conv}} \bigcup m\mathcal{E}_{A,B})$. Then

$$\bigcap_{C \in \mathcal{E}_{A,B}} (C+D') = \bigcap_{C \in m\mathcal{E}_{A,B}} (C+D') = \bigcap_{C \in m\mathcal{E}_{A,B}} (C+D) = \bigcap_{C \in \mathcal{E}_{A,B}} (C+D).$$

Hence by Lemma 4 if $D \in m\mathcal{E}_{B,A}$, then $D = D \cap (-\overline{\text{conv}} \bigcup m\mathcal{E}_{A,B})$, and

$$D \subset (-\overline{\text{conv}} \bigcup m\mathcal{E}_{A,B}) = -(A \overset{D}{-} B).$$

We have just proved the inclusion $B \overset{D}{-} A \subset -(A \overset{D}{-} B)$, which implies the proposition. □

As a consequence from Theorem 5 we obtain:

**Proposition 5.** *Let $X$ be a Hausdorff topological vector space and pairs $(A,B), (C,D) \in \mathcal{K}^2(X) = \mathcal{K}(X) \times \mathcal{K}(X)$. If $[A,B] = [C,D]$, then $A \overset{D}{-} B = C \overset{D}{-} D$.*

**Lemma 5.** *Let $X$ be a Hausdorff topological vector space, $A, B, C \in \mathcal{K}(X)$. Then for every set $M \in \mathcal{K}(X)$ we have*

$$A + M = \bigcap_{D \in \mathcal{E}_{A,B}, E \in \mathcal{E}_{B,C}} (C+D+E+M).$$

*Proof.* Applying Lemma 1 twice we obtain

$$A + M = \bigcap_{D \in \mathcal{E}_{A,B}} (B+D+M) = \bigcap_{D \in \mathcal{E}_{A,B}} \Big( \bigcap_{E \in \mathcal{E}_{B,C}} (C+E+D+M) \Big).$$ □

**Lemma 6.** $A, B, C \in \mathcal{K}(X)$. *The set $F$ belongs to $\mathcal{E}_{C,A}$ if and only if*

$$0 \in \bigcap_{D \in \mathcal{E}_{A,B}, E \in \mathcal{E}_{B,C}} (D+E+F).$$

*Proof.* Let $D \in \mathcal{E}_{A,B}, E \in \mathcal{E}_{B,C}$ and $F \in \mathcal{E}_{C,A}$. Then $C \subset A+F \subset B+D+F \subset C+E+D+F$. By order law of cancellation $0 \in D+E+F$. On the other hand, if $F \in \mathcal{K}(X)$ and $0 \in \bigcap_{D\in\mathcal{E}_{A,B},E\in\mathcal{E}_{B,C}}(D+E+F)$, then by Lemma 5

$$C \subset C + \bigcap_{D\in\mathcal{E}_{A,B},E\in\mathcal{E}_{B,C}} (D+E+F) \subset \bigcap_{D\in\mathcal{E}_{A,B},E\in\mathcal{E}_{B,C}} (C+D+E+F) = A+F.$$

Hence $F \in \mathcal{E}_{C,A}$. □

**Proposition 6.** *Let $A,B,C \in \mathcal{K}(X)$. Then*

$$A \overset{D}{-} C \subset (A \overset{D}{-} B) + (B \overset{D}{-} C).$$

*Proof.* Let $D \in \mathcal{E}_{A,B}, E \in \mathcal{E}_{B,C}$, and $F \in \mathcal{E}_{C,A}$. Denote

$$F' = F \cap ((-\overline{\text{conv}} \bigcup m\mathcal{E}_{A,B}) + (-\overline{\text{conv}} \bigcup m\mathcal{E}_{B,C})).$$

Then by Lemma 6

$$\bigcap_{D\in\mathcal{E}_{A,B},E\in\mathcal{E}_{B,C}} (D+E+F') = \bigcap_{D\in m\mathcal{E}_{A,B},E\in m\mathcal{E}_{B,C}} (D+E+F')$$
$$= \bigcap_{D\in m\mathcal{E}_{A,B},E\in m\mathcal{E}_{B,C}} (D+E+F) = \bigcap_{D\in\mathcal{E}_{A,B},E\in\mathcal{E}_{B,C}} (D+E+F).$$

Hence, also by Lemma 6, if $F \in m\mathcal{E}_{C,A}$, then

$$F = F \cap ((-\overline{\text{conv}} \bigcup m\mathcal{E}_{A,B}) + (-\overline{\text{conv}} \bigcup m\mathcal{E}_{B,C})).$$

Therefore,

$$F \subset ((-\overline{\text{conv}} \bigcup m\mathcal{E}_{A,B}) + (-\overline{\text{conv}} \bigcup m\mathcal{E}_{B,C}))$$
$$= (-(A \overset{D}{-} B)) + (-(B \overset{D}{-} C)).$$

Then $(C \overset{D}{-} A) \subset (-(A \overset{D}{-} B)) + (-(B \overset{D}{-} C))$. By Proposition 4 we obtain our proposition. □

**Theorem 6.** *If $X$ is a locally convex vector space, then $A \ddot{-} B = A \overset{D}{-} B$.*

*Proof.* First, we prove that $A \ddot{-} B \subset A \overset{D}{-} B$. It is enough to show that $\{a-b | a \in A, b \in B, a+b \in \text{ext}(A+B)\} \subset \bigcup m\mathcal{E}_{A,B}$. Let $a+b \in \text{ext}(A+B)$. Decomposition of $a+b$ into a sum of elements of $A$ and $B$ is unique. The same is true of the decomposition of $a = b + (a-b) \in B + (A-b)$. Notice that $A \subset B + (A-b)$. Then $A - b \in \mathcal{E}_{A,B}$. There exist $C \subset A-b$ such that $C \in m\mathcal{E}_{A,B}$. Then $b + (a-b) \subset B + C \subset B + (A-b)$. Therefore, $a-b \in C$.

It is enough to show that if $A,B,C \in \mathcal{K}(X)$ and $A \subset B+C$, then $A \subset B+(C\cap(A\dot{\overline{-}}B))$. Since the set $A\dot{\overline{-}}B$ is a closed subset of the compact $A-B$, the set is compact itself. Hence by translation of intersection of a chain of compact convex sets

$$B+\bigcap_i C_i=\bigcap_i (B+C_i),$$

it is enough to prove that for all $f_1 \in X^\star$ and $C \in \mathcal{E}_{A,B}$ we have $A \subset B+(C\cap(H^-))$ where $H^-=\{x\in X|\, f_1(x)=\sigma=\sup f_1(A\dot{\overline{-}}B)\}$. Let us fix $f_1\in X^\star$, $f_1\neq 0$. If $\sigma\neq 0$, then we can replace $A$ with appropriate translate of $A$ so that in the following we assume that $\sigma=0$.

Let $x\in \mathrm{ext}(A+B)$ such that $f_1(x)=\inf f_1(A+B)$. We have unique decomposition of $x$, i.e. $x=y+z$, where $y\in A$ and $z\in B$. Then

$$\inf f_1(A)-\inf f_1(B)=f_1(y)-f_1(z)=f_1(y-z)\leq \sup(f_1(A\dot{\overline{-}}B))\leq 0.$$

Hence we have

$$\inf f_1(B)+\inf f_1(C)=\inf f_1(B+C)\leq \inf f_1(A)\leq \inf f_1(B).$$

Therefore $\inf f_1(C)\leq 0$, and the set $C\cap(H^-)$ is nonempty.

Let us assume that there exist $w\in A\setminus(B+(C\cap(H^-)))$. Since the set $B+(C\cap(H^-))$ is compact an convex, there exists continuous linear function $f_2\in X^\star$, such that $f_2$ is linearly independent from $f_1$ and such that $f_2(w)>\sup f_2(B+(C\cap(H^-)))$.

By $f:X\longrightarrow\mathbb{R}^2$, we denote a linear function such that $f(x)=(f_1(x),f_2(x))$. Obviously, $f(A)$, $f(B)$, $f(A\dot{\overline{-}}B)$, and $f(C)\in\mathcal{K}(\mathbb{R}^2)$. Consider $a\in f(A)$, $b\in f(B)$ such that $a+b\in \mathrm{ext}(f(A)+f(B))=\mathrm{ext} f(A+B)$. Let us notice that $f^{-1}(a+b)\cap(A+B)$ is an extreme subset of $A+B$, and $\mathrm{ext}(f^{-1}(a+b)\cap(A+B))\subset \mathrm{ext}(A+B)$. By Krein–Milman theorem the set $\mathrm{ext}(f^{-1}(a+b)\cap(A+B))$ is nonempty. Let $x\in \mathrm{ext}(f^{-1}(a+b)\cap(A+B))$. There exist unique $y\in A$ and $z\in B$ such that $x=y+z$. Then $a+b=f(y)+f(z)$. By unique decomposition of $a+b$ in $f(A)+f(B)$, we have $a=f(y)$ and $b=f(z)$. Hence

$$\langle e_1,a\rangle-\langle e_1,b\rangle=f_1(y)-f_1(z)=f_1(y-z)\leq \sup f_1(A\dot{\overline{-}}B)\leq 0$$

for all $a\in f(A)$ and $b\in f(B)$ such that $a+b\in \mathrm{ext} f(A+B)$.

By $h_A,g_A:[\inf f_1(A),\sup f_1(A)]\longrightarrow\mathbb{R}$ we denote such functions that $a=(a_1,a_2)\in f(A)$ if and only if $g_A(a_1)\leq a_2\leq h_A(a_1)$. Obviously, the function $h_A$ is concave and $g_A$ is convex. Both functions are continuous and differentiable almost everywhere. In similar way we denote $h_B,g_B,h_C$, and $g_C$. Let us notice that for almost all $u \in S^1, u_2 \geq 0$ the set $(f(A+B))(u)$ is a singleton, i.e. $(f(A+B))(u) \in \exp f(A+B)$. Then

$$\langle e_1, (f(A))(u)\rangle - \langle e_1, (f(B))(u)\rangle \leq 0.$$

Since $(f(A))(u) = (\alpha, h_A(\alpha))$ and $(f(B))(u) = (\beta, h_B(\beta))$, we have $\alpha - \beta \leq 0$. Moreover, $h'_A(\gamma) \geq -\frac{u_1}{u_2}$ for all $\gamma \leq \alpha$ and $h'_A(\gamma) \leq -\frac{u_1}{u_2}$ for all $\gamma \geq \alpha$. Also $h'_B(\gamma) \geq -\frac{u_1}{u_2}$ for all $\gamma \leq \beta$ and $h'_B(\gamma) \leq -\frac{u_1}{u_2}$ for all $\gamma \geq \beta$. Hence $h'_A(\alpha) = h'_B(\beta)$ implies $\alpha \leq \beta$. Since both functions $h'_A$ and $h'_B$ are non-increasing, we have $h'_A \leq h'_B$. Therefore, the function $h_B - h_A$ is non-decreasing. In other words $h_B$ is increasing faster or decreasing slower than $h_A$.

Let $c \in \mathbb{R}^2, c_1 = 0$ belong to the upper part of the boundary of $f_1(C)$. There exists $u \in S^1, u_2 \geq 0$ such that $c \in (f_1(C))(u)$. Let $\alpha$ be such a number that $(\alpha, h_{B+C}(\alpha)) \in (f(B+C))(u)$. Then for all $\gamma \leq \alpha$ we have

$$h_A(\gamma) \leq h_{B+C}(\gamma) = h_{B+(C\cap H^-)}(\gamma).$$

On the other hand for all $\gamma \geq \alpha$ we have

$$h_B(\gamma) + c_2 = h_{B+(C\cap H^-)}(\gamma).$$

Since $h_B - h_A$ is non-decreasing, we obtain

$$h_{B+(C\cap H^-)}(\gamma) - h_A(\gamma) \geq h_{B+(C\cap H^-)}(\alpha) - h_A(\alpha) \geq 0.$$

By continuity of considered functions $h_A \leq h_{B+(C\cap H^-)}$. Then

$$f_2(w) \leq \sup h_A \leq \sup h_{B+(C\cap H^-)} = \sup f_2(B + (C \cap H^-)) < f_2(w)$$

which contradicts the assumption that $A \not\subset B + (C \cap H^-)$. □

Finally we state the usual properties of the Demyanov difference for locally convex vector spaces:

**Proposition 7.** *Let $X$ be a locally convex vector space and $A, B, C \in \mathcal{K}(X)$. The Demyanov difference has the following properties:*

(D1) *If $A = B + C$, then $C = A \dot{-} B$.*
(D2) $(A \dot{-} B) + B \supseteq A$.
(D3) *If $B \subseteq A$, then $0 \in A \dot{-} B$.*
(D4) $(A \dot{-} B) = -(B \dot{-} A)$
(D5) $A \dot{-} C \subset (A \dot{-} B) + (B \dot{-} C)$.

*Proof.* First, (D4) is proved in Proposition 4 and (D5) in Theorem 6.

Now (D2) $(A \dot{-} B) + B \supseteq A$ follows immediately from (D5) for $C = \{0\}$.

If $B \subseteq A$, then $B + \{0\} \subseteq A \subset (A \stackrel{..}{-} B) + B$ and by the order cancellation law $0 \in A \stackrel{..}{-} B$, which proves (D3).

To prove (D1) note that by (D2) $B + C = A \subseteq (A \stackrel{..}{-} B) + B$ and hence $C \subseteq (A \stackrel{..}{-} B)$. Now we show that $(A \stackrel{..}{-} B) \subseteq C$ holds. Therefore let us consider an extremal point of $z \in \text{ext}(A + B) = \text{ext}(2B + C)$. Hence $z$ has the unique decomposition $z = 2b + c$ with $b \in \text{ext}(B)$ and $c \in \text{ext}(C)$. Now $b + c \in \text{ext}(A) = \text{ext}(B + C)$, because otherwise $b + c = t(b' + c') + (1 - t)(b'' + c'')$ with $b', b'' \in B$ , $c', c'' \in C$, and $0 < t < 1$, which implies that $z = 2b + c = t((b' + b) + c') + (1 - t)((b'' + b) + c'')$ and hence no extremal point of $(2B + C)$. Therefore $z \in \text{ext}(A + B) = \text{ext}(2B + C)$ has the unique decomposition $z = (b + c) + b$ with $b \in \text{ext}(B)$ and $c \in \text{ext}(C)$ and $(b + c) \in \text{ext}(A)$. Now it follows from the definition of the Demyanov difference that $(A \stackrel{..}{-} B) = ((B + C) \stackrel{..}{-} B) \subseteq C$, which completes the proof. □

## References


[1] Clarke, F.H.: Optimization and Nonsmooth Analysis. Wiley, New York (1983)

[2] Demyanov, V.F., Rubinov, A.M.: Quasidifferential Calculus. Springer and Optimization Software Inc., New York (1986)

[3] Hukuhara, M.: Intégration des applications mesurables dont la valeur est un compact convexe (French). Funkcial. Ekvac. **10**, 205–223 (1967)

[4] Husain, T., Tweddle, I.: On the extreme points of the sum of two compact convex sets. Math. Ann. **188**, 113–122 (1970)

[5] Klee, V.: Extremal structure of convex sets II. Math. Zeitschr. **69**, 90–104 (1958)

[6] Pallaschke, D., Przybycień, H., Urbański, R.: On partialy ordered semigroups. J. Set Valued Anal. **16**, 257–265 (2007)

[7] Pallaschke, D., Urbański, R.: Pairs of Compact Convex Sets: Fractional Arithmetic with Convex Sets, Mathematics and Its Applications. Kluwer Academic Publisher, Dortrecht (2002)

[8] Pallaschke, D., Urbański, R.: On the separation and order law of cancellation for bounded sets. Optimization **51**, 487–496 (2002)

[9] Pinsker A.G.: The space of convex sets of a locally convex space. Trudy Leningrad Eng. Econ. Inst. **63**, 13–17 (1966)

[10] Pontryagin L.S.: On linear differential games II (Russian). Dokl. Acad. Nauk SSSR **175**, 764–766 (1967)

[11] Rubinov A.M., Akhundov, I.S.: Differences of compact sets in the sense of Demyanov and its application to non-smooth-analysis. Optimization **23**, 179–189 (1992)

[12] Urbański, R.: A generalization of the Minkowski–Rådström–Hörmander theorem. Bull. Acad. Polon. Sci. Sér. Sci. Math. Astronom. Phys. **24**, 709–715 (1976)

# Separable Reduction of Metric Regularity Properties

**A.D. Ioffe**

**Abstract** We show that for a set-valued mapping $F : X \rightrightarrows Y$ between Banach spaces the property of metric regularity near a point of its graph is separably determined in the sense that it holds, provided for any separable subspaces $L_0 \subset X$ and $M \subset Y$, containing the corresponding components of the point, there is a separable subspace $L \subset X$ containing $L_0$ such that the mapping whose graph is the intersection of the graph of $F$ with $L \times M$ (restriction of $F$ to $L \times M$) is metrically regular near the same point. Moreover, it is shown that the rates of regularity of the mapping near the point can be recovered from the rates of such restrictions.

**Keywords** Set-valued mapping • Linear openness • Metric regularity • Regularity rates • Separable reduction

## 1 Introduction

Separable reduction (if possible) allows to reduce the study of one or another property to objects belonging to or defined on separable spaces. In variational analysis which typically deals with problems involving heavy techniques, the possibility of separable reduction is both attractive and productive.

Speaking a bit more formally, when we say that a certain property *admits separable reduction* or is *separably determined*, we mean that the property is valid on the entire Banach space if and only if it is valid on all or on arbitrarily big separable subspaces. (The latter means that any separable subspace is a part of a bigger separable subspace having the property.) We refer to [3] for a list of such properties. We just mention the role of separable reduction in the calculus

A.D. Ioffe (✉)
Department of Mathematics, Technion, Haifa 32000, Israel
e-mail: ioffe@math.technion.ac.il

V.F. Demyanov et al. (eds.), *Constructive Nonsmooth Analysis and Related Topics*, Springer Optimization and Its Applications 87, DOI 10.1007/978-1-4614-8615-2_3,

of nonconvex subdifferentials, especially the proof by Fabian and Zhivkov in [4] that the Fréchet subdifferentiability is separably determined which made possible extension of the calculus of Fréchet subdifferentials to functions on Asplund spaces (see also [3] and references therein).

This paper deals with three equivalent regularity properties of set-valued mappings between metric spaces: linear openness, metric regularity proper, and the Aubin property of the inverse. It seems reasonable to use the term "regularity" for the unity of these three properties, first to avoid ambiguity with the expression "metric regularity" and secondly to emphasize the direct connection with the classical regularity concept. Needless to say that (metric) regularity is one of the most fundamental concepts of variational analysis.

The main result, Theorem 2, states that regularity is a separably determined property. So the theorem is still another confirmation of a special role played by separable (and more generally, weakly compactly generated[1]) spaces in variational analysis. A practical implication of the theorem is a simplification of computing subdifferential estimates of quantitative measures of regularity.

The central role in the proof of the theorem is played by the new concept of *compact regularity.*[2] In the concluding section we discuss its connection with other compactness properties used in variational analysis in connection with the regularity property. As a result of the discussion we formulate a new regularity criterion in terms of the $G$-coderivatives of restrictions of the mapping at the given point of its graph.

## 2 Preliminaries

We shall start by reminding the definition of two (of the three) equivalent fundamental regularity properties of set-valued mappings. These properties are naturally defined for mappings between arbitrary metric spaces. In this note we deal only with Banach spaces. But passage to the Banach space setting brings about no simplification in the definitions of the regularity properties (and of proofs of many associated results). So at this specific point it looks more natural to talk about metric spaces $X$ and $Y$ and a set-valued mapping $F$ from $X$ into $Y$. (We shall use the standard notation $F: X \rightrightarrows Y$ in what follows.) As usual we denote by Graph $F = \{(x,y) \in X \times Y : y \in F(x)\}$ the *graph* of $F$ and by dom $F = \{x \in X : F(x) \neq \emptyset\}$ the *domain* of $F$.

[1]WCG spaces are distinguished by the property that the limiting versions of all subdifferentials trusted on the space coincide for locally Lipschitz functions.

[2]I am indebted to one of the reviewers for pointing out that compact regularity coincides with "partial cone property up to a compact set" of the inverse map introduced in a just published Penot's monograph [14].

Let $(\bar{x},\bar{y}) \in \text{Graph } F$. It is said that

(a) $F$ is *linearly open near* (or *at* in some publications) $(\bar{x},\bar{y})$ if there is are $\varepsilon > 0$, $r > 0$ such that

$$B(F(x),rt) \cap B(\bar{y},\varepsilon) \subset F(B(\bar{x},t)), \quad \text{if } d(x,\bar{x}) < \varepsilon,\ 0 < t < \varepsilon. \tag{1}$$

The upper bound $\text{sur}F(\bar{x}|\bar{y})$ of such $r$ is the *modulus of surjection* of $F$ at $(\bar{x},\bar{y})$. If no such $\varepsilon$ and $r$ exist, we set $\text{sur}F(\bar{x}|\bar{y}) = 0$.

(b) $F$ is *metrically regular near* (or *at*) $(\bar{x},\bar{y})$ if there are $\varepsilon > 0$, $K \in (0,\infty)$ such that

$$d(x,F^{-1}(y)) \le Kd(y,F(x)), \quad \text{if } d(x,\bar{x}) < \varepsilon,\ d(y,\bar{y}) < \varepsilon. \tag{2}$$

The lower bound $\text{reg}F(\bar{x}|\bar{y})$ of all such $K$ is the *modulus of metric regularity* of $F$ at $(\bar{x},\bar{y})$. If no such $\varepsilon$ and $K$ exist, we set $\text{reg}F(\bar{x}|\bar{y}) = \infty$.

The well-known equivalence theorem states that the equality

$$\text{sur}F(\bar{x}|\bar{y}) \cdot \text{reg}F(\bar{x}|\bar{y}) = 1$$

holds unconditionally (under the convention that $0 \cdot \infty = 1$). The equivalence theorem allows each time to choose just one regularity property to work with—depending on convenience and/or personal preferences. Here we basically deal with linear openness. It is also to be mentioned that there is a third equivalent property called *Aubin* property or *pseudo-Lipschitz* property of the inverse mapping $F^{-1}$. But we do not use it in this note.

The basic infinitesimal mechanism that allows to compute the regularity moduli is associated with the concept of *slope* (see [1, 7]). But in the Banach setting a more convenient (although often less precise) instrument is provided by subdifferentials. For the most recent and the most complete information about subdifferentials we refer to [9, 14]. Here we just mention that there are five nonconvex subdifferentials that are mainly used in variational analysis: Fréchet subdifferential $\partial_F$, Dini–Hadamard subdifferential $\partial_H$, limiting Fréchet subdifferential $\partial_{LF}$, approximate subdifferential $\partial_G$, and Clarke's generalized gradient $\partial_C$. The last two can be effectively used for functions on all Banach spaces. The Fréchet and the limiting Fréchet subdifferentials can be "trusted" only on Asplund spaces and the Dini–Hadamard subdifferentials on Gâteaux smooth spaces (which are space having equivalent norms Gâteaux differentiable off the origin). The word "trusted" means roughly speaking that every l.s.c. function has a nonempty subdifferential at points of a dense subset of its domain and certain embryonic ("fuzzy") calculus rules hold. (For the formal definition of trustworthiness, see, e.g., [7, 9, 14].)

With every subdifferential we associate the concepts of *normal cone* to a set at a point of the set and of *coderivative* of a set-valued mapping at a point of its graph. Namely, given a set $S \subset X$ (usually assumed closed), a point $x \in S$, and a subdifferential $\partial$, the normal cone $N(S,x)$ to $S$ at $x$ (associated with $\partial$) is defined as the subdifferential at $x$ of the *indicator* $i_S$ of $S$ which is the function equal to zero on $S$ and $+\infty$ outside of $S$.

If $F : X \rightrightarrows Y$ is a set-valued mapping and $(x,y) \in \text{Graph}\, F$, then the *coderivative of* $F$ at $(x,y)$ (associated with $\partial$) is the set-valued mapping $D^*F(x,y) : Y^* \rightrightarrows X^*$ defined by

$$D^*F(x,y)(y^*) = \{x^* : (x^*, -y^*) \in N(\text{Graph}\, F, (x,y))\}.$$

The following is the basic coderivative regularity criterion.

**Theorem 1 ([5–7]).** *Let $X$, $Y$ be Banach spaces, and let $\partial$ be a subdifferential that can be trusted on a class of Banach spaces containing both X and Y. Let further $F : X \rightrightarrows Y$ be a set-valued mapping with locally closed graph and $\bar{y} \in F(\bar{x})$. Then*

$$\text{sur} F(\bar{x}|\bar{y}) \geq \lim_{\varepsilon \to +0} \inf\{\|x^*\| : x^* \in D^*F(x,y)(y^*),\ \|y^*\| = 1,$$
$$(x,y) \in (\text{Graph}\, F) \bigcap B((\bar{x},\bar{y}),\varepsilon)\},$$

*or equivalently,*

$$\text{reg} F(\bar{x}|\bar{y}) \leq \lim_{\varepsilon \to +0} \sup\{\|y^*\| : x^* \in D^*F(x,y)(y^*),\ \|x^*\| = 1,$$
$$(x,y) \in (\text{Graph}\, F) \bigcap B((\bar{x},\bar{y}),\varepsilon)\}.$$

For more specific and more precise results relating to the Fréchet and limiting Fréchet subdifferentials we refer to [12–14].

## 3 Main Result

In what follows we denote by $\mathcal{S}(X)$ the collection of closed separable subspaces of a Banach space $X$.

**Lemma 1 ([6]).** *Let $X$ be a Banach space, and let $E_0$ be a separable subspace of $X$. Let further $S_1, S_2, \ldots$ be a countable collection of subsets of $X$. Then there exists a subspace $E \in \mathcal{S}(X)$ containing $E_0$ and such that*

$$d(x,S_i) = d(x,S_i \cap E), \quad \forall\, x \in E, \quad \forall i = 1,2,\ldots.$$

*Moreover, if $X$ is a Cartesian product of Banach spaces $X_1,\ldots,X_k$, then $E$ can have the form $E^1 \times \cdots \times E^k$, where $E^i$ is a subspace of $X_i$.*

*Proof.* To prove the first statement it is sufficient to construct, starting with $E_0$, an increasing sequence of separable subspaces $E_n$ such that $d(x,S_i \cap E_{n+1}) = d(x,S_i)$ for all $x \in E_n$ and all $i$ and then define $E$ as the closure of $\cup E_n$. If we have already $E_n$, then we take a dense countable subset $C_n = \{x_1,x_2,\ldots\} \subset E_n$, for every $i$ and $k$ choose a sequence $(u_{ikm})_{m=1}^{\infty} \subset S_i$ such that $d(x_k,S_i) \geq \|x_k - u_{ikm}\| - (km)^{-1}$

for every $m$ and define $E_{n+1}$ as the space spanned by the union of $E_n$ and all $u_{ikm}$, $i,k,m,=1,2,\dots$. If now $x \in E_n$ and $x_{k_r} \in C_n$ converge to $x$, then

$$\begin{aligned} d(x,S_i) &= \lim_{r\longrightarrow\infty} d(x_{k_r},S_i) = \lim_{r\longrightarrow\infty}\lim_{m\longrightarrow\infty} \|x_{k_r} - u_{ik_rm}\| \\ &\geq \lim_{r\longrightarrow\infty} d(x_{k_r},S_i \cap E_{n+1}) = d(x,S_i \cap E_{n+1}) \geq d(x,S_i) \end{aligned}$$

and the result follows. The second statement is obvious as every $E_n$ can be easily chosen as a Cartesian product of subspaces of $X_i$. □

**Definition 1 (compact regularity).** Let $F : X \rightrightarrows Y$ and $(\bar{x},\bar{y}) \in \text{Graph } F$. We say that $F$ is *compactly regular* near $(\bar{x},\bar{y})$ if there are $r > 0$, $\varepsilon > 0$ and a norm compact set $P \subset Y$ such that

$$\big(F(x) \cap B(\bar{y},\varepsilon)\big) + trB_Y \subset F(B(x,t)) + tP \quad \text{if} \quad \|x-\bar{x}\| < \varepsilon,\ 0<t<\varepsilon. \tag{3}$$

Clearly, (3) reduces to the standard linear openness if $P = \{0\}$.[3] The first property of compact regularity to be emphasized is that it is inherited (up to a closure) by some separable subspaces of $X \times Y$.

**Proposition 1.** *Assume that $F$ is compactly regular near $(\bar{x},\bar{y})$, that is, with a suitable choice of a norm compact set $P$, (3) holds with some positive $\varepsilon$ and $r$. Then for any separable subspaces $L_0 \subset X$ and $M_0 \subset Y$ there are bigger subspaces $L \in \mathcal{S}(X)$ and $M \in \mathcal{S}(Y)$ such that*

*(i)* $d((x,y),\text{Graph } F) = d((x,y),(\text{Graph } F)\bigcap(L\times M))$, $\forall\,(x,y) \in L\times M$ *and for all sufficiently small $t > 0$*

*(ii)* $y + rt(B_Y \bigcap M) \subset \text{cl}F\big(B(x+t(1+\delta)(B_X\bigcap L)\big) + tP$, $\forall\,\delta > 0$ *and all* $(x,y) \in (\text{Graph } F)\bigcap(L\times M)$ *sufficiently close to* $(\bar{x},\bar{y})$

*Proof.* Let a compact set $P \subset Y$, an $r > 0$, and an $\varepsilon > 0$ be such that (3) holds. First we shall prove the following: *for any separable subspaces $L_0 \subset X$ and $M_0 \subset Y$ there is a nondecreasing sequence $(L_n,M_n)$ of separable subspaces of $X$ and $Y$, respectively, such that*

*($i_0$)* $d((x,y),\text{Graph } F) = d((x,y),(\text{Graph } F)\bigcap(L_{n+1}\times M_{n+1}))$ *for all* $(x,y) \in L_n \times M_n$;
*and for all sufficiently small $t > 0$*

*($ii_0$)* $y + rt(B_Y \bigcap M_n) \subset \text{cl}F(B(x+t(1+\delta)(B_X \bigcap L_{n+1})) + tP$, *for all $\delta > 0$ and all* $(x,y) \in (\text{Graph } F)\bigcap(L_n\times M_n)$ *sufficiently close to $(\bar{x},\bar{y})$ and all sufficiently small $t > 0$*

---

[3] We have chosen a slightly different way for writing the inclusion to emphasize close relationship with the popular "compact epi-Lipschitz property" introduced by Borwein and Strojwas in 1986.

Assume that we have already $L_n, M_n$ for some $n$. Take a sequence $(x_i, y_i)$ which is dense in the intersection of (Graph $F)\bigcap(L_n \times M_n)$ with the neighborhood of $(\bar{x}, \bar{y})$ in which (3) is guaranteed, and let the sequences $(v_j)$ and $(t_k)$ be dense in $B_Y \bigcap M_n$. For any $i, j, k = 1, 2, \ldots$ we find from (3) an $h_{ijk} \in B_X$ such that $y_i + rt_k v_j \in F(x_i + t_k h_{ijk}) + t_k P$. Let $\hat{L}_n$ be the subspace of $X$ spanned by the union of $L_n$ and the collection of all $h_{ijk}$.

If now $(x, y) \in$ (Graph $F) \cap (L_n \times M_n)$, $t \in (0, 1)$, $v \in B_Y$ and $(x_{i_m}, y_{i_m})$, $t_{k_m}$, $v_{j_m}$ converge, respectively, to $(x, y)$, $t$, and $v$, then as $x_{i_m} + t_{k_m}(B_X \cap L_n) \subset x + t(1 + \delta)(B_X \cap L_n)$ for sufficiently large $m$, we conclude that $(ii_0)$ holds with $\hat{L}_n$ instead of $L_{n+1}$.

Finally we define $L_{n+1}, M_{n+1}$ using Lemma 1 applied to $S =$ Graph $F$ and $E_0 = \hat{L}_n \times M_n$. Then $(i_0)$ is valid as $L_n \subset \hat{L}_n$ and $(ii_0)$ holds because $\hat{L}_n \subset L_{n+1}$.

Set $L = \mathrm{cl}(\bigcup L_n)$ and $M = \mathrm{cl}(\bigcup M_n)$. We claim that

$$(\text{Graph } F) \cap (L \times M) = \mathrm{cl}((\text{Graph } F) \cap (\bigcup_n (L_n \times M_n)).$$

Indeed, the inclusion $\supset$ is obvious as Graph $F$ is closed. So we need to prove the opposite inclusion. Let $(x, y) \in$ (Graph $F)\bigcap(L \times M)$. Then there is a sequence $(x_m, y_m) \in \bigcup_n (L_n \times M_n)$ converging to $(x, y)$. If $(x_m, y_m) \in L_n \times M_n$, then by $(i_0)$ we can find $(u_m, v_m) \in$ (Graph $F)\bigcap(L_{n+1} \times M_{n+1})$ with, e.g., $\|(u_m, v_m) - (x_m, y_m)\| \leq 2\|(x_m, y_m) - (x, y)\|$ from which the claim easily follows.

Now we can complete the proof. The verification of (i) is like in the proof of Lemma 1: let $(x, y) \in L \times M$, take as above $(x_m, y_m) \in \bigcup_n (L_n \times M_n)$ converging to $(x, y)$ and if $n = n(m)$ is such that $(x_m, y_m) \in L_n \times M_n$, use $(i_0)$ to find a pair $(u_m, v_m) \in$ (Graph $F) \cap (L_{n+1} \times M_{n+1})$ such that

$$\begin{aligned} d((x,y), (\text{Graph } F) \cap (L \times M)) &\leftarrow d((x_m, y_m), (\text{Graph } F) \cap (L \times M)) \\ &\leq \|(u_m, v_m) - (x_m, y_m)\| \longrightarrow d((x,y), \text{Graph } F). \end{aligned}$$

The proof of (ii) is equally straightforward. We have by $(ii_0)$

$$y + rtB_{\bigcup_n M_n} \subset \mathrm{cl} F(x + t(1 + (\delta/2))(B_L)) + tP$$

if $(x, y)$ is close to $(\bar{x}, \bar{y})$ and belongs to (Graph $F)\bigcap(L_n \times M_n)$. But $rtB_M = \mathrm{cl}(rt(B_{\bigcup_n M_n})$ by definition and the set in the right-hand part of the above inclusion is close due to compactness of $P$. Thus (ii) holds for all $(x, y)$ belonging to (Graph $F) \cap (\bigcup_n (L_n \times M_n))$.

If finally $(x, y) \in$ (Graph $F) \cap (L \times M))$, then as we have seen $(x, y)$ is a limit of a sequence $(x_m, y_m)$ belonging to (Graph $F) \cap (\bigcup_n (L_n \times M_n))$. So if $m$ is so big that $\|x - x_m\| < t\delta/2$, we have

$$y_m + rtB_M \subset \mathrm{cl} F(x + t(1 + \delta)(B_L)) + tP$$

and the result follows.

The most important consequence of the above proposition is the following separable reduction principle for regularity.

**Theorem 2 (separable reduction of regularity).** *Let $X$ and $Y$ be Banach spaces. A set-valued mapping $F : X \rightrightarrows Y$ with closed graph is regular at $(\bar{x},\bar{y}) \in \text{Graph}\, F$ if and only if for any separable subspace $M \subset Y$ and any separable subspace $L_0 \subset X$ with $(\bar{x},\bar{y}) \in L_0 \times M$ there exists a bigger separable subspace $L \in \mathcal{S}(X)$ such that the mapping $F_{L\times M} : L \rightrightarrows M$ whose graph is the intersection of* Graph $F$ *with $L \times M$ is regular at $(\bar{x},\bar{y})$.*

*Moreover, if $\text{sur}F(\overline{x}|\overline{y}) > r$, we can choose $L \in \mathcal{S}(X)$ and $M \in \mathcal{S}(Y)$ to make sure that also $\text{sur}F_{L\times M}(\overline{x}|\overline{y}) \geq r$. Conversely, if there is an $r > 0$ such that for any separable $M_0 \subset Y$ and $L_0 \subset X$ there are bigger separable subspaces $M \supset M_0$ and $L \supset L_0$ such that $\text{sur}F_{L\times M}(\overline{x}|\overline{y}) \geq r$, then $F$ is regular at $(\bar{x},\bar{y})$ with $\text{sur}F(\overline{x}|\overline{y}) \geq r$.*

*Proof.* The theorem is connected with the property (ii) of Proposition 1. We note that to prove this property alone in the framework of the proposition we can only deal with subspaces of $X$. In other words, as follows from the proof of (ii) in Proposition 1 , given separable subspaces $M \subset Y$ and $L_0 \subset X$, there is a bigger separable subspace $L \supset L_0$ of $X$ such that the property (ii) of the proposition holds.

Assume that $F$ is regular at $(\bar{x},\bar{y})$ with $\text{sur}F(\overline{x}|\overline{y}) > r$. Applying the proposition with $P = \{0\}$ we see that, given $L_0$ and $M$, we can find a closed separable subspace $L \subset X$ containing $L_0$ such that for any $\delta > 0$, any $(x,y) \in (\text{Graph}\ F) \cap (L \times M)$ sufficiently close to $(\bar{x},\bar{y})$ and any sufficiently small $t > 0$

$$B(y,rt) \cap M \subset \text{cl}F(B(x,(1+\delta)t) \cap L). \tag{4}$$

Application of the density theorem (see, e.g., [8]) allows to drop the closure operation on the right, so that $F_{L\times M}$ is indeed regular near $(\bar{x},\bar{y})$ with $\text{sur}F_{L\times M}(\overline{x}|\overline{y}) \geq (1+\delta)^{-1}r$. As $\delta$ can be arbitrarily small we get the desired estimate for the modulus of surjection of $F_{L\times M}$.

On the other hand, if $F$ were not regular at $(\bar{x},\bar{y})$, then we could find a sequence $(x_n,y_n) \in \text{Graph}\ F$ converging to $(\bar{x},\bar{y})$ such that $y_n + (t_n/n)v_n \notin F(B(x_n,t_n))$ for some $t_n < 1/n$ and $v_n \in B_Y$ (respectively $y_n + t_n(r-\delta)v_n \notin F(B(x_n,t_n))$ for some $\delta > 0$). Clearly this carries over to any closed separable subspace $L \subset X$ and $M \subset Y$ containing, respectively, all $x_n$, all $y_n$, and all $v_n$, so that no such $F_{L\times M}$ cannot be regular at $(\bar{x},\bar{y})$ (with the modulus of surjection $\geq r$) contrary to the assumption.

This theorem effectively reduces the regularity problem to the case of separable spaces. As every separable space has a Gâteaux differentiable renorm, the Dini–Hadamard subdifferential is trusted on the class of separable spaces. So we can use this subdifferential to analyze regularity of restrictions of $F$ to separable subspaces of $X \times Y$. Thus, as an immediate consequence of Theorem 4 and the subdifferential regularity criterion of Theorem 1, we get the following result.

**Theorem 3 (separable reduction of subdifferential criteria).** *Let $X,Y$ be Banach spaces, and let $F : X \rightrightarrows Y$ be a set-valued mapping with closed graph. Let $(\bar{x},\bar{y}) \in$* Graph $F$. *Then $F$ is regular near $(\bar{x},\bar{y})$ with $\text{sur}F(\overline{x}|\overline{y}) > r > 0$ if for any $M \in \mathcal{S}(Y)$*

*and any $L_0 \in \mathcal{S}(X)$, there exist a bigger separable subspace $L \subset X$ containing $L_0$ and an $\varepsilon > 0$ such that*

$$\inf\{\|x^*\| : x^* \in D^*_H F_{L\times M}(x,y)(y^*),\ \|y^*\| = 1,\ \|(x,y) - (\bar{x},\bar{y})\| < \varepsilon\} \geq r.$$

Of course, if $X$ and $Y$ are Asplund, we can use the Fréchet coderivative rather than Dini–Hadamard coderivative and prove (as the Fréchet subdifferential is separably determined) the "only if" statement as well. But in general the last theorem provides the best available coderivative estimate for the surjection modulus that can be applied to set-valued mappings between arbitrary Banach spaces.

## 4 A Few Remarks About Compact Regularity

To place the concept of compact regularity in proper perspective, it is reasonable to mention that there are a number of compactness properties associated with set-valued mappings in general and coderivatives in particular (see, e.g., [6, 10, 11, 13, 14].

Recall (see [7], Theorem 1) that for any subdifferential $\partial$, the property that there is a norm compact $P \subset Y$ such that (7) holds whenever $x^* \in D^*F(x,y)(y^*)$ for $(x,y) \in \text{Graph } F$ close to $(\bar{x},\bar{y})$ (where $D^*$ stands for the coderivative associated with $\partial$) is equivalent to *$\partial$-coderivative compactness* of $F$ at $(\bar{x},\bar{y})$: given a net of quadruples $(x_\alpha, y_\alpha, x^*_\alpha, y^*_\alpha)$ such that $(x_\alpha, y_\alpha) \in \text{Graph } F$ and norm converges to $(\bar{x},\bar{y})$, $x^*_\alpha \in D^*F(x_\alpha,y_\alpha)(y^*_\alpha)$, $\|x^*_\alpha\| \longrightarrow 0$, $\|y^*_\alpha\| = 1$ and $y^*_\alpha$ weak$^*$ converge to zero, then necessarily $\|y^*_\alpha\| \longrightarrow 0$.

If in the definition of $\partial$-coderivative compactness we replace nets by sequences (countable nets), then we shall get the definition of *sequential $\partial$-coderivative compactness*. If $Y$ is a separable Banach space, then coderivative compactness and sequential coderivative compactness coincide (as the weak$^*$-topology on bounded subsets of $Y^*$ is metrizable).

**Proposition 2.** *Let $\partial$ be a subdifferential trusted on a class of spaces containing $X$ and $Y$, and let $F : X \rightrightarrows Y$ be a set-valued mapping with closed graph. Then $F$ is compactly regular at $(\bar{x},\bar{y}) \in$ Graph $F$, provided it is $\partial$-coderivatively compact at the point.*

*Proof.* In the proof we shall consider the norm $X \times Y$: $\|(x,y) - (u,v)\| = \|x - u\| + \|y - v\|$ in $X \times Y$ and compact subsets of $Y$ which are intersections of finite-dimensional subspaces with the unit ball.

So let $P = L \bigcap B_Y$ with $\dim L < \infty$. If $F$ is not compactly regular at $(\bar{x},\bar{y})$, then for any sufficiently small $\varepsilon > 0$ we can find $x,\ y,\ t,\ v$ such that $(x,y) \in \text{Graph } F$, $\|x - \bar{x}\| + \|y - \bar{y}\| < 2\varepsilon$, $0 < t < \varepsilon$, $\|v\| < t\varepsilon$ and $y + v \notin F(B(x,t)) + tP$. This means that $d((u, y+v-tp), \text{Graph } F) > 0$ if $\|u - x\| \leq t$ and $p \in P$.

Consider the function

$$\varphi(u,z,p) = \|z-(y+v-tp)\| + \sqrt{\varepsilon}(\|u-x\| + t\|p\|).$$

This is a nonnegative function and $\varphi(x,y,0) = \|v\| < t\varepsilon$. Applying Ekeland's variational principle, we can find $(u',z',p') \in \text{Graph } F \times P$ such that $\|u'-x\| + \|z'-y\| + \|p'\| \leq \varepsilon$, $\varphi(u',z',p') \leq \varphi(x,y,0) \leq t\varepsilon$ and

$$\psi(u,z,p) := \varphi(u,z,p) + t(\|u'-u\| + \|z'-z\| + \|p'-p\|) \geq \varphi(u',z',p'),$$

for all $(u,z,p) \in \text{Graph } F \times P$. Note that $\|z'-(y+v-tp')\| > 0$. Indeed, otherwise, we would have $\|u'-x\| > t$ and

$$\varphi(u',z',p') \geq \sqrt{\varepsilon}(\|u'-x\| > t\sqrt{\varepsilon} > t\varepsilon \geq \varphi(u',z',p').$$

Recall that by $i_Q$ we denote the *indicator* of $Q$ that is the function equal to zero on $Q$ and $+\infty$ outside of $Q$. If $Q$ is closed, then $i_Q$ is obviously a lower semicontinuous function. Thus the function

$$\begin{aligned}(u,z,p) \mapsto &\|z-(y+v-tp)\| + \sqrt{\varepsilon}(\|u-x\| + t\|p\|)\\ &+t(\|u'-u\| + \|z'-z\| + \|p'-p\|) + i_{\text{Graph } F}(u,z) + i_P(p)\end{aligned}$$

has a local minimum at $(u',z',p')$.

This function is a sum of the indicator of Graph $F$ and a convex lower semicontinuous function which, in turn, is a sum of a convex continuous function and $i_P$. We also observe that $p'$ belongs to the relative interior of $P$ (as $\|p'\| \leq \varepsilon$), so that $(u',z',p')$ is also a local minimum of

$$\begin{aligned}(u,z,p) \mapsto &\|z-(y+v-tp)\| + \sqrt{\varepsilon}(\|u-x\| + t\|p\|)\\ &+t(\|u-u'\| + \|z-z'\| + \|p-p'\|) + i_{\text{Graph } F}(u,z) + i_L(p).\end{aligned}$$

Set

$$\begin{aligned}g(u,z,,p) = &\|z-(y+v-tp)\| + \sqrt{\varepsilon}(\|u-x\| + t\|p\|)\\ &+t(\|u-u'\| + \|z-z'\| + \|p-p'\|) + i_L(p).\end{aligned}$$

By the standard local fuzzy minimization rule (see, e.g., [2,9]) in any neighborhood of $(u',z',p')$ and $(u',p')$ we can find $(u_0,z_0,p_0) \in X \times Y \times L$, $(u_1,z_1) \in \text{Graph } F$, $(u^*,z^*,p^*) \in \partial g(u_0,z_0,p_0)$ and $(x^*,y^*) \in X^* \times Y^*$ such that $x^* \in D^*F(u_1,z_1)(y^*)$ and

$$\|u^* + x^*\| < \varepsilon, \quad \|z^* - y^*\| < \varepsilon, \quad \|p^*\| < t\varepsilon. \tag{5}$$

We have

- $u^* = \sqrt{\varepsilon}u_1^* + tu_2^*$, where $u_1^* \in \partial\|\cdot\|(u_0 - x)$, $u_2^* \in \partial\|\cdot\|(u_0 - u')$, so that $\|u_i^*\| \leq 1$;
- $z^* = z_1^* + tz_2^*$, where $z_1^* \in \partial\|\cdot\|(z_0 - (y + v - tp_0)$ so that $\|z_1^*\| = 1$ (as we can choose the neighborhood of $(u', z', p')$ small enough to guarantee that $\|z_0 - (y + v - tp_0)\| > 0$) and $z_2^* \in \partial\|\cdot\|(z_0 - z')$, so that $\|z_2^*\| \leq 1$;
- $p^* = -tz_1^* + t\sqrt{\varepsilon}p_1^* + tp_2^* + p_3^*$, where $p_1^* \in \partial\|\cdot\|(p_0)$, $p_2^* \in \partial\|\cdot\|(p_0 - p')$, so that $\|p_i^*\| \leq 1$, $i = 1, 2$ and $p_3^* \in L^\perp$.

Thus, by (5), $\|x^*\| \leq 3\varepsilon + \sqrt{\varepsilon}$; $|\|y^*\| - 1| \leq 3\varepsilon$ and (as $t \leq \varepsilon$)

$$y^* \in L^\perp + (3\varepsilon + \sqrt{\varepsilon})B.$$

This can be done for every finite-dimensional $L$ and any $\varepsilon > 0$. Thus for any pair $\alpha = (L, \varepsilon)$ we can find $(x_\alpha, y_\alpha) \in \text{Graph } F$ within $\varepsilon$ of $(\bar{x}, \bar{y})$ and $(x_\alpha^*, y_\alpha^*)$ such that $x_\alpha^* \in D^*F(x_\alpha, y_\alpha)(y_\alpha^*)$, $\|x_\alpha^*\| \leq 3\varepsilon + \sqrt{\varepsilon}$, $y_\alpha^* \in L^\perp + (3\varepsilon + \sqrt{\varepsilon})B$, and $\|y_\alpha^* - 1\| \leq 3\varepsilon$. But the sets $L^\perp + \varepsilon B$ form a basis of neighborhoods of zero in the weak$^*$-topology in $Y^*$ and therefore $y_\alpha^*$ weak$^*$ converge to zero. On the other hand, $\|x_\alpha^*\| \longrightarrow 0$, and coderivative compactness of $F$ implies that $\|y_\alpha^*\| \longrightarrow 0$. Thus, the assumption that $F$ is not compactly regular leads to a contradiction.

**Proposition 3.** *Given $F : X \rightrightarrows Y$, $(\bar{x}, \bar{y}) \in$ Graph $F$ and a compact set $P \subset Y$. Suppose there are $\varepsilon > 0$, $r > 0$ such that*

$$\big(F(x) \cap B(\bar{y}, \varepsilon)\big) + trB_Y \subset \text{cl}F(B(x, t)) + tP, \tag{6}$$

*for any $x \in B(\bar{x}, \varepsilon)$ and $t \in [0, \varepsilon)$. Then the inequality*

$$\|x^*\| + s_P(y^*) \geq r\|y^*\| \tag{7}$$

*holds whenever $(x^*, -y^*) \in \partial_F d(\cdot, \text{Graph } F)(x, y)$ and $\|(x, y) - (\bar{x}, \bar{y})\| < \varepsilon$.*[4]

Here $s_P(y^*) = \sup\{\langle y^*, p\rangle : p \in P\}$ is the *support function* of $P$. We also note that (6) is a weaker property than compact regularity.

*Proof.* Let $\|(u, v) - (\bar{x}, \bar{y})\| < \varepsilon$, and let $(u^*, -v^*) \in \partial_F d(\cdot, \text{Graph } F)(u, v)$. Choose for any $t > 0$ a $(u_t, v_t) \in \text{Graph } F$ such that

$$\|(u_t, v_t) - (u, v)\| = d((u, v), \text{Graph } F) + o(t),$$

[4]It is appropriate to quote here a comment by the other reviewer who states that (6) is equivalent to compact regularity and the closure operation can be harmlessly omitted. This is certainly true in the "full" regularity case when $P = \{0\}$—we have explicitly used this fact in the proof of Proposition 2 So I believe that the reviewer's statement is correct and wish to express my thanks for it.

take a $z_t \in rB_Y$ such that $-\langle v^*, z_t\rangle = r\|v^*\| + o(t)$ and find, using (7), a $p_t \in P$ and a $w_t \in B_X$ such that

$$v_t + tz_t \in F(u_t + tw_t) + tp_t + o(t)B_Y.$$

Then

$$\begin{aligned} d((u+tw_t, v+t(z_t-p_t)), \text{Graph } F) & \\ &\leq \|(u,v)-(u_t,v_t)\| + d((u_t+tw_t, v_t+t(z_t-p_t)), \text{Graph } F) \\ &= d((u,v), \text{Graph } F) + o(t). \end{aligned}$$

It follows that

$$-\|u^*\| + r\|v^*\| \leq \lim_{t \longrightarrow 0} (\langle u^*, w_t\rangle - \langle v^*, z_t\rangle) \leq \lim_{t \longrightarrow 0} \langle v^*, p_t\rangle \leq s_P(v^*)$$

as claimed.

Combining the two propositions and taking into account the well-known fact that the Fréchet normal cone to a set at a certain point is generated by the Fréchet subdifferential of the distance function to the set at the same point (which implies that the property in the conclusion of Proposition 3 is equivalent to $\partial_F$-coderivative compactness), we get

**Corollary 1.** *If $X$ and $Y$ are Asplund spaces, then under the assumptions of Proposition 3, $F$ is compactly regular at $(\bar{x}, \bar{y})$ if and only if it is $\partial_F$-coderivatively compact at $(\bar{x}, \bar{y})$.*

This is a new and intrinsic characterization of coderivative compactness for the Fréchet subdifferential (cf. [6]). A similar characterization for the Dini–Hadamard subdifferential can be obtained under an additional "steadiness" assumption on the set-valued mapping.

**Definition 2.** We shall say that a set $Q \subset X$ is *steady* at $\bar{x} \in S$ if

$$\limsup_{x \longrightarrow \bar{x}} \lim_{t \longrightarrow +0} t^{-1} \sup\{d(u, x+T(Q,x)) : u \in Q,\ d(u,x) \leq t\} = 0.$$

Here $T(Q,x) = \{h \in X : \exists\, h_n \longrightarrow h,\ t_n \longrightarrow 0,\ x + t_n h_n \in Q\}$ is the contingent (Bouligand) tangent cone to $Q$ at $x$. A set-valued mapping $F : X \rightrightarrows Y$ is *steady* at $(\bar{x}, \bar{y}) \in \text{Graph } F$ if so is its graph.

The meaning of the definition is simple: a set is steady at a point if the contingent cone mapping provides a reasonably good upper approximation for the set.

**Proposition 4.** *Suppose that both $X$ and $Y$ have Gâteaux smooth renorms. If $F : X \rightrightarrows Y$ is a set-valued mapping with closed graph which is steady at $(\bar{x}, \bar{y}) \in$*

Graph $F$, *then* $F$ *is compactly regular at* $(\bar{x},\bar{y})$ *if and only if it is* $\partial_H$*-coderivatively compact at* $(\bar{x},\bar{y})$.

*Proof.* If $F$ is $\partial_H$-coderivatively compact, then it is compactly regular by Proposition 2 even without the assumption that $F$ is steady. The proof of the opposite implication is a slight modification of the proof of Proposition 3 and we leave it for the reader.

Observe finally that a simple modification of the above proof allows to include the property of being steady into the list of properties inherited by arbitrarily large separable subspaces. The observation leads to the following separable reduction theorem for the point regularity criterion (cf. [6, 10, 11, 14]) for steady mappings.

**Theorem 4.** *Let* $X$ *and* $Y$ *be Banach spaces, and let* $F : X \rightrightarrows Y$ *be a set-valued mapping with closed graph which is steady and compactly regular at* $(\bar{x},\bar{y}) \in$ Graph $F$. *Then* $F$ *is regular near* $(\bar{x},\bar{y})$ *if for any* $M \in \mathcal{S}(Y)$ *and any* $L_0 \in \mathcal{S}(X)$ *there is a* $L \in \mathcal{S}(X)$ *containing* $L_0$ *such that* $D^*_G F_{L\times M}(\bar{x},\bar{y})$ *is non-singular in the sense that*

$$\inf\{\|x^*\| : x^* \in D^*_G F_{L\times M}(\bar{x},\bar{y}),\ \|y^*\| = 1\} > 0$$

The only consideration that should be taken in addition (to Theorem 3 and Proposition 2) into account while proving the theorem is that the $G$-subdifferential of a Lipschitz function on a separable space coincides with the limiting Dini–Hadamard subdifferential.

*Remark 1.* More thorough arguments (to be published elsewhere) allow to reverse the above theorem and to prove that for a steady and compactly regular set-valued mapping non-singularity of $D^*_G F_E(\bar{x},\bar{y})$ on big separable subspaces *is also necessary for regularity of* $F$ *at* $(\bar{x},\bar{y})$.

**Acknowledgements** I wish to thank the reviewers for valuable comments and many helpful remarks.

## References


[1] Azé, D.: A unified theory for metric regularity of multifunctions. J. Convex Anal. **13**, 225–252 (2006)

[2] Borwein, J.M., Zhu, J.: Techniques of Variational Analysis. CMS Books in Mathematics, vol. 20. Springer, New York (2006)

[3] Fabian, M., Ioffe, A.D.: Separable reduction in the theory of Fréchet subdifferential Set-Valued and Variational Analysis, to appear

[4] Fabian, M., Zhivkov, N.V.: A characterization of Asplund spaces with the help of $\varepsilon$-supports of Ekeland and Lebourg. C.R. Acad. Bulg. Sci. **38**, 671–674 (1985)

[5] Ioffe, A.D.: On the local surjection property. Nonlinear Anal. TMA **11**, 565–592 (1987)

[6] Ioffe, A.D.: Codirectional compactness, metric regularity and subdifferential calculus. In: Thera, M. (ed.) Constructive, Experimental and Nonlinear Analysis. Canadian Mathematical Society Conference Proceedings, vol. 27, pp. 123–163

[7] Ioffe, A.D.: Metric regularity and subdifferential calculus. Uspehi Mat. Nauk **55**(3), 103–162 (2000) (in Russian). English translation: Russian Math. Surveys **55**(3), 501–558 (2000)
[8] Ioffe, A.D.: Regularity on fixed sets. SIAM J. Optimization **21**, 1345–1370 (2011)
[9] Ioffe, A.D.: On the general theory of subdifferentials. Adv. Nonlinear Anal. **1**, 47–120 (2012)
[10] Jourani, A., Thibault, L.: Verifiable conditions for openness and metric regularity in Banach spaces. Trans. Amer. Math. Soc. **347**, 1255–1268 (1995)
[11] Jourani, A., Thibault, L.: Coderivatives of multivalued mappings, locally compact cones and metric regularity. Nonlinear Anal. TMA **35**, 925–945 (1999)
[12] Kruger, A.Y.: A covering theorem for set-valued mappings. Optimization **19**, 763–780 (1988)
[13] Mordukhovich, B.S.: Variational Analysis and Generalized Differentiation, vol. 1–2. Springer, Berlin (2006)
[14] Penot, J.P.: Calculus Without Derivatives. Graduate Texts in Mathematics, vol. 266. Springer, New York (2012)

# Construction of Pairs of Reproducing Kernel Banach Spaces

**Pando G. Georgiev, Luis Sánchez-González, and Panos M. Pardalos**

**Abstract** We extend the idea of reproducing kernel Hilbert spaces (RKHS) to Banach spaces, developing a theory of pairs of reproducing kernel Banach spaces (RKBS) without the requirement of existence of semi-inner product (which requirement is already explored in another construction of RKBS). We present several natural examples, which involve RKBS of functions with supremum norm and with $\ell_p$-norm ($1 \leq p \leq \infty$). Special attention is devoted to the case of a pair of RKBS $(B, B^\sharp)$ in which $B$ has sup-norm and $B^\sharp$ has $\ell_1$-norm. Namely, we show that if $(B, B^\sharp)$ is generated by a universal kernel and $B$ is furnished with the sup-norm, then $B^\sharp$, furnished with the $\ell_1$-norm, is linearly isomorphically embedded in the dual of $B$. We reformulate the classical classification problem (support vector machine classifier) to RKBS and suggest that it will have sparse solutions when the RKBS is furnished with the $\ell_1$-norm.

**Keywords** Reproducing kernel Hilbert space • Banach space • Machine learning

---

P.G. Georgiev (✉)
Center for Applied Optimization, University of Florida, Gainesville, FL, USA
e-mail: pandogeorgiev@ufl.edu

L. Sánchez-González
Department of Mathematical Analysis, Universidad Complutense de Madrid, Madrid, Spain
e-mail: lfsanche@mat.ucm.es

P.M. Pardalos
Department of Industrial and Systems Engineering University of Florida, Gainesville, FL, USA

Laboratory of Algorithms and Technologies for Networks Analysis (LATNA),
Higher School of Economics, Moscow, Russia
e-mail: pardalos@ufl.edu

V.F. Demyanov et al. (eds.), *Constructive Nonsmooth Analysis and Related Topics*,
Springer Optimization and Its Applications 87, DOI 10.1007/978-1-4614-8615-2_4,

# 1 Introduction

The wide applicability of the reproducing kernel Hilbert spaces (RKHS) in applied mathematics, and especially in machine learning problems, comes from the fact that a nonlinear relation defined by a kernel (a positive definite function of two variables) between two vectors can be expressed as a linear one (scalar product) of their substitutes in higher (or infinite) dimensional space. The fundamental ideas allowing this are to embed the vectors into a functional space and to work with functions instead with vectors.

It appears that in applications, the linear operations may not be represented by a scalar product, and still the corresponding algorithms work well. It means that the requirement for learning in a Hilbert space is not obligatory. Instead, it can be a Banach space. Several papers have been already devoted to learning tasks in Banach spaces: [2, 8, 9, 11], for regularization network problems; [2, 5–7, 11], for classifications in Banach (or even metric) spaces; etc. In [12] the authors develop a theory of reproducing kernel Banach spaces (RKBS) based on existence of a semi-inner product and study their basic properties and machine learning tasks.

In the present paper we extend the idea of RKHS to Banach spaces and develop a theory of pairs of RKBS without the requirement of existence of semi-inner product. Several examples illustrate the fruitfulness of the construction. Basic properties of the RKBS are proved and main learning problems are presented to this framework.

The literature on the RKHS in machine learning is so huge that it is impossible to cover even a small number of it. A systematical theory of RKHS was developed by Aronszajn [1]. For applications to statistical learning theory we refer to [4, 18] and references therein.

Specific applications, especially to time series, are given in [3, 10, 13].

We will consider Banach spaces over either the field of real numbers, $\mathbb{R}$, or of complex numbers, $\mathbb{C}$. We will use $\mathbb{K}$ to denote either $\mathbb{R}$ or $\mathbb{C}$.

**Definition 1.** Let $X$ and $Y$ be two sets, let $K : X \times Y \longrightarrow \mathbb{K}$ be a function, and let $V$ and $V^\sharp$ be two Banach spaces composed by functions defined on $Y$ and $X$, respectively. We shall say that the pair $(V, V^\sharp)$ is a pair of **RKBS with the reproducing kernel** $K$ provided that

(1) For all $f \in V$, $||f||_V = 0$ if and only if $f(y) = 0$ for all $y \in Y$.
(2) For all $g \in V^\sharp$, $||g||_{V^\sharp} = 0$ if and only if $g(x) = 0$ for all $x \in X$.
(3) The point evaluation functionals are continuous on $V$ and $V^\sharp$, i.e., for every $x \in X$ and $y \in Y$, the functionals $\delta_y : V \longrightarrow \mathbb{K}$ and $\delta_x : V^\sharp \longrightarrow \mathbb{K}$ defined as $\delta_y(f) = f(y)$ and $\delta_x(g) = g(x)$ for all $f \in V$ and $g \in V^\sharp$ are continuous.
(4) $K(x, \cdot) \in V$ and $K(\cdot, y) \in V^\sharp$ for every $x \in X$ and $y \in Y$.
(5) There exists a bilinear form $< \cdot, \cdot >_K$ in $V \times V^\sharp$ such that

$$< f, K(\cdot, y) >_K = f(y) \qquad \text{for all } y \in Y \text{ and } f \in V \tag{1}$$

and

$$< K(x,\cdot), g >_K = g(x) \qquad \text{for all } x \in X \text{ and } g \in V^\sharp. \tag{2}$$

Let us note that the first difference with RKHS is that we are not making assumptions about the norm, so the RKBS could be nonunique.

Let $(V, V^\sharp)$ be a pair of RKBS of real-valued functions on the sets $X$ and $Y$ with reproducing kernel $K$. Let $W = \{f_1 + if_2 : f_1, f_2 \in V\}$ and $W^\sharp = \{f_1 + if_2 : f_1, f_2 \in V^\sharp\}$, which are vector spaces of complex-valued functions on $X$ and $Y$. If we set the bilinear form

$$< f_1 + if_2, g_1 + ig_2 >_W = (< f_1, g_1 >_K - < f_2, g_2 >_K) + i(< f_1, g_2 >_K + < f_2, g_1 >_K),$$

and the norms

$$||f_1 + if_2||_W = ||f_1||_V + ||f_2||_V \qquad \text{and} \qquad ||g_1 + ig_2||_{W^\sharp} = ||g_1||_{V^\sharp} + ||g_2||_{V^\sharp},$$

we have that the pair $(W, W^\sharp)$ is a pair of RKBS of complex-valued functions on $X$ and $Y$ with reproducing kernel $K$ (which is real-valued). We call $W$ and $W^\sharp$ the complexifications of $V$ and $V^\sharp$. Since every real-valued RKBS can be complexified preserving the reproducing kernel, we shall consider only complex-valued RKBS.

In Sect. 2, we construct a wide class of RKBS under two general assumptions. In this process, we follow and improve the techniques developed in [14], removing some assumptions. In particular, we obtain the following theorem:

**Theorem 1.** *Let $X$ and $Y$ be two sets and $K : X \times Y \longrightarrow \mathbb{C}$ a function. Let us denote $B_0 = \mathrm{span}\{K(x,\cdot) : x \in X\}$ and $B_0^\sharp = \mathrm{span}\{K(\cdot,y) : y \in Y\}$. Let us suppose that there is a norm $||\cdot||_{B_0}$ in $B_0$ satisfying:*

*(HN1) The evaluation functionals are continuous of $B_0$.*
*(HN2) If $\{f_n\}_{n=1}^\infty$ is a Cauchy sequence in $(B_0, ||\cdot||_{B_0})$ such that $f_n(y) \longrightarrow 0$ for all $y \in Y$, then $||f_n||_{B_0} \longrightarrow 0$.*

*Then, there are $(B, ||\cdot||_B)$ and $(B^\sharp, ||\cdot||_{B^\sharp})$—Banach completions of $B_0$ and $B_0^\sharp$, respectively, such that $(B, B^\sharp)$ is a pair of RKBS with the reproducing kernel $K$. Furthermore, if we denote by $< \cdot,\cdot >_K$ the bilinear form in $B \times B^\sharp$, we state*

$$| < f, g >_K | \le ||f||_B ||g||_{B^\sharp} \qquad \text{for all } f \in B \text{ and } g \in B^\sharp, \text{and}$$

$$||g||_{B^\sharp} = \sup_{\substack{f \in B \\ ||f||_B \le 1}} |< f, g >_K| \qquad \text{for any } f \in B \text{ and } g \in B^\sharp.$$

In Sect. 3, we give some examples of RKBS. In particular, we obtain RKBS when the kernel is either a bounded function, continuous and bounded function, or is in $C_0(X)$. Moreover, we obtain a simple condition in the kernel in order to obtain RKBS with $\ell_1$-norm. Lastly, in this section we obtain a class of RKBS with $\ell_p$-norm following the construction given in [14] for RKBS with the $\ell_1$-norm.

In Sect. 4, we sketch the main tasks in statistical learning theory and reformulate support vector machine classification problem for the case of RKBS with $\ell_1$-norm.

## 2 Constructing Reproducing Kernel Banach Spaces

Let $X$ and $Y$ be sets and $K : X \times Y \longrightarrow \mathbb{C}$ a function. Following the ideas given in [14] and the proof of the uniqueness of RKHS in [1], we construct a pair of Banach spaces which is an RKBS with the reproducing kernel $K$.

Let us introduce the vector spaces

$$B_0 := \operatorname{span}\{K(x,\cdot) : x \in X\} \subset \mathbb{C}^Y \qquad \text{and} \qquad B_0^\sharp := \operatorname{span}\{K(\cdot,y) : y \in Y\} \subset \mathbb{C}^X.$$

Let us define the bilinear form $<\cdot,\cdot>_K$ on $B_0 \times B_0^\sharp$ as

$$< f,g >_K = < \sum_{j=1}^{n} \alpha_j K(x_j,\cdot), \sum_{k=1}^{m} \beta_k K(\cdot,y_k) >_K := \sum_{j,k} \alpha_j \beta_k K(x_j,y_k), \tag{3}$$

for every $f = \sum_{j=1}^{n} \alpha_j K(x_j,\cdot) \in B_0$ and $g = \sum_{k=1}^{m} \beta_k K(\cdot,y_k) \in B_0^\sharp$.

**Proposition 1.** *It is straightforward to verify that*

*(i)* $<\cdot,\cdot>_K$ *is a bilinear form on* $B_0 \times B_0^\sharp$.
*(ii)* $K(x,y) = < K(x,\cdot), K(\cdot,y) >_K$ *for all* $x \in X$ *and* $y \in Y$.
*(iii)* $< f, K(\cdot,y) >_K = f(y)$ *and* $< K(x,\cdot), g >_K = g(x)$ *for all* $f \in B_0$, $g \in B_0^\sharp$, $x \in X$, *and* $y \in Y$.

Let us suppose (HN1), i.e., there is a norm $||\cdot||_{B_0}$ on $B_0$ such that the evaluation functionals are continuous on $B_0$. We follow [1,14] to obtain a Banach completion of $B_0$. Let $\{f_n\}_{n=1}^{\infty}$ be a Cauchy sequence in $B_0$; since the point evaluation functionals are continuous, the sequence $\{f_n(y)\}_{n=1}^{\infty}$ is Cauchy for every $y \in Y$. So, we can define $f(y) := \lim_{n \longrightarrow \infty} f_n(y)$, for all $y \in Y$, the space

$$B := \Big\{ f : Y \longrightarrow \mathbb{C} : \text{there exists } \{f_n\}_{n=1}^{\infty} \text{ a Cauchy sequence in } B_0 \text{ such that} \\ f_n(y) \longrightarrow f(y) \text{ for all } y \in Y \Big\},$$

and the function $||\cdot||_B$ on $B$ as

$$||f||_B := \lim_{n \longrightarrow \infty} ||f_n||_{B_0},$$

for all $f \in B$, where $\{f_n\}_{n=1}^{\infty}$ is a Cauchy sequence in $B_0$ such that $f_n(y) \longrightarrow f(y)$ on $Y$.

**Proposition 2.** *If $(B_0, ||\cdot||)$ satisfies the hypothesis (HN2), then $||\cdot||_B$ is well defined, and the pair $(B, ||\cdot||_B)$ is a Banach space.*

*Proof.* The function $||\cdot||_B$ is well defined by (HN2). It is easy to see that $||\cdot||_B$ is a norm in $B$; only note that $f_n(y) \longrightarrow 0$ for all $y \in Y$ whenever $||f_n||_{B_0} \longrightarrow 0$ since the point evaluation functionals are continuous (hypothesis (HN1)). Thus, $||f||_B = 0$ if and only if $f = 0$.

Finally, we shall show that $(B, ||\cdot||_B)$ is a Banach space. Given $\{f_n\}_{n=1}^{\infty}$ a Cauchy sequence in $B$, if there exists $n_0 \in \mathbb{N}$ such that $f_n \in B_0$ for all $n \geq n_0$, then, by definition of $B$, there exists $f \in B$ such that $f_n \longrightarrow f$ in $B$ (i.e., $||f_n - f||_B \longrightarrow 0$). Let us suppose that there are infinitely many functions $f_n \notin B_0$; we can choose $g_n \in B_0$ such that $||g_n - f_n||_B < \frac{1}{n}$ for all $n \in \mathbb{N}$. Then, $\{g_n\}_{n=1}^{\infty}$ is a Cauchy sequence in $B_0$. Therefore, there exists $g \in B$ such that $g_n \longrightarrow g$ in $B$ and, consequently, $f_n \longrightarrow g$.

Let us notice that (HN2) is also a necessary hypothesis. Indeed, if $\{f_n\}_{n=1}^{\infty}$ is a Cauchy sequence in $B_0$ such that $f_n(y) \longrightarrow 0$ for all $y \in Y$, and $(B, ||\cdot||_B)$ is well defined, then

$$0 = ||0||_B = \lim_{n \longrightarrow \infty} ||f_n||_{B_0}.$$

□

**Proposition 3.**

(i) *Every function $f \in B$ such that $||f||_B = 0$ satisfies $f(y) = 0$ for all $y \in Y$.*
(ii) *The point evaluation functionals are continuous on $B$.*

*Proof.* Since (i) is straightforward, we only show the property (ii). Given $f \in B$, there is a Cauchy sequence $\{f_n\}_{n=1}^{\infty}$ in $B_0$ such that $f_n(y) \longrightarrow f(y)$ for all $y \in Y$ and $||f||_B = \lim_{n \longrightarrow \infty} ||f_n||_{B_0}$. Then

$$|\delta_y(f)| = |f(y)| = |\lim_{n \longrightarrow \infty} f_n(y)| = |\lim_{n \longrightarrow \infty} \delta_y(f_n)| \leq \lim_{n \longrightarrow \infty} |\delta_y(f_n)| \leq$$

$$||\delta_y|| \lim_{n \longrightarrow \infty} ||f_n||_{B_0} = ||\delta_y|| ||f||_B.$$

□

Now, let us define the following function on $B_0^{\sharp}$:

$$||g||_{B_0^{\sharp}} := \sup_{\substack{f \in B_0 \\ ||f||_{B_0} \leq 1}} |< f, g >_K| \qquad \text{for all } g \in B_0^{\sharp}.$$

**Proposition 4.** $||\cdot||_{B_0^{\sharp}}$ *is a norm in $B_0^{\sharp}$.*

*Proof.* Let us suppose that there is $g \in B_0^{\sharp}$ with $||g||_{B_0^{\sharp}} = 0$. Then, for every $x \in X$, we have

(i) If $||K(x,\cdot)||_{B_0} \leq 1$, then $|g(x)| = |< K(x,\cdot), g >_K| \leq ||g||_{B_0^{\sharp}} = 0$.

(ii) If $||K(x,\cdot)||_{B_0} > 1$, then $f := \frac{1}{||K(x,\cdot)||_{B_0}} K(x,\cdot) \in B_0$ and $||f||_{B_0} = 1$, so $\frac{1}{||K(x,\cdot)||_{B_0}} |g(x)| = |< f, g >_K| \leq ||g||_{B_0^\sharp} = 0$, and $g(x) = 0$.

It is straightforward to prove the other properties of the norm. □

**Proposition 5.** *For every $f \in B_0$ and $g \in B_0^\sharp$*

$$|< f, g >_K| \leq ||f||_{B_0} ||g||_{B_0^\sharp}. \tag{4}$$

*Proof.* If $||f||_{B_0} = 0$ then $f = 0$ and the inequality is obvious. If $||f||_{B_0} \neq 0$, we define $h = \frac{1}{||f||_{B_0}} f \in B_0$ and $||h||_{B_0} = 1$, then

$$|< f, g >_K| = ||f||_{B_0} |< h, g >_K| \leq ||f||_{B_0} ||g||_{B_0^\sharp}.$$ □

**Corollary 1.** *The point evaluation functionals are continuous on $B_0^\sharp$ with the norm $||\cdot||_{B_0^\sharp}$.*

*Proof.*

$$|\delta_x(g)| = |g(x)| = |< K(x,\cdot), g >_K| \leq ||K(x,\cdot)||_{B_0} ||g||_{B_0^\sharp}.$$ □

In the same way as before, we can define a Banach completion of $B_0^\sharp$ which yields a space of functions, given by

$$B^\sharp := \Big\{ g : X \longrightarrow \mathbb{C} : \text{there exists a Cauchy sequence } \{g_n\}_{n=1}^\infty \text{ in } B_0^\sharp \text{ such that } g_n(x) \longrightarrow g(x) \text{ for all } x \in X \Big\}.$$

**Proposition 6.** *Given $g \in B^\sharp$ and $\{g_n\}_{n=1}^\infty$—any Cauchy sequence in $B_0^\sharp$ such that $g_n(x) \longrightarrow g(x)$ on $X$—the function*

$$||g||_{B^\sharp} := \lim_{n \longrightarrow \infty} ||g_n||_{B_0^\sharp}$$

*is well defined and $(B^\sharp, ||\cdot||_{B^\sharp})$ is a Banach space.*

*Proof.* Let us show that $||\cdot||_{B^\sharp}$ is well defined. First of all, notice that it is sufficient to prove that $||g_n||_{B_0^\sharp} \longrightarrow 0$ whenever $\{g_n\}$ is a Cauchy sequence in $B_0^\sharp$ such that $g_n(x) \longrightarrow 0$ for all $x \in X$. Fixed $\varepsilon > 0$ there is $N_0 \geq 1$ such that

$$||g_m - g_n||_{B_0^\sharp} = \sup_{\substack{f \in B_0 \\ ||f||_{B_0} \leq 1}} |< f, g_m - g_n >_K| < \varepsilon/2 \qquad \text{for all } n, m \geq N_0.$$

Since $g_n(x) \longrightarrow 0$ for all $x \in X$, given $f = \sum_{j=1}^{p} \alpha_j K(x_j, \cdot) \in B_0$ with $||f||_{B_0} \leq 1$, there is $m_0$ $(= m_0(\varepsilon, f))$ such that

$$|< f, g_m >_K| = \left| \sum_{j=1}^{p} \alpha_j g_m(x_j) \right| < \varepsilon/2 \qquad \text{for all } m \geq m_0.$$

Then, for $n \geq N_0$, let us take $m \geq \max\{N_0, m_0\}$; thus,

$$|< f, g_n >_K| \leq |< f, g_n - g_m >_K| + |< f, g_m >_K| < \varepsilon.$$

Hence,

$$||g_n||_{B_0^\sharp} = \sup_{\substack{f \in B_0 \\ ||f||_{B_0} \leq 1}} |< f, g_n >_K| < \varepsilon \qquad \text{for all } n \geq N_0, \text{ and} \qquad ||g_n||_{B_0^\sharp} \longrightarrow 0.$$

□

**Proposition 7.**

*(i) Every function $g \in B^\sharp$ such that $||g||_{B^\sharp} = 0$ satisfies $g(x) = 0$ for all $x \in X$.*
*(ii) The point evaluation functionals are continuous on $B^\sharp$.*

*Proof.* The proofs are the same as Proposition 3; notice that $|\delta_x(g)| \leq ||K(x, \cdot)||_{B_0} ||g||_{B^\sharp}$ for every $x \in X$ and $g \in B^\sharp$. □

Let us extend the bilinear form $< \cdot, \cdot >_K$ from $B_0 \times B_0^\sharp$ to $B \times B^\sharp$. Let us note that there is no Hahn–Banach theorem for bilinear mappings. So, we have to use the "density" of $B_0$ and $B_0^\sharp$ in $B$ and $B^\sharp$, respectively.

- Let us define $< \cdot, \cdot >_K : B \times B_0^\sharp \longrightarrow \mathbb{C}$ by

$$< f, g >_K := \lim_{n \longrightarrow \infty} < f_n, g >_K,$$

where $f \in B$, $g \in B_0^\sharp$, and $\{f_n\}_{n=1}^{\infty}$ is a Cauchy sequence in $B_0$ such that $f_n(y) \longrightarrow f(y)$ for all $y \in Y$. Using inequality (4), $\{< f_n, g >_K\}_{n=1}^{\infty}$ is a Cauchy sequence in $\mathbb{C}$; the limit exits and is well defined. Clearly, the form $< \cdot, \cdot >_K$ is bilinear in $B \times B_0^\sharp$. Finally,

$$|< f, g >_K| = \left| \lim_{n \longrightarrow \infty} < f_n, g >_K \right| \leq \lim_{n \longrightarrow \infty} ||f_n||_{B_0} ||g||_{B_0^\sharp} = ||f||_B ||g||_{B_0^\sharp} \quad (5)$$

for all $f \in B$ and $g \in B_0^\sharp$, where $\{f_n\}_{n=1}^{\infty}$ is a Cauchy sequence in $B_0$ such that $f_n(y) \longrightarrow f(y)$ for all $y \in Y$.

- Let us define $< \cdot,\cdot >_K : B \times B^\sharp \longrightarrow \mathbb{C}$ by

$$< f,g >_K := \lim_{n \longrightarrow \infty} < f,g_n >_K,$$

where $f \in B$, $g \in B^\sharp$, and $\{g_n\}_{n=1}^\infty$ is a Cauchy sequence in $B_0^\sharp$ such that $g_n(x) \longrightarrow g(x)$ for all $x \in X$. Using inequality (5) instead of inequality (4), we prove in a similar way that $< \cdot,\cdot >_K$ is a bilinear form on $B \times B^\sharp \longrightarrow \mathbb{C}$ and

$$|< f,g >_K| \leq ||f||_B ||g||_{B^\sharp} \tag{6}$$

for all $f \in B$ and $g \in B^\sharp$.

**Proposition 8.** *For all* $g \in B^\sharp$

$$||g||_{B^\sharp} = \sup_{\substack{f \in B \\ ||f||_B \leq 1}} |< f,g >_K|.$$

*Proof.* By inequality (6) it is clear that $\sup_{\substack{f \in B \\ ||f||_B \leq 1}} |< f,g >_K| \leq ||g||_{B^\sharp}$. Let us prove the reverse inequality. If $g \in B_0^\sharp$, then

$$||g||_{B^\sharp} = ||g||_{B_0^\sharp} = \sup_{\substack{f \in B_0 \\ ||f||_{B_0} \leq 1}} |< f,g >_K| \leq \sup_{\substack{f \in B \\ ||f||_B \leq 1}} |< f,g >_K|.$$

If $g \in B^\sharp$ and $\{g_n\}_{n=1}^\infty$ is a Cauchy sequence in $B_0^\sharp$ such that $g_n(x) \longrightarrow g(x)$ on $X$, then $||g_n - g||_{B^\sharp} \longrightarrow 0$, and

$$|< f,g_n >_K| \leq |< f,g_n - g >_K| + |< f,g >_K| \leq ||g_n - g||_{B^\sharp} + |< f,g >_K|,$$

for all $f \in B$ with $||f||_B \leq 1$. Thus,

$$||g||_{B^\sharp} = \lim_{n \longrightarrow \infty} ||g_n||_{B_0^\sharp} \leq \lim_{n \longrightarrow \infty} \sup_{\substack{f \in B \\ ||f||_B \leq 1}} |< f,g_n >_K|$$

$$\leq \lim_{n \longrightarrow \infty} \left( ||g_n - g||_{B^\sharp} + \sup_{\substack{f \in B \\ ||f||_B \leq 1}} |< f,g >_K| \right) = \sup_{\substack{f \in B \\ ||f||_B \leq 1}} |< f,g >_K|.$$

□

Let us denote by $< \cdot,\cdot >: B \times B^* \longrightarrow \mathbb{C}$ the evaluation map $< x,y^* >= y^*(x)$ for all $x \in B$ and $y^* \in B^*$. Let us notice that the mapping $\mathcal{L}$ from the Banach space $B^\sharp$ to the dual space $B^*$ of $B$ defined as

$$(\mathcal{L}(g))(f) =< f, \mathcal{L}(g) >:=< f, g >_K \qquad \text{for all } f \in B \text{ and } g \in B^\sharp, \tag{7}$$

is an embedding from $B^\sharp$ to $B^*$, i.e., it is an isometric and linear mapping.

So, we can define $\phi : X \longrightarrow B$ and $\phi^* : Y \longrightarrow B^*$ as

$$\phi(x) = K(x, \cdot) \in B \qquad \text{and} \qquad \phi^*(y) = \mathcal{L}(K(\cdot, y)) \in B^*,$$

and they satisfy

$$K(x,y) =< \phi(x), \phi^*(y) > .$$

Now, we have all ingredients to prove Theorem 1.

*Proof of Theorem 1.* It only remains to check the equalities (1) and (2). Let us prove equality (1) (equality (2) can be proved similarity). Given $f \in B$ and $y_0 \in Y$, there is a Cauchy sequence $\{f_n\}_{n=1}^\infty$ in $B_0$ such that $f_n(y) \longrightarrow f(y)$ for all $y \in Y$. Thus,

$$f(y_0) = \lim_{n \longrightarrow \infty} f_n(y_0) = \lim_{n \longrightarrow \infty} < f_n, K(\cdot, y_0) >_K .$$

By inequality (6), the map $h \in B \longrightarrow < h, K(\cdot, y_0) >_K$ is a bounded linear functional on $B$, then $f(y_0) = \lim_{n \longrightarrow \infty} < f_n, K(\cdot, y_0) >_K =< f, K(\cdot, y_0) >_K$.

## 3 Some Examples of RKBS

### *3.1 Pairs of RKBS Defined on Banach Spaces*

Let $X$ be a Banach space and $X^*$ be its dual space. For a given $p \in \mathbb{N}$, define a function $K : X \times X^* \longrightarrow \mathbb{C}$ such that

$$K(x, y^*) = (< x, y^* >_X + 1)^p,$$

where $< \cdot, \cdot >_X : X \times X^* \longrightarrow \mathbb{C}$ denotes the evaluation map $< x, y^* >_X = y^*(x)$ for all $x \in X$ and $y^* \in X^*$. We can define

$$B_0 := \text{span}\{K(x, \cdot) : x \in X\} \qquad \text{and} \qquad B_0^\sharp := \text{span}\{K(\cdot, y) : y \in X^*\},$$

and the norm in $B_0$ given by

$$||f||_{B_0} := \sup_{y \in X^*} \frac{|f(y)|}{1 + \sum_{j=1}^p ||y||^j} .$$

**Proposition 9.**

*(i) If* $f=\sum_{j=1}^{n}\alpha_j K(x_j,\cdot)\in B_0$, *then* $||f||_{B_0}\le\sum_{j=1}^{n}|\alpha_j|(1+||x_j||)^p<\infty$.
*(ii)* $||\cdot||_{B_0}$ *is a norm on* $B_0$ *and satisfies the hypotheses (HN1) and (HN2).*

*Proof.* It is easy to see that it is a norm; only notice that if $||f||_{B_0}=0$, then $f(y)=0$ for all $y\in X^*$, i.e., $f=0$.

Let us take $y\in X^*$, then

$$|\delta_y(f)|=|f(y)|=\left(1+\sum_{j=1}^{p}||y||^j\right)\frac{|f(y)|}{1+\sum_{j=1}^{p}||y||^j}\le\left(1+\sum_{j=1}^{p}||y||^j\right)||f||_{B_0},$$

for every $f\in B_0$. Then, the evaluation functionals are continuous and $||\cdot||$ satisfies the hypothesis (HN1). Let us prove the hypothesis (HN2). Let us take a Cauchy sequence $\{f_n\}$ in $B_0$ such that $f_n(y)\longrightarrow 0$ for every $y\in X^*$. Since $\{f_n\}$ is a Cauchy sequence in $B_0$, for every $\varepsilon>0$, there is $N_0\ge 1$ such that for every $n,m\ge N_0$,

$$||f_n-f_m||_{B_0}<\varepsilon/2.$$

Fixed $y\in X^*$, since $f_m(y)\longrightarrow 0$, then there is $m_0=m_0(\varepsilon,y)\in\mathbb{N}$ such that for every $m\ge m_0$

$$\frac{|f_m(y)|}{1+\sum_{j=1}^{p}||y||^j}<\varepsilon/2.$$

So, for every $n\ge N_0$, let us take $m\ge\max\{N_0,m_0\}$ and we have that

$$\frac{|f_n(y)|}{1+\sum_{j=1}^{p}||y||^j}\le\frac{|f_n(y)-f_m(y)|}{1+\sum_{j=1}^{p}||y||^j}+\frac{|f_m(y)|}{1+\sum_{j=1}^{p}||y||^j}<||f_n-f_m||_{B_0}+\varepsilon/2<\varepsilon.$$

Then, $||f_n||_{B_0}\longrightarrow 0$. □

Then, according to Theorem 1, we conclude that there are $(B,||\cdot||_B)$ and $(B^\sharp,||\cdot||_{B^\sharp})$ Banach completions of $B_0$ and $B_0^\sharp$, respectively, such the pair $(B,B^\sharp)$ is a pair of RKBS on $X^*$ and $X$ with the reproducing kernel $K$. Furthermore, if we take the map $\mathcal{L}:B^\sharp\longrightarrow B^*$ defined as (7) and denote by $<\cdot,\cdot>:B\times B^*\longrightarrow\mathbb{C}$ the evaluation map $<x,y^*>=y^*(x)$ for all $x\in B$ and $y^*\in B^*$, we can define $\phi:X\longrightarrow B$ and $\phi^*:Y\longrightarrow B^*$ as

$$\phi(x)=K(x,\cdot)=(<x,\cdot>_X+1)^p\in B$$

and

$$\phi^*(y)=\mathcal{L}(K(\cdot,y))=\mathcal{L}((<\cdot,y^*>_X+1)^p)\in B^*,$$

satisfying

$$K(x,y) = < \phi(x), \phi^*(y) > .$$

## 3.2 Bounded, Continuous, and $C_0(X)$ Kernels

Let $X$ be a set and $F(X)$ be one of the following space of functions:

- $\mathcal{B}(X)$, the space of bounded functions on $X$
- $C(X) \cap \mathcal{B}(X)$, the space of continuous and bounded functions on $X$
- $C_0(X)$, the space of continuous functions $f : X \longrightarrow \mathbb{C}$ such that for all $\varepsilon > 0$, the set $\{x \in X : |f(x)| \geq \varepsilon\}$ is compact

Let $K : X \times X \longrightarrow \mathbb{C}$ be a function such that for every $x \in X$ the functions $K(x,\cdot) \in F(X)$.

Consider the vector spaces

$$B_0 := \text{span}\{K(x,\cdot) : x \in X\} \subset F(X) \qquad \text{and} \qquad B_0^\sharp := \text{span}\{K(\cdot,x) : x \in X\},$$

and define the function $||\cdot||_{B_0} : B_0 \longrightarrow \mathbb{C}$ by

$$||f||_{B_0} := \sup\{|f(x)| : x \in X\} \qquad \text{for every } f \in B_0.$$

**Proposition 10.** *$||\cdot||_{B_0}$ is a norm on $B_0$ and satisfies the hypotheses (HN1) and (HN2).*

*Proof.* It is straightforward to verify that $||\cdot||_{B_0}$ is a norm on $B_0$.

Given $x \in X$, let us consider $\delta_x : B_0 \longrightarrow \mathbb{C}$ the point evaluation functional, i.e., $\delta_x(f) = f(x)$ for $f \in B_0$. Then

$$|\delta_x(f)| = |f(x)| \leq \sup\{|f(y)| : y \in X\} = ||f||_{B_0}.$$

Let us show that it satisfies (HN2). We have to prove that $||f_n||_{B_0} \longrightarrow 0$ whenever $\{f_n\}_{n=1}^{\infty}$ is a Cauchy sequence in $B_0$ such that $f_n(x) \longrightarrow 0$ on $X$. Let $\varepsilon > 0$ fixed. On the one hand, there is $N_0 \geq 1$ such that

$$\sup\{|f_n(x) - f_m(x)| : x \in X\} < \frac{\varepsilon}{2} \qquad \text{for every } n,m \geq N_0.$$

On the other hand, for a fixed $x \in X$, there exists $m_0 \in \mathbb{N}$ such that $|f_m(x)| < \varepsilon/2$ for all $m \geq m_0$, since $f_m(x) \longrightarrow 0$. Thus, let us take $m \geq \max\{N_0, m_0\}$ and

$$|f_n(x)| \leq |f_n(x) - f_m(x)| + |f_m(x)| < \varepsilon$$

for all $n \geq N_0$. Then $||f_n||_{B_0} \longrightarrow 0$. □

Then, by Theorem 1, we conclude that there are $(B,||\cdot||_B)$ and $(B^\sharp,||\cdot||_{B^\sharp})$ Banach completions of $B_0$ and $B_0^\sharp$, respectively, such that the pair $(B,B^\sharp)$ is a pair of RKBS on $X$ with the reproducing kernel $K$. In fact, the completion of $B_0$ is given by

$$B := \Big\{ f : X \longrightarrow \mathbb{C} : \textit{there exists } \{f_n\}_{n=1}^{\infty} \textit{ a Cauchy sequence in } B_0 \text{ such that}$$
$$f_n(x) \longrightarrow f(x) \text{ for all } x \in X \Big\},$$

and the norm $||\cdot||_B$ on $B$ by

$$||f||_B := \lim_{n \longrightarrow \infty} ||f_n||_{B_0},$$

for all $f \in B$, where $\{f_n\}_{n=1}^{\infty}$ is a Cauchy sequence in $B_0$ such that $f_n(x) \longrightarrow f(x)$ for all $x \in X$.

**Proposition 11.** $B \subset F(X)$.

*Proof.* If $f \in B$, then there exists a Cauchy sequence $\{f_n\}_{n=1}^{\infty}$ on $B_0$ such that $f_n(x) \longrightarrow f(x)$ for all $x \in X$. Recall that $||\cdot||_{B_0} = ||\cdot||_\infty$:

- $B \subset \mathcal{B}(X)$. There is $n_0 \in \mathbb{N}$ such that $||f_n - f_m||_{B_0} < 1$ for all $n,m \geq n_0$. Then,

$$|f(x)| \leq \lim_n |f_n(x)| \leq \lim_n ||f_n||_{B_0} \leq ||f_{n_0}||_{B_0} + 1, \qquad \text{for all } x \in X.$$

- If $\{f_n\} \subset C(X)$ then $f \in C(X)$, since if a sequence of continuous functions uniform converges to a function $f$, then $f$ is continuous.
- If $\{f_n\} \subset C_0(X)$, then $f \in C_0(X)$. For any $\varepsilon > 0$, there is $n_0 \in \mathbb{N}$ such that $||f_n - f_m||_{B_0} < \varepsilon/4$ for all $n,m \geq n_0$. Since $K = \{x \in X : |f_{n_0}(x)| \geq \varepsilon/2\}$ is compact, the closed set $\{x \in X : |f(x)| \geq \varepsilon\} \subset K$ is also compact. Indeed, let us take $x \in X$ such that $|f(x)| \geq \varepsilon$. Then there is $m_0 \in \mathbb{N}$ such that $|f(x) - f_m(x)| < \varepsilon/4$ for every $m \geq m_0$. Let us fix $m \geq n_0, m_0$. Then we have

$$\begin{aligned}|f_{n_0}(x)| &\geq |f(x)| - |f(x) - f_{n_0}(x)| \geq \varepsilon - (|f(x) - f_m(x)| + |f_m(x) - f_{n_0}(x)|)\\ &\geq \varepsilon - (\varepsilon/4 + \varepsilon/4) = \varepsilon/2,\end{aligned}$$

showing that $\{x \in X : |f(x)| \geq \varepsilon\} \subset K$, and the proof is completed. □

**Proposition 12.** *The norm* $||\cdot||_B$ *satisfies*

$$||f||_B = \sup\{|f(x)| : x \in X\} \qquad \textit{for all } f \in B.$$

*Proof.* Firstly, let us show that $\sup\{|f(x)| : x \in X\} \leq ||f||_B$. Let $f \in B$ and $\{f_n\}_{n=1}^{\infty}$ be a Cauchy sequence of $B_0$ with $f_n(x) \longrightarrow f(x)$ on $X$. Then, for every $x \in X$,

$$|f(x)| \leq \lim_{n \longrightarrow \infty} |f_n(x)| \leq \lim_{n \longrightarrow \infty} ||f_n||_{B_0} = ||f||_B.$$

Let us see the reverse inequality. If $f \in B$, then there exists a Cauchy sequence $\{f_n\}_{n=1}^{\infty}$ on $B_0$ such that $f_n(x) \longrightarrow f(x)$ for all $x \in X$. For any $\varepsilon > 0$, there is $n_0 \in \mathbb{N}$ such that $||f_n - f_m||_{B_0} < \varepsilon/3$ for all $n, m \geq n_0$. Furthermore, there are $x_n \in X$ for all $n$ such that $|f_n(x_n)| > ||f_n||_{B_0} - \varepsilon/3$. Since $|f_m(x_{n_0}) - f_{n_0}(x_{n_0})| \leq ||f_m - f_{n_0}||_{B_0} < \varepsilon/3$, then

$$||f_m||_{B_0} \leq ||f_{n_0}||_{B_0} + \varepsilon/3 \leq |f_{n_0}(x_{n_0})| + 2\varepsilon/3 \leq |f_m(x_{n_0})| + \varepsilon \qquad \text{for every } m \geq n_0.$$

Thus,

$$||f||_B = \lim_m ||f_m||_{B_0} \leq \lim_m |f_m(x_{n_0})| + \varepsilon = |f(x_{n_0})| + \varepsilon \leq \sup\{|f(x)| : x \in X\} + \varepsilon.$$

Since the above inequality holds for every $\varepsilon > 0$, then $||f||_B \leq \sup\{|f(x)| : x \in X\}$ for every $f \in B$. □

As in the general case, we have that for all $g \in B^\sharp$

$$||g||_{B^\sharp} = \sup_{\substack{f \in B \\ ||f||_B \leq 1}} |< f, g >_K|.$$

## 3.3 *RKBS with $\ell_1$-Norm*

Now, we can obtain an RKBS with $\ell_1$-norm in an easier way than [15]. We suppose the following condition, which is weaker that the given condition in [15].

($H_1$) For every sequence $\{x_i\}_{i=1}^n \subset X$ and $\varepsilon_i = +1$ or $\varepsilon_i = -1$ for $1 \leq i \leq n$, there is $f \in B$ with $||f||_B = 1$ such that $f(x_i) = \varepsilon_i$.

This assumption is satisfied if the kernel $K$ is *universal*, i.e., span$\{K(x,.) : x \in X\}$ is dense in $C_0(X)$ (see [16]).

In [15] it has been proved that the following kernels satisfy the hypothesis ($H_1$):

- The exponential kernel, $K(s,t) = e^{-|s-t|}$ for $s,t \in \mathbb{R}$
- The Brownian bridge kernel, $K(s,t) = \min\{s,t\} - st$ for $s,t \in (0,1)$

**Proposition 13.** *If $K$ satisfies hypothesis ($H_1$), then, for every $g = \sum_{j=1}^m \beta_j K(\cdot, y_j) \in B_0^\sharp$, the norm $||\cdot||_{B^\sharp}$ satisfies*

$$||g||_{B^\sharp} = \sum_{j=1}^m |\beta_j| = ||\{\beta_j\}||_1.$$

*Proof.* For every $f \in B$ with $||f||_B \leq 1$, we have

$$| < f, g >_K | \leq \sum_{j=1}^{m} |\beta_j||f(y_j)| \leq \sum_{j=1}^{m} |\beta_j|||f||_B \leq \sum_{j=1}^{m} |\beta_j|.$$

Now, if we chose $\varepsilon_i = \text{sign}(\beta_i)$, then there is $f \in B$ with $||f||_B = 1$ such that $f(y_i) = \varepsilon_i$. Thus

$$||g||_{B^\sharp} \geq | < f, g >_K | = |\sum_{j=1}^{m} \beta_j f(y_j)| = \sum_{j=1}^{m} |\beta_j|.$$

□

**Proposition 14.** *If K satisfies hypothesis* $(H_1)$*, then*

$$B^\sharp \subset \left\{ \sum_{n=1}^{\infty} \beta_n K(\cdot, y_n) : y_n \in X \text{ and } \{\beta_n\}_{n=1}^{\infty} \in \ell_1 \right\}.$$

*Furthermore,*

$$||g||_{B^\sharp} = \sum_{n=1}^{\infty} |\beta_n| \qquad \text{for every } g = \sum_{n}^{\infty} \beta_n K(\cdot, y_n) \in B^\sharp.$$

*Proof.* First of all, let us note that the function $K(x, \cdot)$ is bounded for every $x \in X$, and let us denote its bound by $M_x$.

If $g \in B^\sharp$, then there exists a Cauchy sequence $\{g_n\}_{n=1}^{\infty}$ on $B_0^\sharp$ such that $g_n(x) \longrightarrow g(x)$ for all $x \in X$ and $||g_n||_{B^\sharp} \longrightarrow ||g||_{B^\sharp}$. Since $g_n = \sum_{j=1}^{m_n} \beta_j^n K(\cdot, y_j^n)$, by Proposition 13 $||g_n||_{B^\sharp} = ||\{\beta_j^n\}_j||_1$ and $\{g_n\}$ is a Cauchy sequence in $B_0^\sharp$, the set of points $\{y_j^n\}_{j,n}$ can be ordered and denoted by $\{y_n\}_n$, and there is $\beta = \{\beta_n\}_n \in \ell_1$ such that $\{\beta_j^n\}_j \longrightarrow \beta$ when $n \longrightarrow \infty$.

Let us define $G = \sum_{n=1}^{\infty} \beta_n K(\cdot, y_n)$ on $X$. $G$ is well defined since $|G(x)| \leq \sum_n |\beta_n||K(x, y_n)| \leq M_x ||\beta||_1$ for every $x \in X$. Moreover, since $g_n(x) \longrightarrow g(x)$ for every $x \in X$ and $K(x, \cdot)$ is bounded, we conclude that $g_n(x) \longrightarrow G(x)$ and $G(x) = g(x)$ for all $x \in X$.

Finally,

$$||g||_{B^\sharp} = \lim_n ||g_n||_{B^\sharp} = \lim_n ||\{\beta_j^n\}_j||_1 = ||\beta||_1.$$

□

Summarizing, we obtain the following theorem.

**Theorem 2.** *Let X be a set and* $K : X \times X \longrightarrow \mathbb{C}$ *a function such that for every* $x \in X$ *the functions* $K(x, \cdot) \in F(X)$*. Let us denote* $B_0 = \text{span}\{K(x, \cdot) : x \in X\}$ *and* $B_0^\sharp = \text{span}\{K(\cdot, x) : x \in X\}$*. Then, there are* $(B, ||\cdot||_B)$ *and* $(B^\sharp, ||\cdot||_{B^\sharp})$ *Banach completions of* $B_0$ *and* $B_0^\sharp$*, respectively, such that they are RKBS on X with the reproducing kernel K. Furthermore,*

$$|<f,g>_K| \leq ||f||_B ||g||_{B^\sharp} \qquad \textit{for all } f \in B \textit{ and } g \in B^\sharp,$$

$$||f||_B = \sup\{|f(x)| : x \in X\},\ ||g||_{B^\sharp} = \sup_{\substack{f \in B \\ ||f||_B \leq 1}} |<f,g>_K| \textit{ for any } f \in B \textit{ and } g \in B^\sharp.$$

*Moreover, if $K$ satisfies hypothesis ($H_1$), then, for every $g \in B^\sharp$, there are $\beta = \{\beta_n\}_n \in \ell_1$ and $\{y_n\}_n$ a sequence of points in $X$ such that $g = \sum_{n=1}^\infty \beta_n K(\cdot, y_n)$ and the norm $||\cdot||_{B^\sharp}$ satisfies*

$$||g||_{B^\sharp} = \sum_{n=1}^\infty |\beta_n| = ||\beta||_1.$$

## 3.4 RKBS with the Norm $\ell_p$

Here we follow the construction of RKBS with $\ell_1$-norm given in [14]. Let $X$ be a set, $1 \leq p, q \leq +\infty$ such that $\frac{1}{p} + \frac{1}{q} = 1$. Let us denote by

$$\ell_q(X) = \{\alpha = \{\alpha(x)\}_{x \in X} \in \mathbb{C}^X : \sum_{x \in X} |\alpha(x)|^q < +\infty\},$$

with the norm $||\alpha||_q = (\sum_{x \in X} |\alpha(x)|^q)^{\frac{1}{q}}$ for all $\alpha \in \ell_q(X)$.

Suppose that $K : X \times X \longrightarrow \mathbb{C}$ is a function which satisfies:

(Hp1) $\{K(x,y)\}_{x \in X} \in \ell_q(X)$ for every $y \in X$.

(Hp2) If $\{\alpha_j\}_{j=1}^\infty \in \ell_p$ and $\{x_j\}_{j=1}^\infty$ is a sequence of $X$ such that $\sum_{j=1}^\infty \alpha_j K(x_j, \cdot) = 0$, then $\alpha_j = 0$ for any $j$.

**Fact**

*(i) The hypothesis (Hp1) implies that for every $y \in X$, the set $\{x \in X : K(x,y) \neq 0\}$ is countable.*

*(ii) When $X = \{x_j\}_{j=1}^\infty$ is a countable set, $\ell_q(X) = \ell_q(\mathbb{N})$.*

Let us introduce the vector spaces

$$B_0 := \text{span}\{K(x,\cdot) : x \in X\} \qquad \text{and} \qquad B_0^\sharp := \text{span}\{K(\cdot,x) : x \in X\},$$

and define the function $||\cdot||_{B_0} : B_0 \longrightarrow \mathbb{C}$ by

$$||f||_{B_0} = ||\sum_{j=1}^n \alpha_j K(x_j,\cdot)||_{B_0} := \left(\sum_{j=1}^n |\alpha_j|^p\right)^{\frac{1}{p}}, \quad \text{for every } f = \sum_{j=1}^n \alpha_j K(x_j,\cdot) \in B_0.$$

**Proposition 15.** $||\cdot||_{B_0}$ *is a **norm** on $B_0$ and satisfies the hypothesis (HN1) and (HN2).*

*Proof.* Using hypothesis (Hp2), it is straightforward to verify that $||\cdot||_{B_0}$ is a norm on $B_0$.

Let us prove that the evaluation functionals are continuous. Given $y \in X$, let us denote $[y]_K := (\sum_{x\in X} |K(x,y)|^q)^{\frac{1}{q}}$. Then

$$|\delta_y(f)| = |f(y)| = |\sum_{j=1}^{n} \alpha_j K(x_j,y)| \leq \left(\sum_{j=1}^{n} |\alpha_j|^p\right)^{\frac{1}{p}} \left(\sum_{j=1}^{n} |K(x_j,y)|^q\right)^{\frac{1}{q}} \leq ||f||_{B_0}[y]_K$$

Let us show that it satisfies (HN2). For every $n \in \mathbb{N}$ we can write

$$f_n(\cdot) = \sum_{j=1}^{\infty} \alpha_j^n K(x_j,\cdot),$$

where $\alpha^n = \{\alpha_j^n\}_{j=1}^{\infty}$ has finitely nonzero components, so $\alpha^n \in \ell_p(\mathbb{N})$. Since $\{f_n\}_{n=1}^{\infty}$ is a Cauchy sequence in $B_0$, then $\{\alpha^n\}_{n=1}^{\infty}$ is a Cauchy sequence in $\ell_p(\mathbb{N})$, so there exists $\alpha \in \ell_p(\mathbb{N})$ such that $\alpha^n \longrightarrow \alpha$ in $\ell_p(\mathbb{N})$. Let us define

$$f(\cdot) = \sum_{j=1}^{\infty} \alpha_j K(x_j,\cdot).$$

Hence, for all $x \in X$,

$$|f(x) - f_n(x)| = |\sum_{j=1}^{\infty} (\alpha_j - \alpha_j^n) K(x_j,x)| \leq ||\alpha - \alpha^n||_p [x]_K,$$

and, then, $f_n(x) \longrightarrow f(x)$ for all $x \in X$. Since $f_n(x) \longrightarrow 0$, then $f(x) = 0$ for all $x \in X$; by hypothesis (Hp2) we obtain that $\alpha_j = 0$ for any $j \in \mathbb{N}$ and

$$\lim_n ||f_n||_{B_0} = \lim_n ||\alpha^n||_p = ||\alpha||_p = 0.$$

Let us notice that the hypothesis (Hp2) is also necessary to prove that $||\cdot||_B$ is well defined. □

Then, by Theorem 1, we can conclude that there are $(B, ||\cdot||_B)$ and $(B^\sharp, ||\cdot||_{B^\sharp})$ Banach completions of $B_0$ and $B_0^\sharp$, respectively, such that the pair $(B, B^\sharp)$ is a pair of RKBS on $X$ with the reproducing kernel $K$. In fact, the completion of $B_0$ is given by

$$B := \Big\{ f : X \longrightarrow \mathbb{C} : \text{there exists } \{f_n\}_{n=1}^{\infty} \text{ a Cauchy sequence in } B_0 \text{ such that}$$
$$f_n(x) \longrightarrow f(x) \text{ for all } x \in X \Big\},$$

and the norm $||\cdot||_B$ on $B$ by

$$||f||_B := \lim_{n\longrightarrow\infty} ||f_n||_{B_0},$$

for all $f \in B$, where $\{f_n\}_{n=1}^{\infty}$ is a Cauchy sequence in $B_0$ such that $f_n(x) \longrightarrow f(x)$ for all $x \in X$.

In fact, in the same way as the proof of the above proposition, it can be proved that

$$B = \{\sum_{x \in X} c_x K(x,\cdot) : \{c_x\}_{x \in X} \in \ell_p(X)\} \qquad \text{and} \qquad ||\sum_{x \in X} c_x K(x,\cdot)||_B = ||\{c_x\}_{x \in X}||_p.$$

**Theorem 3.** *Let $X$ be a set and $K : X \times X \longrightarrow \mathbb{C}$ a function which satisfies the hypothesis (Hp1) and (Hp2). Let us denote $B_0 = \text{span}\{K(x,\cdot) : x \in X\}$ and $B_0^{\sharp} = \text{span}\{K(\cdot,x) : x \in X\}$. Then, there are $(B, ||\cdot||_B)$ and $(B^{\sharp}, ||\cdot||_{B^{\sharp}})$ Banach completions of $B_0$ and $B_0^{\sharp}$, respectively, such that they are RKBS on $X$ with the reproducing kernel $K$. Furthermore,*

$$| < f,g >_K | \leq ||f||_B ||g||_{B^{\sharp}} \qquad \text{for all } f \in B \text{ and } g \in B^{\sharp},$$

$$B = \{\sum_{x \in X} c_x K(x,\cdot) : \{c_x\}_{x \in X} \in \ell_p(X)\}, \qquad ||\sum_{x \in X} c_x K(x,\cdot)||_B = ||\{c_x\}_{x \in X}||_p,$$

*and*

$$||g||_{B^{\sharp}} = \sup_{\substack{f \in B \\ ||f||_B \leq 1}} |< f,g >_K| \quad \text{for any } g \in B^{\sharp}.$$

Several common kernels satisfy the hypothesis (Hp1) and (Hp2): the Gaussian kernel, the exponential kernel, etc.

## 4 Main Problems in Statistical Learning Theory

The statistical learning theory originates in the paper of Vapnik and Chervonenkis [17] and is systematically developed in [18]. A rigorous and comprehensive description of this theory can be found in [4]. There is a huge number of publications devoted to various branches of this theory, increasing rapidly, showing the importance and applicability of this theory in practical problems. Here we sketch briefly some of its main tasks.

We are given input-output data $(\mathbf{x}, \mathbf{y})$, $\mathbf{x} = (x_1, \ldots, x_N), \mathbf{y} = (y_1, \ldots, y_N)$, and cost functional $L_{\mathbf{x},\mathbf{y}}$. The task is to find a function from an admissible class, usually RKHS $H$, that minimizes the perturbed functional

$$L_{\mathbf{x},\mathbf{y}}(f) + \lambda \Phi(\|f\|_H)$$

where $\Phi$ is a regularization functional.

Some of the main problems in statistical learning theory are:

- Regularization networks

$$Y = \mathbb{R},\ L_{\mathbf{x},\mathbf{y}}(f) = \sum_{j\in\mathbb{N}_N} |f(x_j) - y_j|^2,\ \Phi(t) = t^2$$

- Support vector machine regression

$$Y = \mathbb{R},\ L_{\mathbf{x},\mathbf{y}}(f) = \sum_{j\in\mathbb{N}_N} |f(x_j) - y_j|_\varepsilon,\ \Phi(t) = t^2,$$

where $|t|_\varepsilon = \max(|t| - \varepsilon, 0)$ is called Vapnik $\varepsilon$-insensitive norm
- Support vector machine classification

$$Y = \{-1,1\},\ L_{\mathbf{x},\mathbf{y}}(f) = \sum_{j\in\mathbb{N}_N} \max(1 - y_j f(x_j), 0),\ \Phi(t) = t^2$$

The support vector machine classification can be reformulated equivalently as

$$\max_{\gamma,w,b}\{\gamma : y_n(< x_n, w > +b) \geq \gamma :\ 1 \leq n \leq N \text{ and } ||w|| = 1\}$$

(linear SVM), and

$$\text{(RKHS)} \qquad \max_{\gamma,\{\alpha_j\}_j,b}\{\gamma : y_n(\sum_{j=1}^{N} \alpha_j K(x_n,x_j) + b) \geq \gamma :\ \sum_{i,j=1}^{N} \alpha_i\alpha_j K(x_i,x_j) = 1\} \tag{8}$$

(nonlinear SVM)

Our contribution to the above model is to replace RKHS with RKBS with $\ell_1$-norm, i.e., to solve the problem

$$\text{(RKBS)} \qquad \max_{\gamma,\{\alpha_j\}_j,b}\{\gamma : y_n(\sum_{j=1}^{N} \alpha_j K(x_n,x_j) + b) \geq \gamma :\ \sum_{j=1}^{N} |\alpha_j| = 1\} \tag{9}$$

We expect sparse solutions of this problem. For the case of regularization network, sparse solutions in RKBS with $\ell_1$-norm are obtained in [14].

## 5 Conclusion

We defined a pair of RKBS without the requirement of existence of semi-inner product. We show that several natural examples of pairs of RKBS can be constructed on bases of an arbitrary kernel function. The classical machine learning problems involving kernel extension can be formulated in our abstract setting of pairs of RKBS. Further exploration and testing is needed to show expected advantages, as sparse solutions, in pairs of RKBS involving $l_1$-norm. Future work may include also additional properties of pairs of RKBS specific to concrete problems and their algorithmic implementations.

**Acknowledgements** The work is partially supported by LATNA Laboratory, NRU HSE, RF government grant, ag. 11.G34.31.0057.

Luis Sánchez-González conducted part of this research while at the Center for Applied Optimization. This author is indebted to the members of the Center and very especially to Pando Georgiev and Panos Pardalos for their kind hospitality and for many useful discussions.

## References

[1] Aronszajn, N.: Theory of reproducing kernels. Trans. Amer. Math. Soc. **68**, 337–404 (1950)
[2] Bennett, K.P., Bredensteiner, E.J.: Duality and geometry in SVM classifiers. In: Proceeding of the Seventeenth International Conference on Machine Learning, pp. 57–64. Morgan Kaufmann, San Francisco (2000)
[3] Chaovalitwobgse, W., Pardalos, P.: On the time series support vector machine using dynamic time warping kernel for brain activity classification. Cybernet. Syst. Anal. **44**(1), 125–138 (2008)
[4] Cucker, F., Smale, S.: On the mathematical foundations of learning. Bull. Amer. Math. Soc. **39**, 1–49 (2002)
[5] Der, R., Lee, D.: Large-margin classification in Banach spaces. In: JMLR Workshop and Conference Proceedings: AISTATS, vol. 2, pp. 91–98 (2007)
[6] Hein, M., Bousquet, O., Schölkopf, B.: Maximal margin classification for metric spaces. J. Comput. System Sci. **71**, 333–359 (2005)
[7] von Luxburg, U., Bousquet, O.: Distance-based classification with Lipschitz functions. J. Mach. Learn. Res. **5**, 669–695 (2004)
[8] Micchelli, C.A., Pontil, M.: A function representation for learning in Banach spaces. In: Learning Theory. Lecture Notes in Computer Science, vol. 3120, pp. 255–269. Springer, Berlin (2004)
[9] Micchelli, C.A., Xu, Y., Ye, P.: Cucker Smale learning theory in Besov spaces. In: Advances in Learning Theory: Methods, Models and Applications, pp. 47–68. IOS Press, Amsterdam, The Netherlands (2003)
[10] Seref, O., Erhun Kundakcioglu, O., Prokopyev, O.A., Pardalos, P.: Selective support vector machines. J. Combinat. Optimizat. **17**(1), 3–30 (2009)
[11] Zhang, T.: On the dual formulation of regularized linear systems with convex risks. Mach. Learn. **46**, 91–129 (2002)
[12] Zhang, H., Xu, Y., Zhang, J.: Reproducing kernel Banach spaces for machine learning J. Machile Learn. Res. **10**, 2741–2775 (2009)
[13] Pyrgiotakis, G., Kundakcioglu, E., Finton, K., Powers, K., Moudgil, B., Pardalos, P.: Cell death discrimination with Raman spectroscopy and support vector machines. Ann.Biomed. Eng. **37**(7), 1464–1473 (2009)
[14] Song, G., Zhang, H., Hickernell, F.J.: Reproducing kernel Banach spaces with the $\ell_1$ norm. Appl. and Comp. Harmonic Anal. **34**(1), 96–116 (2013)
[15] Song, G., Zhang, H.: Reproducing kernel Banach spaces with the $\ell_1$ norm II: Error analysis for regularized least square regression. Neural comp. **23**(10), 2713–2729 (2011)
[16] Steinwart, I.: On the influence of the kernel on the consistency of support vector machines. J. Mach. Learn. Res. **2**, 67–93 (2001)
[17] Vapnik, V.N., Chervonenkis, A.Y.: On the uniform convergence of relative frequencies of events to their probabilities. Theor. Probab. Appl. **16**, 264–280 (1971)
[18] Vapnik, V.: Statistical Learning Theory. Wiley, New York (1998)

# On the Regularization Method in Nondifferentiable Optimization Applied to Hemivariational Inequalities

**N. Ovcharova and J. Gwinner**

**Abstract** In this paper we present the regularization method in nondifferentiable optimization in a unified way using the smoothing approximation of the plus function. We show how this method can be applied to hemivariational inequalities. To illustrate our results we consider bilateral contact between elastic bodies with a nonmonotone friction law on the contact boundary and present some numerical results.

**Keywords** Regularization method • Nondifferentiable optimization • Smoothing approximation • Plus function • Hemivariational inequalities

## 1 Introduction

The motivation of this paper comes from the numerical treatment of nonlinear nonsmooth variational problems of continuum mechanics involving nonmonotone contact of elastic bodies. These contact problems lead to nonmonotone and multivalued laws which can be expressed by means of the Clarke subdifferential of a nonconvex, nonsmooth but locally Lipschitz function, the so-called superpotential. The variational formulation of these problems involving such laws gives rise to hemivariational inequalities introduced for the first time by Panagiotopoulos in the 1980s; see [14,15]. For the mathematical background of hemivariational inequalities we refer to Naniewicz and Panagiotopoulos [12]. For more recent works on the mathematical analysis of nonsmooth variational problems and contact problems, see also the monographs [3,8,11,18]. Numerical methods for such problems can be

---

N. Ovcharova (✉) • J. Gwinner
Institute of Mathematics, Department of Aerospace Engineering,
Universität der Bundeswehr, München, Germany
e-mail: nina.ovcharova@unibw.de; joachim.gwinner@unibw.de

V.F. Demyanov et al. (eds.), *Constructive Nonsmooth Analysis and Related Topics*,
Springer Optimization and Its Applications 87, DOI 10.1007/978-1-4614-8615-2_5,

found in the classical book of Haslinger et al. [9] as well as in the recent papers of Baniotopoulos et al. [2], Hintermüller et al. [10], etc.

Since in most applications the nonconvex superpotential can be modelled by means of maximum or minimum functions, we turn our attention to their regularizations. The idea of regularization goes back to Sobolev and is based more or less on convolution. However, regularizations via convolution are not easily applicable in practice, since it generally involves a calculation of a multivariate integral. But for the special class of maximum and minimum functions considered here, using regularization by a specified, e.g., piecewisely defined kernel, we can compute the smoothing function explicitly; see, e.g., [16, 20] and the recent survey in [13]. Moreover, since all nonsmooth functions under consideration can be reformulated by using the plus function, we can present the regularization method on nondifferentiable optimization (NDO) in a unified way. A large class of smoothing functions for the plus function can be found, e.g., in [4, 5, 7, 16, 19, 20].

In Sect. 2 we present a smoothing approximation of the maximum function based on the approximation of the plus function via convolution. We analyze some approximability property of the gradients of the smoothing function and show that the Clarke subdifferential of the nonsmooth but locally Lipschitz maximum function coincides with the subdifferential associated with the smoothing function.

Finally, in Sect. 3 we sketch how the regularization procedure from Sect. 2 can be used to solve nonmonotone contact problems. Here we focus on an elastic structure supported by a rigid foundation with a nonmonotone friction law. For further details concerning the regularization methods for hemivariational inequalities and their numerical realization by finite element methods, we refer to [13].

## 2 A Unified Approach to Regularization in NDO

Consider

$$f(x) = \max\{g_1(x), g_2(x)\}, \tag{1}$$

where $g_i : \mathbb{R}^n \longrightarrow \mathbb{R}$, $i = 1, 2$. Obviously, the maximum function (1) can be expressed by means of the plus function $\mathrm{p}(x) = x^+ = \max(x, 0)$ as

$$f(x) = \max\{g_1(x), g_2(x)\} = g_1(x) + \mathrm{p}[g_2(x) - g_1(x)]. \tag{2}$$

Replacing now the plus function by its approximation via convolution, we present the following smoothing function $S : \mathbb{R}^n \times \mathbb{R}_{++} \longrightarrow \mathbb{R}$ for the maximum function (see also [6]):

$$S(x, \varepsilon) := g_1(x) + P(\varepsilon, g_2(x) - g_1(x)). \tag{3}$$

Here, $P : \mathbb{R}_{++} \times \mathbb{R} \longrightarrow \mathbb{R}$ is the smoothing function via convolution for the plus function p defined by

$$P(\varepsilon,t) = \int_{-\infty}^{\frac{t}{\varepsilon}} (t - \varepsilon s)\rho(s)\,\mathrm{d}s. \tag{4}$$

*Remark 1.* The representation formula (2) can be extended to the maximum of finite number of arbitrary functions due to Bertsekas [1]:

$$f(x) = g_1(x) + \mathrm{p}\left[g_2(x) - g_1(x) + \cdots + \mathrm{p}\left[g_m(x) - g_{m-1}(x)\right]\right]. \tag{5}$$

Therefore the smoothing of the plus function gives a unified approach to regularization in NDO.

We restrict $\rho : \mathbb{R} \longrightarrow \mathbb{R}_+$ to be a density function of finite absolute mean; that is,

$$k := \int_{\mathbb{R}} |s|\rho(s)\,\mathrm{d}s < \infty.$$

From [16] we know that $P$ is continuously differentiable on $\mathbb{R}_{++} \times \mathbb{R}$ with

$$P_t(\varepsilon,t) = \int_{-\infty}^{\frac{t}{\varepsilon}} \rho(s)\,\mathrm{d}s \tag{6}$$

and satisfies

$$|P(\varepsilon,t) - \mathrm{p}(t)| \leq k\varepsilon \quad \forall \varepsilon > 0, \forall t \in \mathbb{R}. \tag{7}$$

The inequalities in (7) imply

$$\lim_{t_k \longrightarrow t, \varepsilon \longrightarrow 0^+} P(\varepsilon, t_k) = \mathrm{p}(t) \quad \forall t \in \mathbb{R}.$$

Moreover, $P(\varepsilon,\cdot)$ is twice continuously differentiable on $\mathbb{R}$ and we compute

$$P_{tt}(\varepsilon,t) = \varepsilon^{-1}\rho\left(\frac{t}{\varepsilon}\right). \tag{8}$$

Due to this formula we can also get the smoothing function (4) by twice integrating the density function; see [5].

In what follows, we suppose that all the functions $g_i$ are continuously differentiable. The major properties of $S$ (see [16]) inherit the properties of the function $P$ (see, e.g., [5, 7, 16]) and are collected in the following lemma:

**Lemma 1.**

*(i) For any $\varepsilon > 0$ and for all $x \in \mathbb{R}^n$,*

$$|S(x,\varepsilon) - f(x)| \leq k\varepsilon. \tag{9}$$

*(ii) The function S is continuously differentiable on* $\mathbb{R}^n \times \mathbb{R}_{++}$ *and for any* $x \in \mathbb{R}^n$ *and* $\varepsilon > 0$ *there exist* $\Lambda_i \geq 0$ *such that* $\sum_{i=1}^{2} \Lambda_i = 1$ *and*

$$\nabla_x S(x,\varepsilon) = \sum_{i=1}^{2} \Lambda_i \nabla g_i(x). \tag{10}$$

*Moreover,*

$$co\{\xi \in \mathbb{R}^n : \xi = \lim_{k \longrightarrow \infty} \nabla_x S(x_k, \varepsilon_k),\ x_k \longrightarrow x,\ \varepsilon_k \longrightarrow 0^+\} \subseteq \partial f(x), \tag{11}$$

*where "co" denotes the convex hull and* $\partial f(x)$ *is the Clarke subdifferential.*

We recall that the Clarke subdifferential of a locally Lipschitz function $f$ at a point $x \in \mathbb{R}^n$ can be expressed by

$$\partial f(x) = co\{\xi \in \mathbb{R}^n : \xi = \lim_{k \longrightarrow \infty} \nabla f(x_k),\ x_k \longrightarrow x,\ f \text{ is differentiable at } x_k\},$$

since in finite-dimensional case, according to Rademacher's theorem, $f$ is differentiable almost everywhere.

The maximum function given by (1) is clearly locally Lipschitz continuous and the Clarke subdifferential can be written as

$$\partial f(x) = co\{\nabla g_i(x) : i \in I(x)\}$$

with

$$I(x) := \{i : f(x) = g_i(x)\}.$$

In particular, if $x \in \mathbb{R}^n$ is a point such that $f(x) = g_i(x)$ then $\partial f(x) = \{\nabla g_i(x)\}$. For such a point $x \in \mathbb{R}^n$ we show later on that

$$\lim_{z \longrightarrow x, \varepsilon \longrightarrow 0^+} \nabla_x S(z,\varepsilon) = \nabla g_i(x).$$

Note that the set on the left-hand side in (11) goes back to Rockafellar [17]. In [4], this set is denoted by $G_S(x)$ and is called there the subdifferential associated with the smoothing function. The inclusion (11) shows in fact that $G_S(x) \subseteq \partial f(x)$. Moreover, according to the part $(b)$ of Corollary 8.47 in [17], $\partial f(x) \subseteq G_S(x)$. Thus, $\partial f(x) = G_S(x)$.

*Remark 2.* Note that $S$ is a smoothing approximation of $f$ in the sense that

$$\lim_{z \longrightarrow x, \varepsilon \longrightarrow 0^+} S(z,\varepsilon) = f(x) \quad \forall x \in \mathbb{R}^n.$$

This is immediate from (9).

*Remark 3.* The regularization procedure (3) can be also applied to a minimum function by

$$\begin{aligned} \min\{g_1(x), g_2(x)\} &= -\max\{-g_1(x), -g_2(x)\} = -\{-g_1(x) + \mathrm{p}[-g_2(x) + g_1(x)]\} \\ &\approx g_1(x) - P(\varepsilon, g_1(x) - g_2(x)) =: \tilde{S}(x,\varepsilon) \end{aligned} \tag{12}$$

Since all the nonsmooth functions considered in this paper can be reformulated by using the plus function, all our regularizations are based in fact on a class of smoothing approximations for the plus function. Some examples from [7] and the references therein are in order.

*Example 1.*

$$P(\varepsilon,t) = \int_{-\infty}^{\frac{t}{\varepsilon}} (t - \varepsilon s)\,\rho_1(s)\,\mathrm{d}s = t + \varepsilon \ln(1 + \mathrm{e}^{-\frac{t}{\varepsilon}}) = \varepsilon \ln(1 + \mathrm{e}^{\frac{t}{\varepsilon}}) \tag{13}$$

where $\rho_1(s) = \frac{\mathrm{e}^{-s}}{(1+\mathrm{e}^{-s})^2}$. Due to (8) the smoothing function (13) is obtained by integrating twice the function $\varepsilon^{-1}\rho_1(\frac{s}{\varepsilon})$.

*Example 2.*

$$P(\varepsilon,t) = \int_{-\infty}^{\frac{t}{\varepsilon}} (t - \varepsilon s)\,\rho_2(s)\,\mathrm{d}s = \frac{\sqrt{t^2 + 4\varepsilon^2} + t}{2}, \tag{14}$$

where $\rho_2(s) = \frac{2}{(s^2+4)^{3/2}}$. The formula (14) is similarly obtained as formula (13) via (8).

*Example 3.*

$$P(\varepsilon,t) = \int_{-\infty}^{\frac{t}{\varepsilon}} (t - \varepsilon s)\,\rho_3(s)\,\mathrm{d}s = \begin{cases} 0 & \text{if } t < -\frac{\varepsilon}{2} \\ \frac{1}{2\varepsilon}(t + \frac{\varepsilon}{2})^2 & \text{if } -\frac{\varepsilon}{2} \le t \le \frac{\varepsilon}{2} \\ t & \text{if } t > \frac{\varepsilon}{2}, \end{cases} \tag{15}$$

where $\rho_3(s) = \begin{cases} 1 \text{ if } -\frac{1}{2} \le s \le \frac{1}{2} \\ 0 \text{ otherwise.} \end{cases}$

*Example 4.*

$$P(\varepsilon,t) = \int_{-\infty}^{\frac{t}{\varepsilon}} (t-\varepsilon s)\,\rho_4(s)\,\mathrm{d}s = \begin{cases} 0 & \text{if } t < 0 \\ \frac{t^2}{2\varepsilon} & \text{if } 0 \leq t \leq \varepsilon \\ t - \frac{\varepsilon}{2} & \text{if } t > \varepsilon, \end{cases} \tag{16}$$

where $\rho_4(s) = \begin{cases} 1 \text{ if } 0 \leq s \leq 1 \\ 0 \text{ otherwise.} \end{cases}$

In the following, we denote

$$A_1 = \{x \in \mathbb{R}^n : g_1(x) > g_2(x)\} \quad \text{and} \quad A_2 = \{x \in \mathbb{R}^n : g_2(x) > g_1(x)\}.$$

**Lemma 2.** *The following properties hold:*

*(a) If* $x \in A_1$ *then* $\lim_{z \longrightarrow x, \varepsilon \longrightarrow 0^+} P_t(\varepsilon, g_2(z) - g_1(z)) = 0.$

*(b) if* $x \in A_2$ *then* $\lim_{z \longrightarrow x, \varepsilon \longrightarrow 0^+} P_t(\varepsilon, g_2(z) - g_1(z)) = 1.$

*Proof.* The proof is straightforward and is based on formula (6). We provide only the proof of (a). The proof of (b) is analogous.

Let $x \in A_1$, i.e., $g_1(x) > g_2(x)$. Using (6), it follows that

$$P_t(\varepsilon, g_2(z) - g_1(z)) = \int_{-\infty}^{\frac{g_2(z)-g_1(z)}{\varepsilon}} \rho(s)\,\mathrm{d}s \longrightarrow 0 \quad \text{as } z \longrightarrow x, \varepsilon \longrightarrow 0^+$$

and (a) is verified. □

Now we show that the gradient of the given function $g_i$ on $A_i$ can be approximated by the gradients of the smoothing function.

**Theorem 1.** *For any* $x \in A_i$, $i = 1,2$,

$$\lim_{z \longrightarrow x, \varepsilon \longrightarrow 0^+} \nabla_x S(z,\varepsilon) = \nabla g_i(x).$$

*Proof.* From (3), by direct differentiation with respect to $x$ [see also (10)], it follows that

$$\nabla_x S(z,\varepsilon) = \big(1 - P_t\big(\varepsilon, g_2(z) - g_1(z)\big)\big)\nabla g_1(z) + P_t\big(\varepsilon, g_2(z) - g_1(z)\big)\nabla g_2(z).$$

First, we take $x \in A_1$. From Lemma 2(a) we have

$$\lim_{z \longrightarrow x, \varepsilon \longrightarrow 0^+} P_t\big(\varepsilon, g_2(z) - g_1(z)\big) = 0$$

and therefore $\lim_{z \longrightarrow x, \varepsilon \longrightarrow 0^+} \nabla_x S(z,\varepsilon) = \nabla g_1(x)$. Let now $x \in A_2$. Then, from Lemma 2(b), it follows that

$$\lim_{z \longrightarrow x, \varepsilon \longrightarrow 0^+} P_t\big(\varepsilon, g_2(z) - g_1(z)\big) = 1$$

and consequently, $\lim_{z \longrightarrow x, \varepsilon \longrightarrow 0^+} \nabla_x S(z,\varepsilon) \longrightarrow \nabla g_2(x)$. The proof of the theorem is complete. □

*Remark 4.* Note that if $x \in \mathbb{R}^n$ is a point such that $g_1(x) = g_2(x)$ then for any sequences $\{x_k\} \subset \mathbb{R}^n$, $\{\varepsilon_k\} \subset \mathbb{R}_{++}$ such that $x_k \longrightarrow x$ and $\varepsilon_k \longrightarrow 0^+$ we have

$$\lim_{k \longrightarrow \infty} \nabla_x S(x_k, \varepsilon_k) \in \partial f(x).$$

# 3 Bilateral Contact with Nonmonotone Friction: A 2D Benchmark Problem

## 3.1 Statement of the Problem

In this section we sketch how our regularization method presented in Sect. 2 can be applied to numerical solution of nonmonotone contact problems that can be formulated as hemivariational inequality with maximum or minimum superpotential. As a model example we consider the bilateral contact of an elastic body with a rigid foundation under given forces and a nonmonotone friction law on the contact boundary. Here the linear elastic body $\Omega$ is the unit square $1m \times 1m$ (see Fig. 1) with modulus of elasticity $E = 2.15 \times 10^{11}\,\mathrm{N/m^2}$ and Poisson's ration $\nu = 0.29$ (steel). Then the linear Hooke's law is given by

$$\sigma_{ij}(\mathbf{u}) = \frac{E\nu}{1-\nu^2}\delta_{ij}\,tr\big(\varepsilon(\mathbf{u})\big) + \frac{E}{1+\nu}\varepsilon_{ij}(\mathbf{u}), \quad i,j = 1,2, \tag{17}$$

where $\delta_{ij}$ is the Kronecker symbol and

$$tr\big(\varepsilon(\mathbf{u})\big) := \varepsilon_{11}(\mathbf{u}) + \varepsilon_{22}(\mathbf{u}).$$

The boundary $\partial\Omega$ of $\Omega$ consists of four disjoint parts $\Gamma_u$, $\Gamma_c$, $\Gamma_F^1$, and $\Gamma_F^2$. On $\Gamma_u$ the body is fixed, i.e., we have

$$u_i = 0 \quad \text{on } \Gamma_u,\ i = 1,2.$$

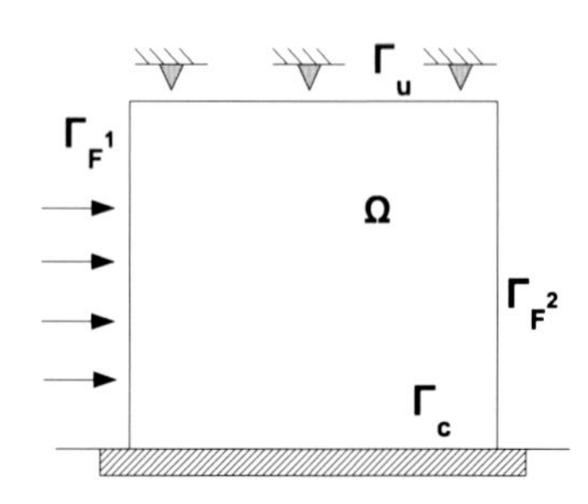


**Fig. 1** A 2D benchmark with force distribution and boundary decomposition

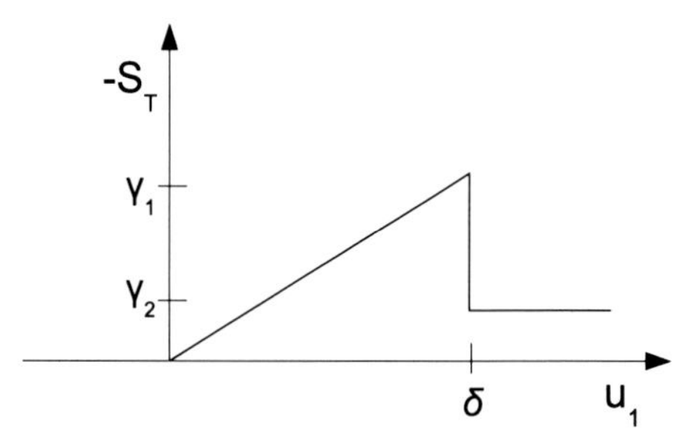


**Fig. 2** A nonmonotone friction law

The body is loaded with horizontal forces, i.e., $\mathbf{F} = (P,0)$ on $\Gamma_F^1$, where $P = 1.2 \times 10^6\,\mathrm{N/m^2}$, $\mathbf{F} = (0,0)$ on $\Gamma_F^2$. Further, we assume that

$$\begin{cases} u_2(s) = 0 & s \in \Gamma_c \\ -S_T(s) \in \partial j(u_1(s)) & \text{for a.a. } s \in \Gamma_c\,. \end{cases}$$

Note that $S_T$ denotes the tangential component of the stress vector on the boundary. The assumed nonmonotone multivalued law $\partial j$ holding in the tangential direction is depicted in Fig. 2 with parameters $\delta = 9.0 \times 10^{-6}\,\mathrm{m}$, $\gamma_1 = 1.0 \times 10^3\,\mathrm{N/m^2}$ and $\gamma_2 = 0.5 \times 10^3\,\mathrm{N/m^2}$. Notice that here $j$ is a minimum of a convex quadratic and a linear function, for instance, $j(\xi) = \min\{\frac{1}{2}\alpha\xi^2, \beta\xi\}$ for some $\alpha, \beta > 0$. Let

$$V = \{\mathbf{v} \in H^1(\Omega;\mathbb{R}^2) \,:\, v_i = 0 \text{ on } \Gamma_u,\ i = 1,2,\ v_2 = 0 \text{ on } \Gamma_c\}$$

be the linear subspace of all admissible displacements. The weak formulation of this bilateral contact problem leads to the following hemivariational inequality: find $\mathbf{u} \in V$ such that

$$a(\mathbf{u}, \mathbf{v} - \mathbf{u}) + \int_{\Gamma_c} j^0(u_1(s); v_1(s) - u_1(s))\,\mathrm{d}s \geq \langle \mathbf{g}, \mathbf{v} - \mathbf{u} \rangle \tag{18}$$

for all $\mathbf{v} \in V$. Here, $a(\mathbf{u}, \mathbf{v})$ is the energy bilinear form of linear elasticity

$$a(\mathbf{u}, \mathbf{v}) = \int_{\Omega} \sigma_{ij}(\mathbf{u})\varepsilon_{ij}(\mathbf{v})\,\mathrm{d}x \quad \mathbf{u}, \mathbf{v} \in V \tag{19}$$

with $\sigma$, $\varepsilon$ related by means of (17) and the linear form $\langle \mathbf{g}, \cdot \rangle$ defined by

$$\langle \mathbf{g}, \mathbf{v} \rangle = P \int_{\Gamma_{F1}} v_1 \, \mathrm{d}s. \tag{20}$$

## 3.2 Numerical Solution

We solve this problem numerically by first regularizing the hemivariational inequality (18) and then discretizing by the Finite Element Method. More precisely, we regularize $j(\xi)$ by $\tilde{S}(\xi, \varepsilon)$ defined by (12) and using (15) from Example 3 as a smoothing approximation of the plus function. Then, we introduce the functional $J_\varepsilon : V \longrightarrow \mathbb{R}$

$$J_\varepsilon(\mathbf{v}) = \int_{\Gamma_c} \tilde{S}(v_1(s), \varepsilon) \, \mathrm{d}s.$$

Since $\tilde{S}(\cdot, \varepsilon)$ is continuously differentiable for all $\varepsilon > 0$ the functional $J_\varepsilon$ is everywhere Gâteaux differentiable with continuous Gâteaux derivative $DJ_\varepsilon : V \longrightarrow V^*$ given by

$$\langle DJ_\varepsilon(\mathbf{u}), \mathbf{v} \rangle = \int_{\Gamma_c} \tilde{S}'_\xi(u_1(s), \varepsilon) v_1(s) \, \mathrm{d}s.$$

Notice that for $\mathbf{v} \in V$ the trace on $\Gamma_c$ is well defined, so $J_\varepsilon$ and $DJ_\varepsilon$ make sense.

The regularized problem of (18) now reads as follows: find $u_\varepsilon \in V$ such that

$$a(u_\varepsilon, \mathbf{v} - u_\varepsilon) + \langle DJ_\varepsilon(u_\varepsilon), \mathbf{v} - u_\varepsilon \rangle \geq \langle g, \mathbf{v} - u_\varepsilon \rangle \quad \forall v \in V. \tag{21}$$

Further, we consider a triangulation $\{\mathcal{T}_h\}$ of $\Omega$. Let $\Sigma_h$ be the set $\{x_i\}$ of all vertices of the triangles of $\{\mathcal{T}_h\}$ and $\mathcal{P}_h^c$ the set of all nodes on $\overline{\Gamma}_c$, i.e.,

$$\mathcal{P}_h^c = \{x_i \in \Sigma_h : x_i \in \overline{\Gamma}_c\}.$$

Using continuous piecewise linear functions we approximate the subspace of all admissible displacements $V$ by

$$V_h = \{v_h \in C(\overline{\Omega}; \mathbb{R}^2) : v_{h|_T} \in (\mathbb{P}_1)^2, \forall T \in \mathcal{T}_h, v_{hi} = 0 \text{ on } \Gamma_U, \, i = 1, 2, \\ v_{h2}(x_i) = 0 \, \forall x_i \in \mathcal{P}_h^c\}.$$

The discretization of the regularized problem (21) is defined now as follows: find $u_h \in V_h$ such that

$$a(u_h, v_h - u_h) + \langle DJ_h(u_h), v_h - u_h \rangle \geq P \int_{\Gamma_F^1} (v_{h1} - u_{h1}) \, \mathrm{d}x_2 \quad \forall v_h \in V_h,$$

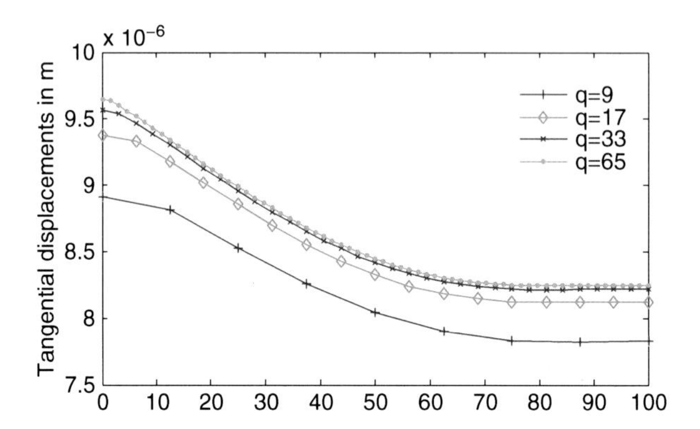


**Fig. 3** The tangential component of the displacement vector on $\Gamma_c$

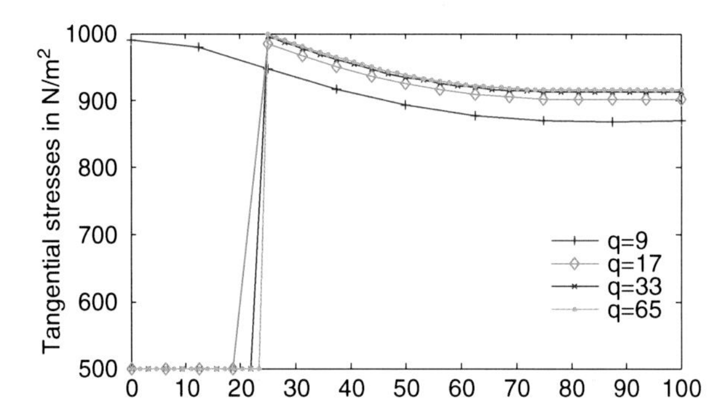


**Fig. 4** The distribution of the tangential stresses along $\Gamma_c$

where

$$\langle DJ_h(u_h), v_h\rangle = \frac{1}{2}\sum |P_iP_{i+1}| \left[\frac{\partial \tilde{S}}{\partial \xi}(u_{h1}(P_i), \varepsilon)v_{h1}(P_i) + \frac{\partial \tilde{S}}{\partial \xi}(u_{h1}(P_{i+1}), \varepsilon)v_{h1}(P_{i+1})\right].$$

Further, using the condensation technique, we pass to a reduced finite-dimensional variational inequality problem formulated only in terms of the contact displacements. To solve this problem numerically we use the equivalent KKT system, which is further reformulated as a smooth, unconstrained minimization problem by using an appropriate merit function. Finally, the merit function is minimized by applying an algorithm based on trust region methods. We did numerical experiments for different mesh sizes $h = 1/8, 1/16$, $1/32$, and $1/64$ m. The number of the contact nodes is $q = 9, 17, 33$, and 65, respectively. The obtained results are collected in the pictures below. Figures 3 and 4 show the behavior of the tangential displacements $u_1$ and the distribution of $-S_T$ along the contact boundary $\Gamma_c$ for the different number of the contact nodes ($q = 9, 17, 33, 65$) and for the constant load $P = 1.2 \times 10^6\,\mathrm{N/m^2}$. From Fig. 4 we can see that the computed stresses indeed follow the law depicted in Fig. 2. It is easy to see that with a finer discretization (e.g., $q = 17, 33, 65$) some of the computed displacements are larger than $\delta = 9.0 \times 10^{-6}\,m$ and the computed tangential stresses jump down to the parallel branch $-S_T = 500\,\mathrm{N/m^2}$ as described by a nonmonotone friction law in Fig. 2. All computations are made with regularization parameter $\varepsilon$ fixed to 0.1.

## References


[1] Bertsekas, D.P.: Nondifferentiable optimization via approximation. Math. Program. Study **3**, 1–25 (1975)

[2] Baniotopoulos, C.C., Haslinger, J., Morávková, Z.: Contact problems with nonmonotone friction: discretization and numerical realization. Comput. Mech. **40**, 157–165 (2007)

[3] Carl, S., Le, V.K., Motreanu, D.: Nonsmooth Variational Problems and Their Inequalities. Springer, New York (2007)

[4] Chen, X.: Smoothing methods for nonsmooth, nonconvex minimization. Math. Program. Ser. B **134**, 71–99 (2012)

[5] Chen, C., Mangasarian, O.L.: A class of smoothing functions for nonlinear and mixed complementarity problems. Comp. Optim. Appl. **5**, 97–138 (1996)

[6] Chen, X., Qi, L., Sun, D.: Global and superlinear convergence of the smoothing Newton method and its application to general box constrained variational inequalities. Math. Comp. **67**, 519–540 (1998)

[7] Facchinei, F., Pang, J.-S.: Finite-Dimensional Variational Inequalities and Complementarity Problems, vol. I, II. Springer, New York (2003)

[8] Goeleven, D., Motreanu, D., Dumont, Y., Rochdi, M.: Variational and hemivariational inequalities: Theory, methods and applications. In: Unilateral Problems, vol. II. Kluwer, Dordrecht (2003)

[9] Haslinger, J., Miettinen, M., Panagiotopoulos, P.D.: Finite Element Methods for Hemivariational Inequalities. Kluwer Academic Publishers, Dordrecht (1999)

[10] Hintermüller, M., Kovtunenko, V.A., Kunisch K.: Obstacle problems with cohesion: A hemivariational inequality approach and its efficient numerical solution. SIAM J. Optim. **21**(2), 491–516 (2011)

[11] Migórski, S., Ochal, A., Sofonea, M.: Nonlinear Inclusions and Hemivariational Inequalities: Models and Analysis of Contact Problems. Springer, New York (2013)

[12] Naniewicz, Z., Panagiotopoulos, P.D.: Mathematical Theory of Hemivariational Inequalities and Applications. Dekker, New York (1995)

[13] Ovcharova, N.: Regularization methods and finite element approximation of hemivariational inequalities with applications to nonmonotone contact problems. Ph.D. thesis, Universität der Bundeswehr München, Cuvillier, Göttingen (2012)

[14] Panagiotopoulos, P.D.: Hemivariational Inequalities: Applications in Mechanics and Engineering. Springer, Berlin (1993)

[15] Panagiotopoulos, P.D.: Inequality Problems in Mechanics and Application: Convex and Nonconvex Energy Functions. Birkhäuser, Basel (1998)

[16] Qi, L., Sun, D.: Smoothing functions and a smoothing Newton method for complementarity and variational inequality problems. J. Optim. Theory Appl. **113**(1), 121–147 (2002)

[17] Rockafellar, R.T., Wets, R.J.-B.: Variational Analysis. Springer, New York (1998)

[18] Sofonea, M., Matei, A.: Variational Inequalities with Applications. Springer, New York (2009)
[19] Sun, D., Qi, L.: Solving variational inequality problems via smoothing-nonsmooth reformulations. J. Comput. Appl. Math. **129**, 37–62 (2001)
[20] Zang, I.: A smoothing-out technique for min-max optimization, Math. Program. **19**, 61–77 (1980)

# Dynamics and Optimization of Multibody Systems in the Presence of Dry Friction

**F.L. Chernousko**

**Abstract** Motions of multibody mechanical systems over a horizontal plane in the presence of dry friction forces are considered. Since the dry friction force is defined by Coulomb's law as a nonsmooth function of the velocity of the moving body, the dynamics of systems under consideration is described by differential equations with nonsmooth right-hand sides. Two kinds of multibody systems are analyzed: snakelike multilink systems with actuators placed at the joints and vibro-robots containing movable internal masses controlled by actuators. It is shown that both types of systems can perform progressive locomotion caused by periodic relative motions of the bodies. The average speed of the systems is evaluated. Optimal values of the system parameters and optimal controls are found that correspond to the maximum locomotion speed. Experimental data confirm the obtained theoretical results. Principles of motions considered are of interest for biomechanics and robotics, especially, for mobile robots moving in various environments and inside tubes.

**Keywords** Multibody systems • Dry friction • Locomotion • Dynamics of systems • Optimal control

## 1 Introduction

In the paper, motions of various multibody mechanical systems over a horizontal plane are considered in the presence of dry friction forces acting between the system and the plane. We assume that the dry friction force obeys Coulomb's law. For a point mass $m$ moving with the velocity $v$ along the horizontal plane, the friction

---

F.L. Chernousko (✉)
Institute for Problems in Mechanics, Russian Academy of Sciences, Moscow Institute of Physics and Technology, pr.Vernadskogo 101-1, Moscow, 119526, Russia
e-mail: chern@ipmnet.ru

V.F. Demyanov et al. (eds.), *Constructive Nonsmooth Analysis and Related Topics*, Springer Optimization and Its Applications 87, DOI 10.1007/978-1-4614-8615-2_6,

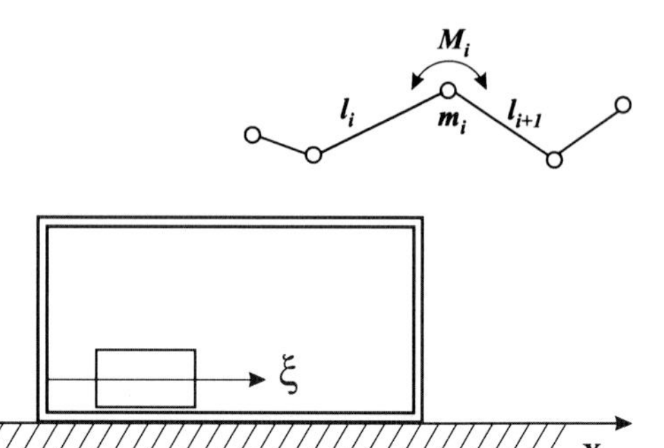


**Fig. 1** Snakelike system

**Fig. 2** System with an internal mass

force **F** acting upon the mass is defined as

$$\begin{aligned} \mathbf{F} &= -kmg\mathbf{v}/|\mathbf{v}|, \text{ if } \quad \mathbf{v} \neq 0, \\ |\mathbf{F}| &\leq kmg, \quad \text{ if } \quad \mathbf{v} = 0. \end{aligned} \tag{1}$$

Here, $k$ is the coefficient of friction and $g$ is the gravity acceleration. Thus, the friction force is a nonsmooth function of the velocity $\mathbf{v}$, and, in the case of rest ($\mathbf{v}$=0), it can take any value within the bounds indicated in (1).

In this case, the friction force can be defined as follows:

$$\begin{aligned} \mathbf{F} = -\min(kmg, |\mathbf{F}_1|)\mathbf{F}_1/|\mathbf{F}_1|, &\text{ if } \quad \mathbf{v} = \mathbf{0}, \quad \mathbf{F}_1 \neq \mathbf{0}, \\ \mathbf{F} = \mathbf{0}, \qquad\qquad &\text{ if } \quad \mathbf{v} = \mathbf{0}, \quad \mathbf{F}_1 = \mathbf{0}, \end{aligned}$$

where $\mathbf{F}_1$ is the resultant of the forces, other than frictional ones, applied to the mass $m$ and acting in the horizontal plane.

Thus, the differential equations describing the dynamics of systems subjected to dry friction forces have nonsmooth right-hand sides.

We consider below two kinds of multibody systems moving along a horizontal plane: snakelike systems and systems containing movable internal masses.

Snakelike systems consist of several links lying on the horizontal plane and connected consecutively by cylindrical joints where the actuators are placed (Fig. 1). The motion of the snakelike system is a result of twisting at the joints caused by torques generated by the actuators.

The other kind of multibody systems considered in the paper consists of systems containing movable internal masses and lying on the horizontal plane (Fig. 2). The controlled relative motion of these masses results in the displacement of the system as a whole.

Principles of motion considered in the paper are of interest for both biomechanics [1] and robotics.

Motions of snakelike systems were considered in a number of papers; see, e.g., [1–4]. In [4], the survey of these papers is given. Mostly, in these papers, the locomotion of snakelike systems was considered under additional assumptions, in the presence of obstacles or wheels attached to the snake's links; as a result, sideways sliding of the mechanism was bounded or eliminated.

The analysis of the locomotion of snakelike systems over a plane in the presence of the dry friction obeying Coulomb's law as well as the optimization of the locomotion were first carried out in [5–10]; see also [11–13].

Mechanical systems containing internal moving masses with applications to robotics were considered in [14–22]. The detailed analysis and optimization of these systems were performed in [23–27].

In this paper, we give the overview of our previous publications and also present certain new results.

For the snakelike systems, both quasistatic and dynamic modes of motions are considered. In quasistatic motions, the dynamic terms in the equations of motion are omitted, and these slow motions are indeed a sequence of equilibrium positions. In dynamic motions, full equations of motions are taken into account.

Wavelike quasistatic progressive motions of multilink systems with more than four links are proposed. For two-link and three-link systems, progressive periodic motions are designed that consist of alternating slow and fast phases.

Progressive periodic motions of systems with internal moving masses are proposed. Different possible controls of internal masses are considered: either velocities or accelerations of internal masses can play the role of controls.

The most important characteristic of a progressive motion is its average velocity. Hence, it is natural to maximize this velocity with respect to the parameters of the system and controls applied.

A number of optimization problems are solved. The average speed of the progressive motions for two-link and three-link systems is maximized with respect to their geometrical and mechanical parameters (lengths of links, masses, etc.) as well as to certain control parameters. Similar problems are solved for the systems with internal moving masses. As a result, optimal parameters and controls are obtained that correspond to the maximum locomotion speed of the systems.

Computer simulation and experimental data illustrate and confirm the obtained theoretical results. These results are useful for mobile robots that can move in various environment and inside tubes.

## 2 Multilink System

Consider a multilink system shown in Fig. 1. We assume that the system lies on a horizontal plane and consists of straight rigid links connected consecutively by cylindrical joints where actuators are placed. Denote by $l_i$ the length of the $i$th link, by $m_i$ the mass of the $i$th joint, and by $M_i$ the torque created by the $i$th actuator. There are also point masses at both ends of the linkage. Suppose that the masses of the links are negligible compared with the masses of the joints and end points.

Under the assumptions made, the only external forces acting upon our multilink system are the gravity, the normal reaction of the plane, and the dry friction forces applied to the joints. Internal forces include torques $M_i$ created by the actuators and interaction forces acting between neighboring links and joints.

Since the friction forces obeying Coulomb's law (1) are not defined at the state of rest, it is not possible to follow the routine approach of dynamics, namely, to write down the differential equations of motion and then to solve them analytically or numerically. Here, we must first decide which masses move and which stay at rest; the equations of motion will depend on this decision.

Our goal is, first, to determine possible progressive motions of the multilink system and, second, to maximize the average speed of these motions with respect to the parameters of the system.

## 3 Quasistatic Motions

If the motion of the system is sufficiently slow so that the inertia forces are negligible compared to the friction forces, then the inertia terms can be omitted in the equations of motion. In this case of quasistatic motions, we can regard the motion as a succession of equilibrium positions.

Possible quasistatic motions were studied for multilink systems with different number of links. It was shown [28] that the two-link system cannot move progressively in the quasistatic mode: its positions are always situated within a bounded domain in the horizontal plane. For the three-link system, progressive quasistatic motions are possible, at least, under certain bounds imposed on the system parameters [29].

Consider a multilink system consisting of $N$ identical links assumed to be rigid straight rods of length $a$. The mass of the system is concentrated at the joints and end points $P_i$, $i = 0, 1, \ldots, N$, which are equal point masses $m$. The actuators installed at the joints $P_1, .., P_{N-1}$ can create torques about vertical axes.

Two types of wavelike quasistatic motions of the system (with three and four moving links) along the horizontal plane $Oxy$ are proposed [6].

At the beginning and at the end of the cycle of motions, the linkage is straight and placed along the $x$-axis. Its successive configurations for two types of motion are shown in Figs. 3 and 4.

In the first stage of the motion with *three moving links* (Fig. 3), the end mass $P_0$ advances along the $x$-axis while the points $P_i$, $i \geq 2$, remain fixed. The angle $\alpha$ between the $x$-axis and link $P_0P_1$ grows monotonically from zero to a certain given value $\alpha_0$; see states $a$, $b$, $c$ in Fig. 3. At the next stage, links $P_0P_1$, $P_1P_2$, and $P_2P_3$ are moving. The angle $\alpha$ decreases monotonically from $\alpha_0$ to zero, while the angle $\beta$ between link $P_2P_3$ and the $x$-axis grows from $O$ to $\alpha_0$; see state $d$ in Fig. 3. At the end of this stage, the system will be in state $e$ in which links $P_1P_2$ and $P_2P_3$ form an isosceles triangle congruent to the triangle $P_0P_1P_2$ in state $c$ but with its apex pointing in the opposite direction. Here, all points except $P_2$ lie on the $x$-axis. Next, the motion involves links $P_1P_2$, $P_2P_3$, and $P_3P_4$ and is identical with the preceding stage, apart from a displacement along the $x$-axis and a mirror reflection in the axis.

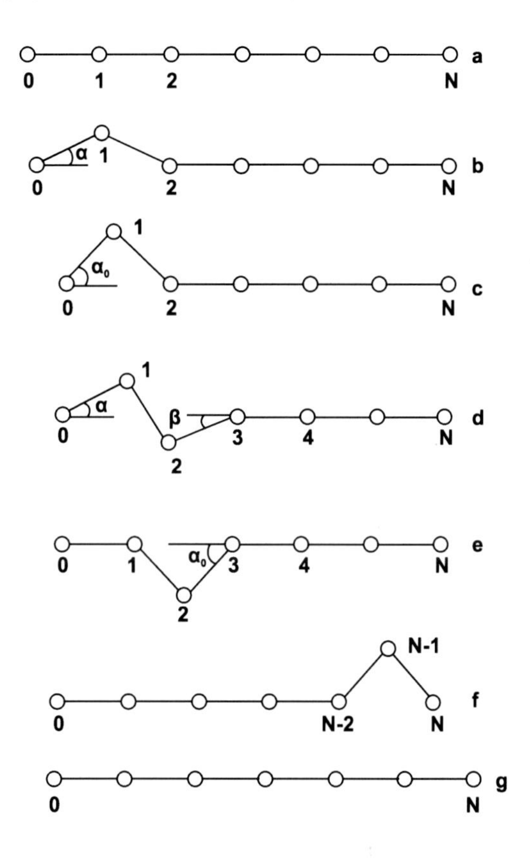


**Fig. 3** Wavelike motion with three moving links

Continuing this process, we see that after each stage all joints except one lie on the $x$-axis and that one joint is the apex of the isosceles triangle with angle $\alpha_0$ at the base. The triangle gradually moves towards the right. Finally, the point $P_{N-1}$ becomes the apex of such a triangle; see state $f$ in Fig. 3. At the last stage of the motion, the point $P_N$ advances to the right along the $x$-axis. The angle at the base of the triangle $P_{N-2}P_{N-1}P_N$ decreases from $\alpha_0$ to zero, and the linkage becomes straight again; see state $g$ in Fig. 3.

As a result of the entire cycle of motions, the linkage advances along the $x$-axis for a distance

$$L = 2a(1 - \cos\alpha_0). \tag{2}$$

In the motion with *four moving links*, the first stage proceeds exactly as in the previous case (states $a$–$c$ are the same in Figs. 3 and 4). At the next stage, links $P_0P_1$, $P_1P_2$, $P_2P_3$, and $P_3P_4$ are involved. The point $P_2$ moves to the right along the $x$-axis. As this happens, the angle $\alpha$ at the base of the isosceles triangle $P_0P_1P_2$ decreases from $\alpha_0$ to zero, while the angle $\beta$ at the base of the triangle $P_2P_3P_4$ grows from zero to $\alpha_0$; see state $d$ in Fig. 4. At the end of this stage, all joints of the linkage except $P_3$ lie on the $x$-axis; see state $e$ in Fig. 4. Continuing this process, we come to

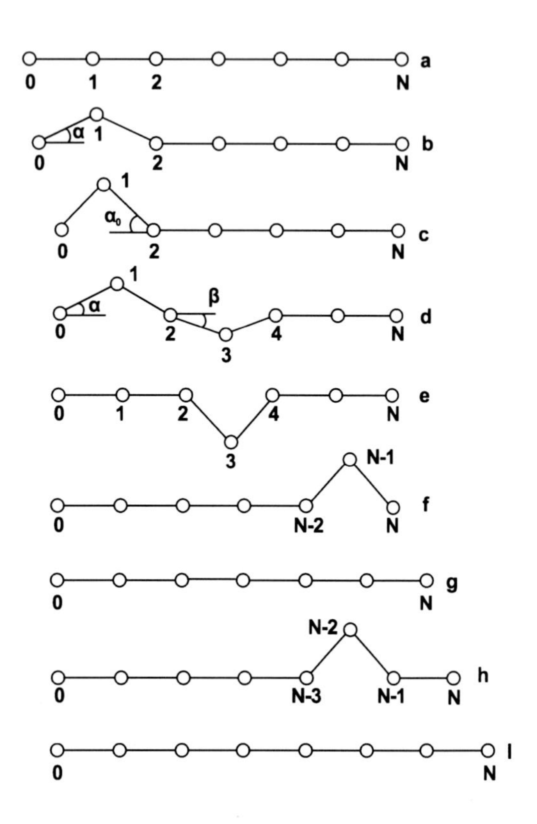


**Fig. 4** Wavelike motion with four moving links

the right end of the linkage. The final stages are somewhat different for the cases of even and odd $N$; see states $f$–$i$ in Fig. 4. The resultant displacement of the linkage is again given by formula (2).

Comparing the two types of wavelike motions, we see that the motion with three moving links is somewhat simpler. However, it requires larger angles between links, or more intensive twisting, for the same values of $\alpha$ and $L$ [6].

The wavelike motions whose kinematics is described above were analyzed in a quasistatic formulation [6]. We assume that all velocities and accelerations are extremely small, and hence the external forces applied to the system must balance out. In other words, the friction forces must satisfy three equilibrium conditions (two for forces and one for moments). The friction forces applied to the moving points are readily evaluated, whereas for the points at rest they are unknown but bounded by the inequalities (1). Our equilibrium problem is statically indeterminate and can have nonunique solution. The simplest distributions of the friction forces are found [6] for which the equilibrium is attained with the participation of the least possible number of points adjacent to the moving ones. It is shown that the wavelike motion with three moving links is feasible, if the linkage has at least five links ($N \geq 5$), whereas the motion with four moving links is possible, if $N \geq 6$. Also, the magnitude $M$ of the required control torques at the joints of the linkage is estimated [6]. It is shown that $M \leq 2mgka$, where $k$ is the friction coefficient.

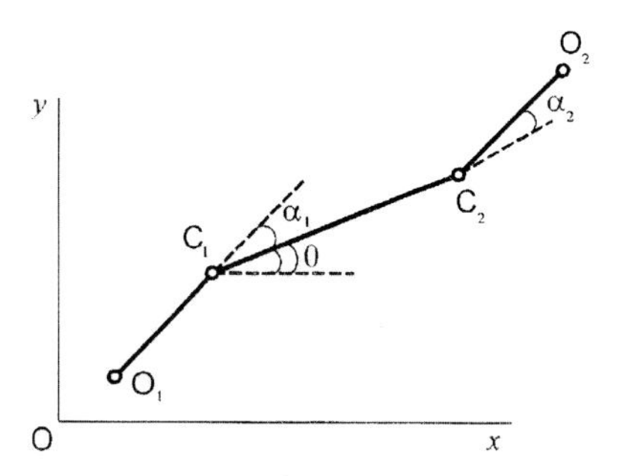


**Fig. 5** Three-link system

## 4 Three-Link System

Dynamical modes of locomotion were analyzed in [5,7] for a three-link system and in [8] for a two-link system.

Consider first a plane three-member linkage $O_1C_1C_2O_2$ moving along a fixed horizontal plane $Oxy$ (Fig. 5). For the sake of simplicity, we assume that the links $O_1C_1$, $C_1C_2$, and $C_2O_2$ are rigid and massless bars, and the entire mass of the linkage is concentrated at the end points $O_1$, $O_2$ and joints $C_1$, $C_2$ which can slide along the plane. The mass of each of the end points is denoted by $m_0$, and the mass of each of the joints is denoted by $m_1$. Thus, the total mass of the linkage is $m = 2(m_0 + m_1)$. The length of the central link $C_1C_2$ is denoted by $2a$, and the lengths of the end links are equal to $l$.

Denote by $x$, $y$ the Cartesian coordinates of the middle of the central link, by $\theta$ the angle between this link and the $x$-axis, and by $\alpha_i$ the angles between the central link $C_1C_2$ and the end links $O_iC_i$, $i = 1,2$ (Fig. 5).

We assume that the dry friction forces acting between the masses $O_i$, $C_i$, $i = 1,2$, and the horizontal plane obey Coulomb's law (1). If a point mass $m_0$ (or $m_1$) moves, the friction force is directed opposite to the point velocity and equal to its weight $m_0g$ (or $m_1g$) multiplied by the friction coefficient $k_0$ (or $k_1$). If the point mass $m_0$ (or $m_1$) is at rest, the friction force does not exceed $m_0gk_0$ (or $m_1gk_1$), and its direction can be arbitrary.

The control torques $M_1$ and $M_2$ about the vertical axes are created by the actuators installed at the joints $C_1$ and $C_2$. We assume that these torques can produce some prescribed time-history of angles $\alpha_i(t)$, $i = 1,2$.

## 5 Elementary Motions

We will construct various motions of the linkage as a combination of more simple motions, which we call elementary [5]. All elementary motions begin and end at the states of rest of the linkage. In each elementary motion, the angles $\alpha_i(t)$, $i = 1,2$, change within the interval $(-\pi, \pi)$ between the prescribed initial value $\alpha_i^0$ and terminal value $\alpha_i^1$; the time-histories $\alpha_i(t)$ can be more or less arbitrary. Either one or both angles $\alpha_1,\alpha_2$ can change in the elementary motion. In the latter case, they must change synchronously so that $\dot{\alpha}_2(t) = \pm\dot{\alpha}_1(t)$. Thus, the end links can rotate either in the same direction or in the opposite directions.

Elementary motions are divided into slow and fast ones.

In *slow* motions, the values of the angular velocity $\omega(t) = |\dot{\alpha}_i(t)|$ and angular acceleration $\varepsilon(t) = \dot{\omega}(t)$ are small enough, so that the central link stays at rest during these motions. To derive conditions under which the central link does not move in slow motions, one should write down the equations of motion of the end links and determine forces and torques with which these links act upon the central one. Then one should consider the equations of equilibrium for the central link $C_1C_2$. This link will stay at rest, if the friction forces acting upon the points $C_1$ and $C_2$ exist which satisfy Coulomb's law and ensure the equilibrium of the central link. According to this plan, sufficient conditions were derived which guarantee that the central link does not move during the slow motions [5, 7]. To write down these conditions, we denote by $\omega_0$ and $\varepsilon_0$, respectively, the maximal values of the angular velocity and acceleration of the end links during the slow motion:

$$\omega_0 = \max|\dot{\alpha}_i(t)|, \quad \varepsilon_0 = \max|\ddot{\alpha}_i(t)|. \tag{3}$$

Here, the maxima are taken along the whole slow motion. Since $\dot{\alpha}_2 = \pm\dot{\alpha}_1$, they do not depend on $i = 1,2$.

If both end links rotate in the same direction in the slow motion, then the sufficient condition, which ensures that the central link stays at rest, can be expressed as follows [5, 7]:

$$m_0 l \left\{ \left[\omega_0^4 + \left(\varepsilon_0 + g k_0 l^{-1}\right)^2\right]^{1/2} + \left(\varepsilon_0 + g k_0 l^{-1}\right) l a^{-1} \right\} \le m_1 g k_1. \tag{4}$$

This condition is also true for the case where only one end link rotates during the slow motion.

Note that condition (4) holds, if the motion is slow enough (i.e., if $\omega_0$ and $\varepsilon_0$ in (3) are sufficiently small) and $m_0 k_0 (a + l) < m_1 k_1 a$. The latter inequality can be easily satisfied by the choice of lengths $a, l$ and masses $m_0, m_1$.

If the end links rotate in the opposite directions, condition (4) can be replaced by a weaker one [5, 7]:

$$m_0 l \left[\omega_0^4 + \left(\varepsilon_0 + g k_0 l^{-1}\right)^2\right]^{1/2} \le m_1 g k_1. \tag{5}$$

This condition holds, if $\omega_0$ and $\varepsilon_0$ in (3) are small enough and $m_0 k_0 < m_1 k_1$.

In *fast* motions, the angular velocities and accelerations of the end links are sufficiently high, and the duration $\tau$ of this motion is much less than the duration $T$ of the slow motion: $\tau \ll T$. The magnitudes of the control torques $M_1$ and $M_2$ during the fast motion are high compared to the torques due to the friction forces:

$$|M_i| \gg m^* g k^* l^*, \quad i = 1,2, \quad m^* = \max(m_0, m_1),$$
$$k^* = \max(k_0, k_1), \quad l^* = \max(a, l). \tag{6}$$

Hence, the friction can be neglected during the fast motion, and the conservation laws for the momentum and angular momentum hold in this motion. Therefore, the center of mass $C$ of the linkage stays at rest, and its angular momentum is zero during the fast motion. Using these conservation laws, one can evaluate the terminal state of the linkage after the fast motion.

## 6 Locomotion of the Three-Link System

Let us show how longitudinal (or lengthwise), lateral (or sideways), and rotational motions of the three-member linkage can be composed from elementary ones. Suppose that at the initial instant of time the linkage is at rest, and all its links are parallel to the $x$-axis. We have $\theta = \alpha_1 = \alpha_2 = 0$ in this state.

For the sake of brevity, we denote slow and fast motions by the capital letters $S$ and $F$, respectively. Let us indicate the initial and terminal values of the angles $\alpha_1$, $i = 1,2$, in each elementary motion by the respective superscripts $^0$ and $^1$. Thus, we will describe the limits between which the angle $\alpha_i$ changes in an elementary motion as follows: $\alpha_i$: $\alpha_i^0 \longrightarrow \alpha_i^1$, $i = 1,2$.

*Longitudinal motion* (Fig. 6).
First, the auxiliary slow motion is carried out: $S, \alpha_1 : 0 \longrightarrow \gamma, \alpha_2(t) \equiv 0$ where $\gamma \in (-\pi, \pi)$ is a fixed angle. Then, the following four motions are performed:

$$(1)\ F, \alpha_1 : \gamma \longrightarrow 0, \alpha_2 : 0 \longrightarrow \gamma; \qquad (2)\ S, \alpha_1 : 0 \longrightarrow -\gamma, \alpha_2 : \gamma \longrightarrow 0;$$
$$(3)\ F, \alpha_1 : -\gamma \longrightarrow 0, \alpha_2 : 0 \longrightarrow -\gamma; \ (4)\ S, \alpha_1 : 0 \longrightarrow \gamma, \alpha_2 : -\gamma \longrightarrow 0.$$

After stage 4, the configuration of the linkage coincides with its configuration before stage 1, so that stages 1–4 can be repeated any number of times. Thus, we obtain a periodic motion consisting of two fast and two slow phases. To return the

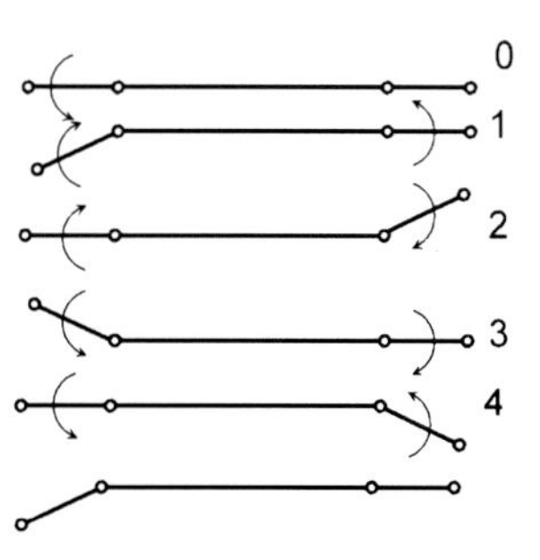


**Fig. 6** Longitudinal motion

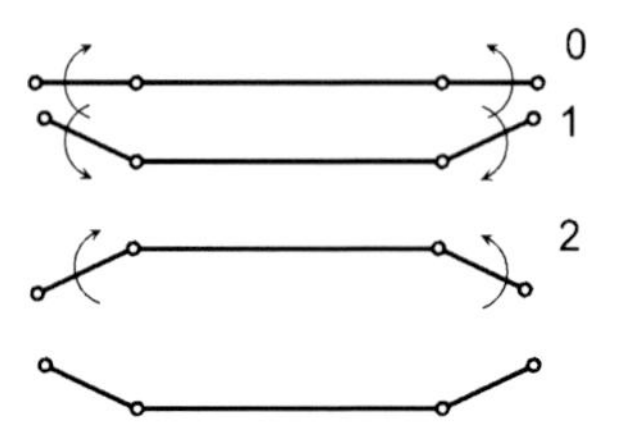

**Fig. 7** Lateral motion

linkage to its initial rectilinear configuration $\alpha_1 = \alpha_2 = 0$, one should perform the motion $S, \alpha_1 : \gamma \longrightarrow 0, \alpha_2 \equiv 0$.

Let us evaluate the displacements of the linkage during the cycle of the periodic motion described above. During the slow phases, the central link stays at rest, whereas the center of mass $C$ of the linkage moves. One can see from Fig. 6 that the $x$-displacements of both end masses are negative in both slow phases, whereas their $y$-displacements are of the opposite signs. Hence, the $x$-coordinate $x_C$ of the center of mass increases in both slow phases, whereas its $y$- coordinate $y_C$ does not change. During the fast phases, the center of mass $C$ stays at rest, while the central link moves. Since the $x$-displacements of both end masses are negative in both fast phases, the central link moves forward along the $x$-axis in these phases. The $y$-displacements of both end masses are positive in the first fast phase and negative in the other one. As a result, the total displacement of the central link along the $y$-axis during the whole cycle is zero. Since both end links rotate in the opposite directions in both fast phases, the central link, by virtue of the conservation law for the angular momentum, does not rotate at all.

The considerations presented above and the calculations performed in [5, 7] give the following total increments for the coordinates $x$, $y$ of the middle of the central link and for the angle $\theta$ during the cycle of the longitudinal motion:

$$\Delta x = 8m_0 m^{-1} l \sin^2(\gamma/2), \quad \Delta y = 0, \quad \Delta\theta = 0. \tag{7}$$

Since the duration of the fast phases $\tau$ is negligible compared to that of the slow ones, the average speed $v_1$ of the longitudinal motion is evaluated by

$$v_1 = \Delta x (2T)^{-1}.$$

*Lateral motion* (Fig. 7).

First, the auxiliary slow motion $S, \alpha_1 : 0 \longrightarrow -\gamma,\ \alpha_2 : 0 \longrightarrow \gamma$ is performed. Then, the following two motions are carried out:

(1) $F, \alpha_1 : -\gamma \longrightarrow \gamma, \quad \alpha_2 : \gamma \longrightarrow -\gamma$; (2) $S, \alpha_1 : \gamma \longrightarrow -\gamma, \quad \alpha_2 : -\gamma \longrightarrow \gamma$.

Note that the configuration of the linkage after stage 2 coincides with its configuration before stage 1. To return to the initial configuration, it is sufficient to perform the motion $S, \alpha_1 : -\gamma \longrightarrow 0, \alpha_2 : \gamma \longrightarrow 0$.

Due to the symmetry of the motions of the end links, the displacement of the central link along the $y$-axis and its angle of rotation stay equal to zero during the

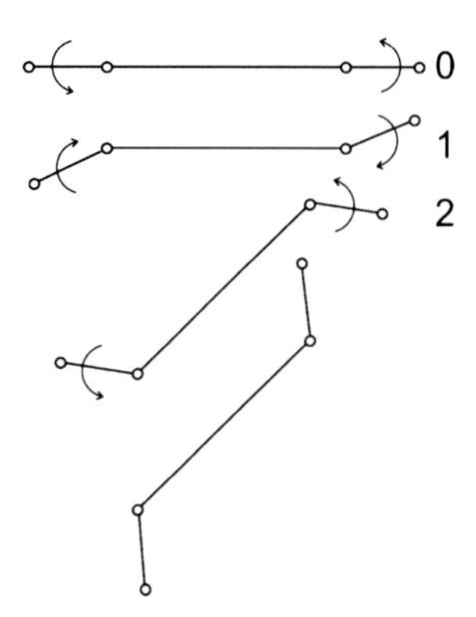


**Fig. 8** Rotation

whole cycle of motion. The total displacement of the middle of the central link along the $y$-axis and the average speed of the lateral motion are given by [6]:

$$\Delta y = 4m_0 m^{-1} l \sin\gamma, \quad v_2 = \Delta y T^{-1}. \tag{8}$$

*Rotation* (Fig. 8).
First, the auxiliary motion $S, \alpha_1 : 0 \longrightarrow \gamma, \alpha_2 : 0 \longrightarrow \gamma$ is carried out. Then, the following two motions are performed:

(1) $F, \alpha_1 : \gamma \longrightarrow -\gamma, \quad \alpha_2 : \gamma \longrightarrow -\gamma;$
(2) $S, \alpha_1 : -\gamma \longrightarrow \gamma, \quad \alpha_2 : -\gamma \longrightarrow \gamma.$

Stages 1 and 2 can be repeated any number of times. To bring the linkage to its initial configuration, one should perform the motion $S, \alpha_1 : \gamma \longrightarrow 0, \alpha_2 : \gamma \longrightarrow 0$.

Using the conservation laws, we can calculate the displacements of the linkage during the cycle of the rotational motion. We obtain [10]:

$$\Delta x = \Delta y = 0,$$
$$\Delta\theta = \gamma + 2\frac{[m_0 l^2 - (m_0 + m_1)a^2]}{R} \arctan\left[\frac{R \tan(\gamma/2)}{m_0(a+l)^2 + m_1 a^2}\right], \tag{9}$$
$$R = \left[m_0^2(l^2 - a^2)^2 + m_1 a^2(2m_0 l^2 + 2m_0 a^2 + m_1 a^2)\right]^{1/2}.$$

Kinematics of basic motions listed above can be also described as follows.

With respect to the perpendicular axis passing through the central link:

- *Longitudinal* motion is produced by an alternating sequence of symmetric and antisymmetric elementary motions, corresponding to a double alternating sequence of clockwise and anticlockwise rotations of the left end link (and associated rotations of the right end link).
- *Lateral* motion is produced by a sequence of symmetric elementary motions corresponding to an alternating sequence of anticlockwise and clockwise rotations of the left end link (and associated rotations of the right end link).

**Table 1** Basic motions of the three-link system

| | Elementary motion | w.r.t. to $C_1C_2$ center | $O_1C_1$ rotation | $C_2O_2$ rotation |
|---|---|---|---|---|
| Longitudinal | Fast 1 | Symmetric | Clockwise | Anticlockwise |
| motion | Slow 2 | Antisymmetric | Clockwise | Clockwise |
| | Fast 3 | Symmetric | Anticlockwise | Clockwise |
| | Slow 4 | Antisymmetric | Anticlockwise | Anticlockwise |
| Lateral | Fast 1 | Symmetric | Anticlockwise | Clockwise |
| motion | Slow 2 | Symmetric | Clockwise | Anticlockwise |
| Rotation | Fast 1 | Antisymmetric | Clockwise | Clockwise |
| | Slow 2 | Antisymmetric | Anticlockwise | Anticlockwise |

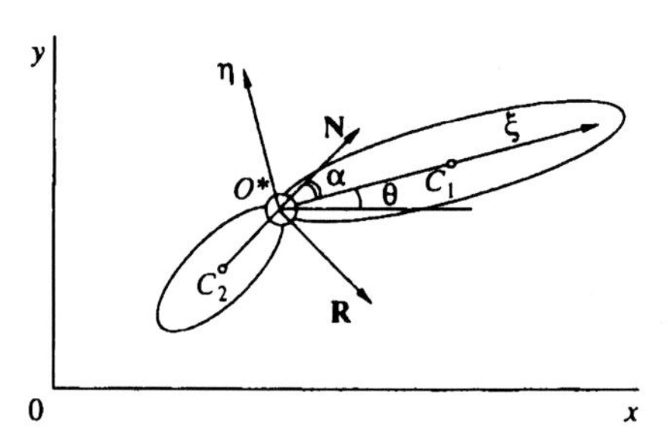


**Fig. 9** Two-link system

- *Rotation* is produced by a sequence of antisymmetric elementary motions corresponding to an alternating sequence of clockwise and anticlockwise rotations of the left end link (and associated rotations of the right end link).

The whole matter is summarized in the (Table 1).

The displacements and the average speed of the longitudinal, lateral, and rotational motions are evaluated by (7)–(9). Combining these three types of motions, the linkage can move from any initial position and configuration in the horizontal plane to any prescribed terminal position and configuration in this plane.

## 7 Two-Link System

Consider now a system of two rigid bodies connected by a joint $O^*$ which can move along the horizontal plane. Denote the masses of the bodies by $m_1$ and $m_2$, their centers of mass by $C_1$ and $C_2$, their moments of inertia about the vertical axis passing through the point $O^*$ by $J_1$ and $J_2$, the distances $O^*C_1$ and $O^*C_2$ by $a_1$ and $a_2$, and the friction coefficients for these bodies by $k_1$ and $k_2$, respectively. The body with index 1 will be referred to as the body, whereas the body with index 2 will be called the tail (Fig. 9).

The joint $O^*$ is treated as a point mass $m_0$ with the friction coefficient $k_0$.

We again consider the motion of the linkage along the horizontal plane $Oxy$. The actuator is installed at the point $O^*$ and creates the torque $M$. The coordinates of the joint $O^*$ are denoted by $x$, $y$, the angle between $O^*C_1$ and the $x$-axis by $\theta$, and the angle between $C_2O^*$ and $O^*C_1$ by $\alpha$ (Fig. 9).

Again, we introduce the notion of slow and fast motions. These motions begin and end at the states of rest of the linkage, and the angle $\alpha(t)$ changes between $\alpha^0$ and $\alpha^1$ in each of these motions, where $\alpha^0$, $\alpha^1 \in (-\pi, \pi)$. The durations of slow and fast motions are denoted by $\tau$ and $T$, respectively, $\tau \ll T$. In the slow motions, the tail rotates slowly enough, so that the body stays at rest. In the fast motions, the friction can be neglected, so that the conservation laws for the momentum and angular momentum of the linkage hold.

Denote by $\omega_0$ and $\varepsilon_0$, respectively, the maximal values of the angular velocity and acceleration of the tail during the slow motion:

$$\omega_0 = \max|\dot{\alpha}(t)|, \quad \varepsilon_0 = \max|\ddot{\alpha}(t)|. \tag{10}$$

Here, the maxima are taken along the whole slow motion.

As in the case of the three-member linkage, we are to obtain sufficient conditions which ensure that the body does not move during the slow motion. Here, however, we do not assume that the mass of the system is concentrated at certain points. Therefore, the distribution of the normal reactions and, hence, of the friction forces is not known a priori. In other words, we are to deal with the statically indeterminate case. For this case, the conditions of equilibrium for the rigid body in the presence of dry friction forces were analyzed in [30], where the notion of the guaranteed equilibrium conditions was introduced. The guaranteed equilibrium conditions ensure that the equilibrium holds under any admissible distribution of the normal reactions.

For our two-member linkage, the following two inequalities make up a sufficient condition which ensures that the body stays at rest during the slow motions of the tail [8]:

$$\begin{gathered} J_2\varepsilon_0 + m_2gk_2a_2 \le m_1gk_1a_1, \\ J_2\varepsilon_0 + m_2gk_2a_2 + m_2a_1a_2[\omega_0^4 + (\varepsilon_0 + gk_2a_2^{-1})^2]^{1/2} \le m_0gk_0a_1. \end{gathered} \tag{11}$$

These inequalities hold, if $\omega_0$ and $\varepsilon_0$ in (10) are sufficiently small and the following two simpler inequalities are true:

$$m_2k_2a_2 < m_1k_1a_1, \quad m_2k_2(a_1 + a_2) < m_0k_0a_1. \tag{12}$$

The magnitude of the control torque $M$ during the fast motion must satisfy the following condition similar to (6):

$$\begin{gathered} |M| \gg m^*gk^*a^*, \qquad m^* = \max(m_0, m_1, m_2), \\ k^* = \max(k_0, k_1, k_2), \quad a^* = \max(a_1, a_2). \end{gathered} \tag{13}$$

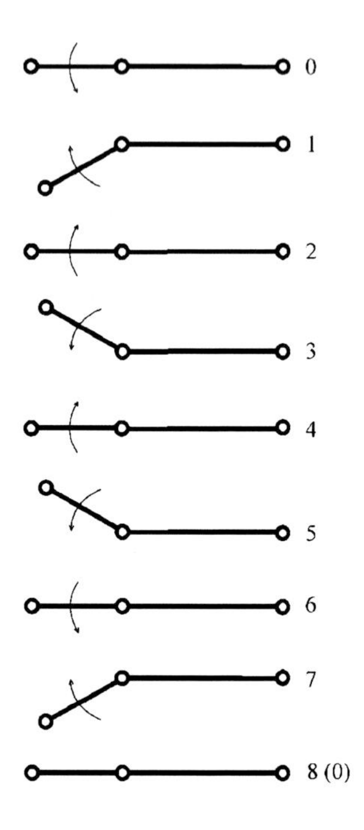

**Fig. 10** Motion of two-link system

The following sequences of slow ($S$) and fast ($F$) motions form the longitudinal motion of the two-member linkage (Fig. 10):

$$
\begin{array}{ll}
(1)\ S,\alpha:0\longrightarrow\beta; & (2)\ F,\alpha:\beta\longrightarrow0;\\
(3)\ S,\alpha:0\longrightarrow-\beta; & (4)\ F,\alpha:-\beta\longrightarrow0;\\
(5)\ S,\alpha:0\longrightarrow-\beta; & (6)\ F,\alpha:-\beta\longrightarrow0;\\
(7)\ S,\alpha:0\longrightarrow\beta; & (8)\ F,\alpha:\beta\longrightarrow0.
\end{array}
$$

Here, $\beta$ is a fixed angle, $\beta\in(-\pi,\pi)$. As a result of this sequence of motions, the linkage moves along itself by the distance [8]

$$\Delta x = 8m_2m^{-1}a_2\sin(\beta/2)\cos(\gamma/2)\sin[(\beta-\gamma)/2]. \tag{14}$$

The total lateral displacement and rotation are equal to zero. In (14), the following denotations are used:

$$
\begin{gathered}
m = m_0+m_1+m_2,\ \gamma=\beta/2+A_0A_1^{-1}A_2^{-1}\arctan\left[A_1A_2^{-1}\tan(\beta/2)\right],\\
A_0 = m(J_2-J_1)+m_1^2a_1^2-m_2^2a_2^2,\\
A_{1,2} = \left[m(J_1+J_2)-(m_1a_1\pm m_2a_2)^2\right]^{1/2}.
\end{gathered}
\tag{15}
$$

The average speed of the longitudinal motion is $v_3=\Delta x/(4T)$.

Note that the longitudinal motion of the two-link system is more complicated than the respective motion of the three-link system. This can be easily explained by the fact that the two-link system has only one actuator, and, to perform the periodic longitudinal motion, the lateral and angular displacements of the body must be compensated by motions of one link (the tail) relative to the body.

It is shown [8] that the two-member linkage satisfying conditions (12) can, starting from any given position and configuration in the horizontal plane and combining slow and fast motions, reach any prescribed position and configuration.

## 8 Optimization of Locomotion

The average speed of motion of the linkages depends on their geometrical and mechanical parameters such as the lengths and masses of links and coefficients of friction, as well as on the parameters of the motion itself. This dependence was analyzed in [9, 10] where also the parametric optimization of the average speed with respect to the parameters of the linkages and their motions was performed.

To specify the problem, let us consider the longitudinal motions of the three-member linkage. We assume that the angular velocity $\omega(t)$ of the end links in slow motions first increases linearly from 0 to its maximal value $\omega_0$ and then decreases linearly from $\omega_0$ to 0. Then we have for the slow stages of the longitudinal motion

$$\begin{gathered} \omega(t)=\varepsilon_0 t, \quad t\in[0,T/2], \\ \omega(t)=\varepsilon_0(T-t), \quad t\in[T/2,T], \\ \omega_0=|\dot{\alpha}_i(t)|, \quad i=1,2, \quad \omega_0=2\gamma T^{-1}, \quad \varepsilon_0=4\gamma T^{-2}. \end{gathered} \tag{16}$$

The sufficient condition for the longitudinal motion (4) for our case takes the form

$$m_0 l\left\{\left[(2\gamma T^{-1})^4+P^2\right]^{1/2}+Pla^{-1}\right\}\leq m_1 g k_1, \quad P=4\gamma T^{-2}+gk_0 l^{-1}. \tag{17}$$

Let us fix the mass $m_1$ of the joints, the length $2a$ of the central link, and the angle of rotation $\gamma$. The mass $m_0$, the length $l$ of the end links, the duration $T$ of the slow motion, and the friction coefficients $k_0$ and $k_1$ are to be chosen in order to maximize the speed $v_1=\Delta x(2T)^{-1}$; see (7). The imposed constraints comprise the inequality (17) and the bounds on the friction coefficients:

$$k^-\leq k_0\leq k^+, \quad k^-\leq k_1\leq k^+. \tag{18}$$

Here, $k^-$ and $k^+$ are given positive constants.

It is shown [9, 10] that the desired maximum of $v_1$ is reached in case of the equality sign in (17), and the optimal values of $k_0$ and $k_1$ are equal to $k^-$ and $k^+$, respectively. Let us introduce the characteristic speed and time given by

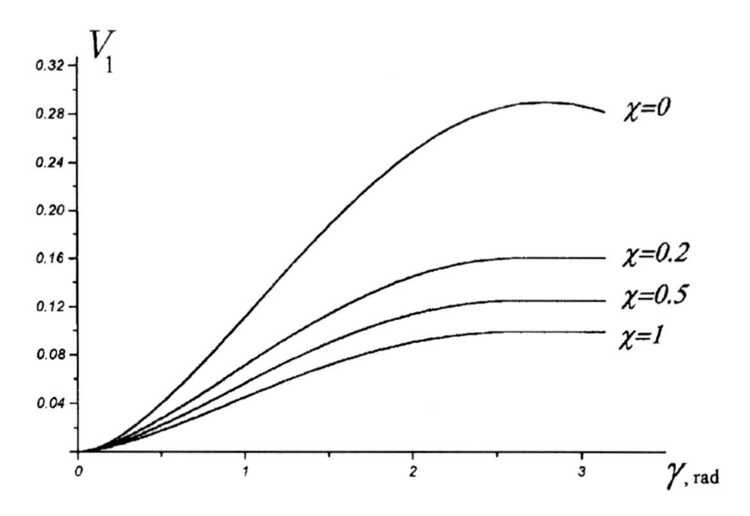


**Fig. 11** Maximal speed

$$v_0 = (gak^+)^{1/2}, \quad T_0 = a^{1/2}(gk^+)^{-1/2}, \tag{19}$$

and the following dimensionless quantities:

$$\lambda = l/a, \quad \mu = m_1/m_0, \quad T_1 = T/T_0,$$
$$\chi = k^-/k^+, \quad V_1 = v_1/v_0. \tag{20}$$

Then our optimization problem is reduced to maximizing the dimensionless speed $V_1$ given by

$$V_1 = 2\lambda(\mu+1)^{-1}T_1^{-1}\sin^2(\gamma/2) \tag{21}$$

over positive $\lambda$, $\mu$, and $T_1$ under the constraint

$$[16\gamma^4 T_1^{-4}\lambda^2 + (4\gamma T_1^{-2}\lambda + \chi)^2]^{1/2} + (4\gamma T_1^{-2}\lambda + \chi)\lambda = \mu. \tag{22}$$

Equation (21) stems from (7), whereas constraint (22) follows from (17) where the inequality sign is replaced by the equality. Also, (16), (19), and (20) are taken into account.

The optimization problem stated above was solved numerically in [9, 10]. Some results are presented in Figs. 11–14 where the maximal speed $V_1$ as well as optimal values of $\lambda$, $\mu$, and $T_1$ are shown as functions of the angle $\gamma$ given in radians. These data correspond to the following values of $\chi = k^-/k^+ = 0;\ 0.2;\ 0.5;\ 1$ which are shown in the figures.

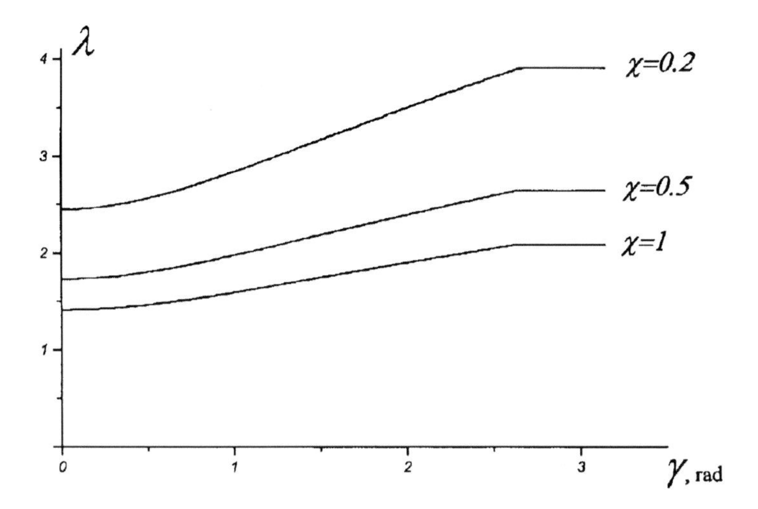


**Fig. 12** Optimal $\lambda$

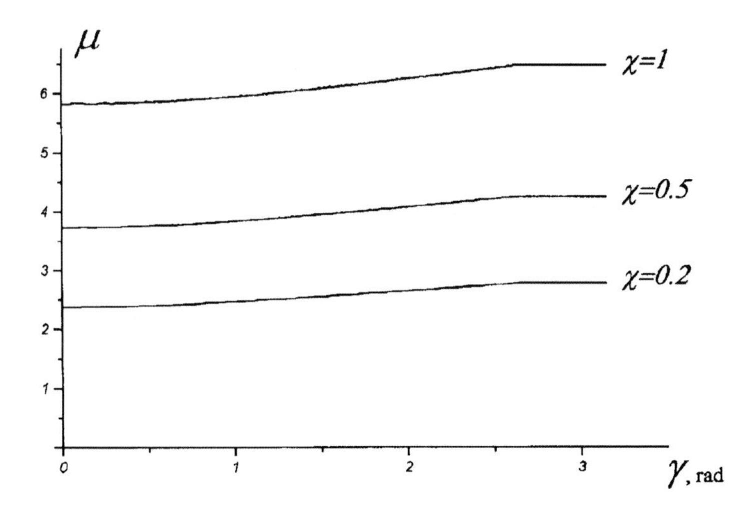


**Fig. 13** Optimal $\mu$

The obtained numerical results reveal the following properties of the optimal solutions. The maximal speed and the duration $T_1$ depend significantly on the angle of rotation $\gamma$ and increase with $\gamma$. The optimal value of $\lambda$ also increases with $\gamma$, whereas the optimal value of $\mu$ does not depend significantly on $\gamma$.

All optimal parameters depend considerably on the ratio $\chi = k^-/k^+$ of the friction coefficients: the speed $V_1$ as well as the optimal values of $\lambda$ and $T_1$ decrease with $\chi$, whereas $\mu$ increases with $\chi$.

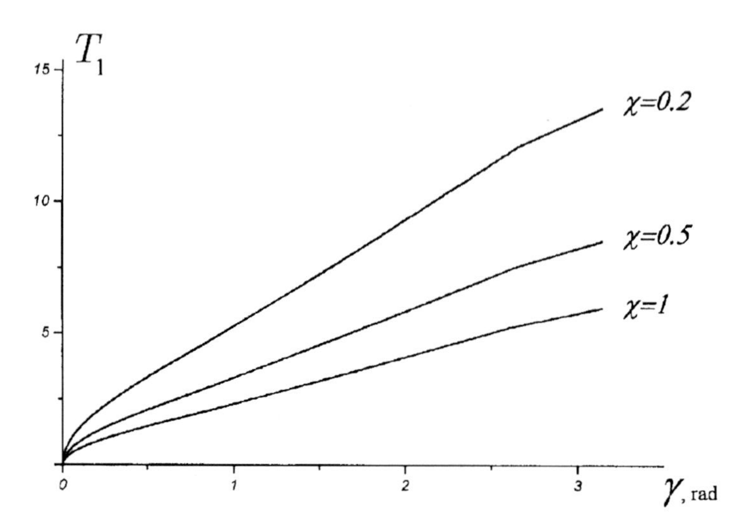


**Fig. 14** Optimal $T_1$

**Table 2** Optimal parameters for the longitudinal motion

| $k^-/k^+$ | 0.2/0.2 | 0.1/0.5 | 0.2/1 | 0.5/1 | 0/1 |
|---|---|---|---|---|---|
| $\gamma$, deg | 60 | 60 | 90 | 60 | 90 |
| $l, m$ | 0.32 | 0.57 | 0.64 | 0.4 | $\infty$ |
| $m_0$, kg | 0.17 | 0.40 | 0.39 | 0.26 | 1 |
| $m = 2(m_0 + m_1)$, kg | 2.34 | 2.80 | 2.78 | 2.52 | 4 |
| $T_1, s$ | 0.77 | 1.10 | 1.08 | 0.49 | $\infty$ |
| $V_1, m/s$ | 0.030 | 0.075 | 0.167 | 0.084 | 0.28 |

The required magnitude of the control torques, according to (6), can be now estimated as follows:

$$|M| \sim 10 m_1 g k^+ l. \tag{23}$$

Note that we used the sufficient condition (4) as the constraint in our optimization problem. Strictly speaking, we should use, instead of (4), the necessary and sufficient condition, which is not available in the explicit form for our nonlinear system. However, the results of computer simulations based on the numerical integration of the complete set of nonlinear differential equations show that our sufficient condition is rather close to the necessary one for a wide range of parameters. Therefore, the obtained results are close to the optimal ones.

As an example, we consider the three-member linkage having the parameters $a = 0.2\,\text{m}$, $m_1 = 1\,\text{kg}$. Optimal dimensional parameters of the longitudinal motion for some values of the friction coefficients and angle $\gamma$ are presented in Table 2. Similar results are obtained also for the optimal lateral motion of the three-member linkage [10] as well as for the two-member linkage [9].

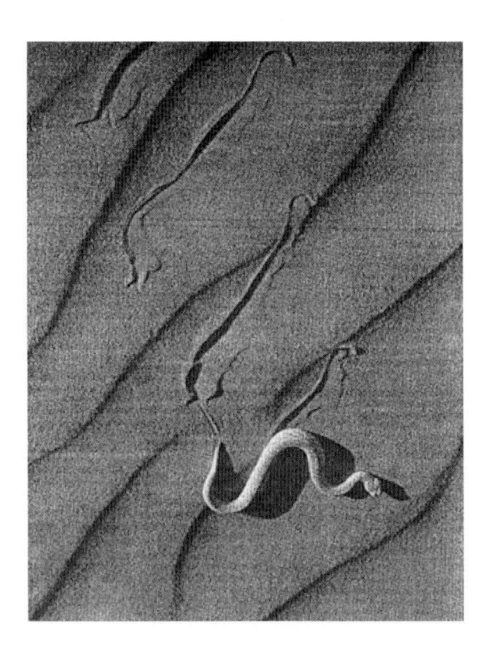

**Fig. 15** Snake in a desert

Note some differences between the three-member and two-member optimal linkages. For the three-member linkage, the optimal length $l$ of the end links is greater than the half-length $a$ of the central link: $l > a$; see Fig. 12. By contrast, the optimal length $a_2$ of the tail of the two-member linkage is smaller than the half-length of the body: $a_2 < a_1/2$. The gain in speed due to the optimization of lengths and masses is rather essential, up to 50 % and more.

It follows from the obtained results that the maximal longitudinal speed of the three-member linkage can be estimated by the formula

$$v_1 \approx 0.1(gak^+)^{1/2}. \tag{24}$$

The maximal speed of the lateral motion for this linkage is several times higher.

Note that snakes often prefer the lateral mode of motion to the longitudinal one, especially, if the speed is essential (Fig. 15).

As for the two-member linkage, its maximal longitudinal speed is about two times smaller than the speed $v_1$ for the three-member linkage of approximately the same size; see (24). This difference can be easily explained by the fact that the three-member linkage is equipped with two actuators, whereas the two-member linkage has only one.

Let us compare the obtained results for the dynamical locomotion with those for the quasistatic one from Sect. 3. For the quasistatic motion, we have $M \sim 2mgka$, whereas for the dynamical mode, according to (23), we find $M \sim 10mgka$, where $a$ is a characteristic length of links and $m$ is a mass of the joints. Thus, the dynamical mode requires higher values of the control torques, but it provides a higher speed of locomotion.

Apart from the average velocity of motion, another important characteristic of the locomotion is the energy consumption per unit of path. The analysis of the snakelike locomotion taking account of the two optimality criteria, average speed and energy consumption, is performed in [13]. It was shown that the increase of speed leads to the increase in the energy consumption, but there exists such maximal speed that cannot be surpassed by the increase of the energy consumption.

## 9 System with a Movable Internal Mass

Consider now a two-mass mechanical system consisting of two rigid bodies: a main body (a container) of mass $M$ and an internal body of mass $m$ (Fig. 2). For brevity, we will call these bodies "body $M$" and "mass $m$," respectively. Both bodies move translationally along the horizontal $x$-axis. Body $M$ is subject to the dry friction force obeying Coulomb's law (1).

Denote by $x$ and $v$ the absolute coordinate and velocity of body $M$ and by $\xi$, $u$, and $w$ the displacement, velocity, and acceleration of mass $m$ with respect to body $M$. The kinematic equations of mass $m$ with respect to body $M$ have the form

$$\dot{\xi} = u, \quad \dot{u} = w. \tag{25}$$

The dynamics of the system is described by equations

$$\dot{x} = v, \quad M\dot{v} + (\dot{v} + w) = F. \tag{26}$$

where the friction force is defined by (1) as

$$\begin{aligned} F &= -k(M+m)g \text{ sign } v, \quad \text{if } v \neq 0, \\ | F | &\leq k(M+m)g, \quad \text{if } v = 0. \end{aligned} \tag{27}$$

Introducing the notation

$$\mu = m/(M+m) \tag{28}$$

and using expressions (27), we transform (26) to the form

$$\dot{x} = v, \quad \dot{v} = -\mu v - r(v), \tag{29}$$

where

$$\begin{aligned} r(v) &= kg, \quad \text{if} \quad v \neq 0, \\ |r(v)| &\leq \quad kg, \quad \text{if} \quad v = 0. \end{aligned} \tag{30}$$

The motion of mass $m$ relative to body $M$ is assumed to be periodic with period $T$ and confined within a certain range, i.e.,

$$0 \leq \xi(t) \leq L. \tag{31}$$

Without loss of generality, we can assume that at the beginning and at the end of the period mass $m$ is at the left end of interval (31). The periodicity condition then takes the form

$$\xi(0) = \xi(T) = 0, \quad u(0) = u(T) = 0. \tag{32}$$

At a certain time instant $\theta \in (0,T)$, mass $m$ reaches the right end of the interval $[0,L]$, so that $\xi(\theta) = L$.

The motion of the system as a whole is controlled by relative motion of mass $m$, so that the functions $\xi(t)$, $u(t)$, and $w(t)$ play the role of controls. These functions are related by (25) and satisfy constraints (31) and periodicity conditions (32).

We will seek those relative periodic motions of mass $m$ for which the absolute velocity of body $M$ is also periodic, i.e.,

$$v(0) = v(T) = v_0, \tag{33}$$

and the average velocity of the system as a whole, i.e.,

$$V = \triangle x/T, \quad \triangle x = x(T) - x(0), \tag{34}$$

is a maximum. Here, $v_0$ is a constant.

Two different assumptions about this constant were considered. In [24], this constant was supposed to be zero: $v_0 = 0$. In [26], such value of $v_0$ was determined that maximizes the average velocity $V$ from (34). It is natural that the optimal average velocity in the second case is higher than in the first one.

## 10 Relative Motions

Let us confine ourselves to examining two simple types of periodic relative motions of mass $m$ that satisfy conditions (31)–(33). We will call them two-phase and three-phase motions. In *two-phase* motion, the relative velocity $u(t)$ of mass $m$ is piecewise constant, and there are two segments of constant velocity in a period of motion. In *three-phase* motion, the relative acceleration $w(t)$ of mass $m$ is piecewise constant, and there are three segments of constant acceleration in a period of motion. It can be shown easily that the two-phase and three-phase motions have the smallest possible number of segments of constant velocity and acceleration, respectively, compatible with the periodicity conditions imposed.

**Fig. 16** Two-phase motion

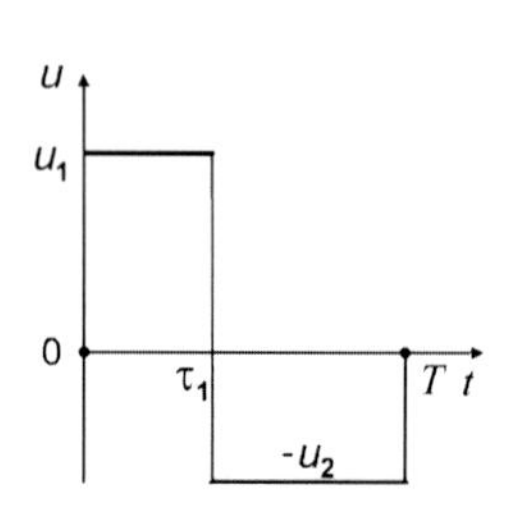


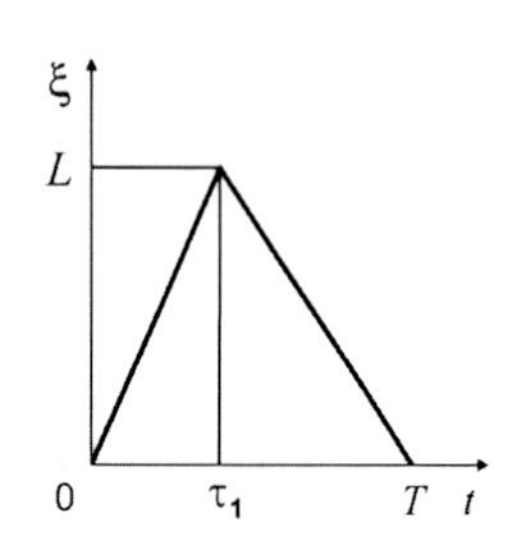


The two-phase and three-phase motions are the simple models that correspond to different possibilities for the actuators that control mass $m$. The two-phase motion corresponds to the case of a high possible acceleration and bounded relative velocity. The three-phase motion corresponds to a bounded relative acceleration.

For the two-phase motion, we denote by $\tau_1$ and $\tau_2$ the durations of the intervals of constant velocity and by $u_1$ and $u_2$ the magnitudes of the velocity in these intervals, respectively. We have

$$u(t) = u_1, \quad t \in (0, \tau_1); \quad u(t) = -u_2, \quad t \in (\tau_1, T), \\ \tau_1 + \tau_2 = T. \tag{35}$$

Here, $u_1$ and $u_2$ are positive quantities that are assumed to be bounded by a given constant:

$$0 < u_i \leq U, \quad i = 1, 2. \tag{36}$$

For the three-phase motion, we denote the durations of the intervals of constant relative acceleration by $\tau_1$, $\tau_2$, and $\tau_3$. The magnitudes of the acceleration in these time intervals are denoted by $w_1$, $w_2$, and $w_3$, respectively. We have

$$w(t) = w_1, \quad t \in (0, \tau_1); \quad w(t) = -w_2, \quad t \in (\tau_1,\ \tau_1 + \tau_2); \\ w(t) = w_3, \quad t \in (\tau_1 + \tau_2,\ T); \quad \tau_1 + \tau_2 + \tau_3 = T. \tag{37}$$

The constants $w_1$, $w_2$, and $w_3$ are bounded, similarly to (36):

$$0 < w_i \leq W, \quad i = 1, 2, 3, \tag{38}$$

where $W$ is a given positive constant.

The two-phase and three-phase motions are illustrated by Figs. 16 and 17.

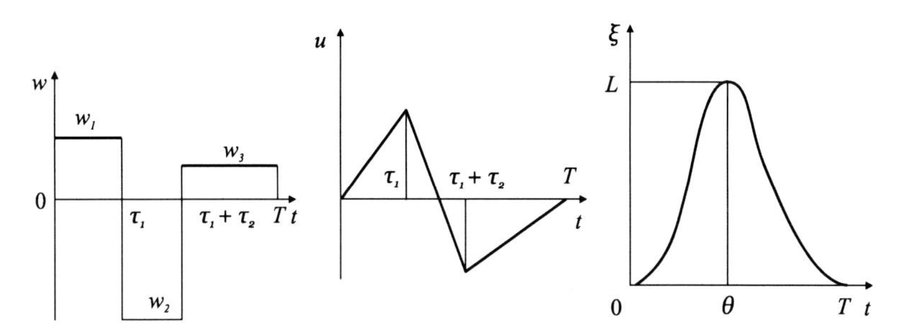


**Fig. 17** Three-phase motion

## 11 Optimization Results

In papers [24, 26], the locomotion of the two-body system controlled by the motion of the internal mass is considered under the assumption more general than Coulomb's law (1), namely, under the anisotropic friction. Instead of (30), the anisotropic friction force is defined as follows:

$$\begin{aligned} &r(v) = k_+ g, \quad \text{if} \quad v > 0, \\ &r(v) = k_- g, \quad \text{if} \quad v < 0, \\ &-k_- \, g \leq r(v) \leq k_+ g, \quad \text{if} \quad v = 0. \end{aligned} \tag{39}$$

Here, $k_+$ and $k_-$ are the coefficients of friction for the forward and backward motions, respectively.

Complete solution of the optimization problem formulated at the end of Sect. 9 is given in [24] both for the two-phase and three-phase motions in the case where $v_0 = 0$. The case where $v_0$ can be chosen arbitrarily is considered in [26].

Here, we restrict ourselves with formulating the final results only for the simplest case of the isotropic friction where $k_+ = k_- = k$ in (39)

1. The optimal two-phase motion for the case where $v_0 = 0$ [24] is given by (35) with

$$\begin{aligned} &u_1 = u_2 = (La/\mu)^{1/2}, \quad \tau_1 = \tau_2 = (\mu L/a)^{1/2}, \quad a = kg, \\ &T = 2\tau_1, \quad V = (\mu La)^{1/2}/2. \end{aligned} \tag{40}$$

This solution is valid, if the admissible relative velocity $U$ of the internal body in (36) is sufficiently high, i.e., under the condition

$$U \geq (La/\mu)^{1/2}. \tag{41}$$

If condition (41) is violated, then the motion described by (40) is impossible for the fixed value of $L$. However, this motion can be realized under the smaller value of $L$. Let us take

$$L' = \mu U^2/a < L, \quad \text{if} \quad U < (L\,a/\mu)^{1/2} \tag{42}$$

and replace $L$ by $L'$ in (40). Thus, we obtain finally

$$\begin{aligned} &u_1 = u_2 = (L'a/\mu)^{1/2}, \quad \tau_1 = \tau_2 = (\mu L'/a)^{1/2}, \quad a = kg, \\ &T = 2\tau_1, \quad V = (\mu L'a)^{1/2}/2, \\ &L' = \min(L, \mu U^2/a). \end{aligned} \tag{43}$$

2. The optimal two-phase motion for the case where $v_0$ can be chosen arbitrarily is described by formulas [26]:

$$\begin{aligned} &u_1 = u_2 = U, \quad \tau_1 = \tau_2 = L/U, \quad T = \tau_1 + \tau_2, \\ &V = \frac{2U^2 - u_0^2}{2u_0 U}, \quad v_0 = \frac{U^2 - u_0^2}{u_0 U}, \quad u_0 = (La/\mu)^{1/2}, \quad a = kg, \\ &\text{for} \quad U \geq u_0/\sqrt{2}. \end{aligned} \tag{44}$$

Note that, by decreasing $L$ in (44), $u_0$ can be made very small. Then the inequality in (44) will be satisfied, and the average velocity $V$ can be made very high; however, this situation will happen due to the increase of the initial velocity $v_0$ and the decrease of the period $T$.

3. The optimal three-phase motion for the case where $v_0 = 0$ is defined by the following formulas [24]:

$$w_i = ay_i/\mu, \quad i = 1,2,3,$$

$$\tau_1 = \left[\frac{2w_2 L}{w_1(w_1 + w_2)}\right]^{1/2}, \quad \tau_2 = \left(\frac{2L}{w_2}\right)^{1/2}\left[\left(\frac{w_1}{w_1 + w_2}\right)^{1/2} + \left(\frac{w_3}{w_2 + w_3}\right)^{1/2}\right],$$

$$\tau_3 = \left[\frac{2w_2 L}{w_3(w_2 + w_3)}\right]^{1/2}, \quad T = \left(\frac{2L}{w_2}\right)^{1/2}\left[\left(\frac{w_1 + w_2}{w_1}\right)^{1/2} + \left(\frac{w_2 + w_3}{w_3}\right)^{1/2}\right],$$

$$a = kg, \quad Y = \mu W/a, \quad V = (\mu La/2)F,$$

$$y_1 = y_3 = 1, \quad y_2 = Y, \quad F = \frac{Y - 1}{Y^{1/2}(Y + 1)}^{1/2}, \quad \text{if} \quad 1 < Y \leq 2 + \sqrt{5}$$

$$y_1 = 1, \quad y_2 = Y, \quad y_3 = \frac{Y + 1}{Y - 3}, \quad F = \left(\frac{Y}{Y + 1}\right)^{1/2}, \quad \text{if} \quad Y > 2 + \sqrt{5}.$$

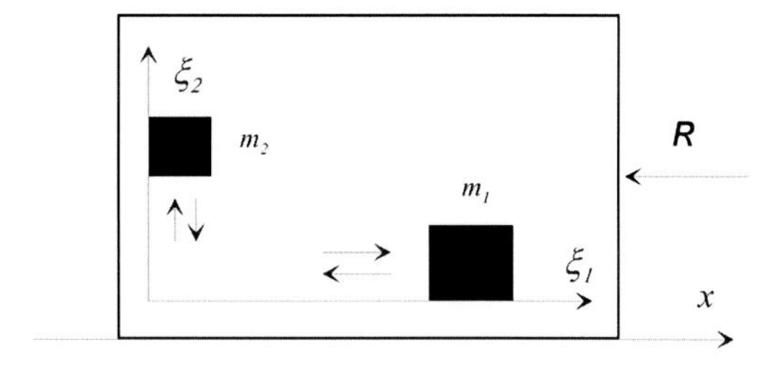


**Fig. 18** Body with two internal masses

If $Y < 1$, then this type of motion under the given values of $\mu$, $w$, and $a$ is impossible: the main body $M$ will not move, if the internal mass moves with the relative acceleration bounded by condition (38).

Note that if the two-phase mode of locomotion of our system is performed, then the velocity of the main body is not continuous and has two jumps within the period of motion: at $t = 0$ and $t = \tau_1$. Though the total displacement $\triangle x$ of the main body (34) during the period is positive, this body moves also backwards for the part of the period.

By contrast, in the three-phase motion the velocity of the body is a continuous function of time, and there are no intervals of backward motion during the period. Here, the motion consists only of intervals of forward motion and intervals of rest.

Since the energy consumption in the case of dry friction is proportional to the path of the body $M$, it is clear that, from the point of view of energy consumption, the three-phase motion is preferable to the two-phase one.

The problem of optimal control for a body containing a moving internal mass in the presence of the isotropic dry friction is considered in [30] under the constraint $|w(t)| \leq W$ imposed on the relative acceleration of the internal mass. The obtained optimal acceleration occurs to be piecewise constant with three intervals of constancy.

Up till now, we considered the case where there is only one internal mass moving horizontally inside the main body. Let us discuss briefly the case where, besides the mass $m_1$ moving horizontally, there is another internal mass $m_2$ moving vertically within the main body $M$ (Fig. 18). The vertical motion of mass $m_2$ changes the normal reaction of the horizontal plane and, as a result, changes the dry friction force that is proportional to this reaction. This circumstance can be used in order to increase the average velocity of the system as a whole. The corresponding optimal control problem is considered in [31].

## 12 Experimental Results

The principles of motion described above are implemented in several experimental models.

**Fig. 19** Three-phase motion

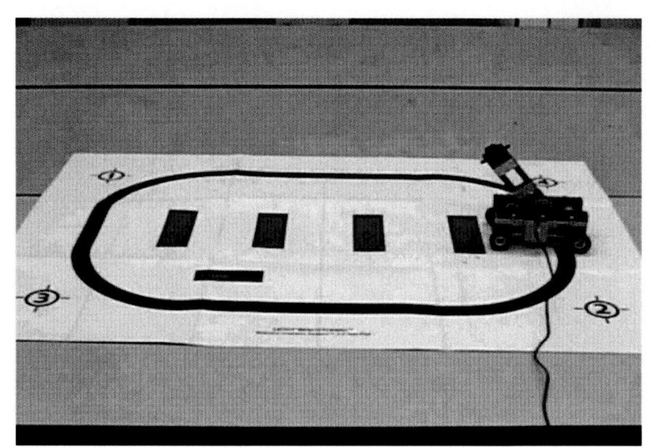

**Fig. 20** Inverted pendulum on a cart

The three-member snakelike mechanism is shown in Fig. 19. Experiments carried out with this mechanism [12] confirmed the theoretical results presented and showed that this mechanism can perform longitudinal, lateral, and rotational motions described in Sect. 6.

Mechanical systems with movable internal masses were modelled by several experimental devices [32, 33]. In Fig. 20, the internal motion is performed by an inverted pendulum. This system is described in [34].

In the capsubot shown in Fig. 21, the internal motions are performed by an electromagnetic actuator [35]. In Fig. 22, the cart carries eccentric rotating wheels.

Their rotation results in the vibration of an equivalent mass both in the horizontal and vertical directions. Thus, the case of the two internal masses is modelled in this experiment.

Mini-robots that utilize the same principle of the moving internal mass and can move inside tubes have been designed [19]. Such vibro-robots consist of two parts which vibrate with respect to each other with the frequency $20 \div 40$ Hz and can move inside tubes, both straight and curved, of the diameter $4 \div 70$ mm with a speed $10 \div 30$ mm/s (Fig. 23).

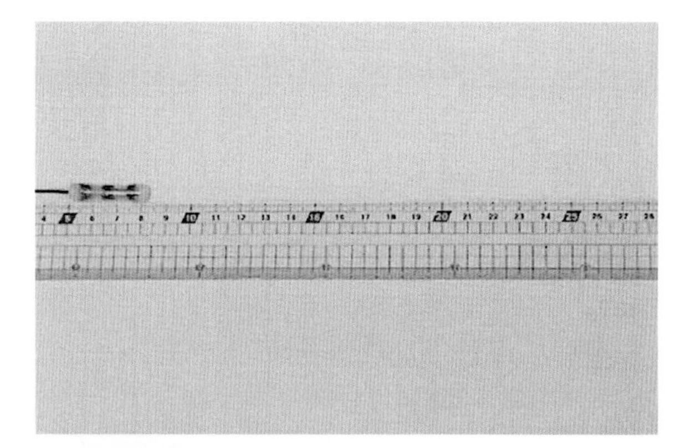

**Fig. 21** Capsubot

**Fig. 22** Cart with rotating wheels

# 13 Conclusions

Snakelike multibody linkage and bodies containing moving internal masses are mechanical systems that can, by changing their configuration, move progressively along a plane. Their locomotion is a result of relative motions of bodies that make up these systems and is possible only in the presence of external forces such as resistance of the environment.

Note that whereas the multilink systems are analogous to snakes and some other animals, the system with moving internal masses has no direct biological prototypes.

In the paper, we have considered the dry friction forces obeying Coulomb's law as a resistance forces. The influence of other resistance forces such as linear, piecewise linear, and quadratic forces depending on the velocity of motion was analyzed in [26, 36, 37].

Various modes of locomotion have been considered above. The average velocity of motion has been estimated. The optimal values of the geometrical and mechanical parameters of the systems under consideration have been obtained that correspond

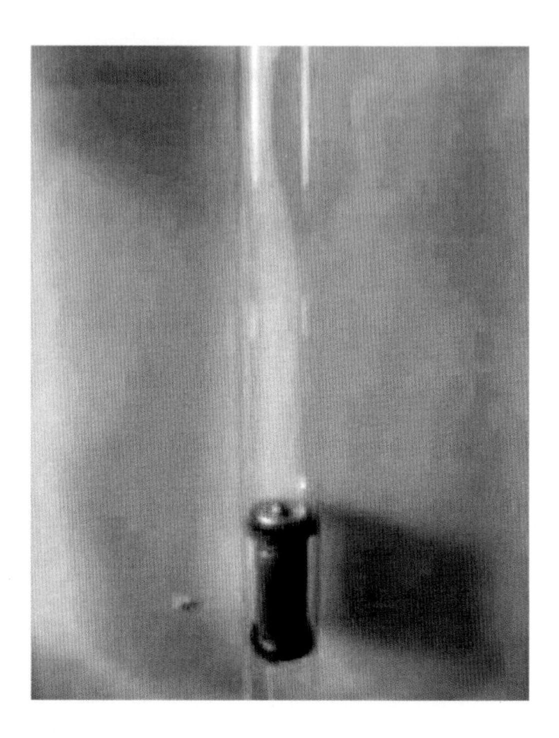

**Fig. 23** Vibro-robot in a tube

to the maximum velocity of locomotion. The results of experiments confirm the theoretical estimates and conclusions.

The systems under consideration are of interest for robotics. They can be regarded as prototypes of mobile robots that can move and fulfil certain tasks such as inspection and repair in complicated and hazardous environment, in pipelines and also in medical applications. Such robotic systems already exist.

## 14 Acknowledgements

The work was supported by the Russian Foundation for Basic Research, Projects 11-01-00513-a, 11-01-12110-ofi, and by the Grant of the President of the Russian Federation for leading scientific schools No. 369.2012.1.

## References

[1] Gray, J.: Animal Locomotion. Norton, New York (1968)
[2] Hirose, S.: Biologically Inspired Robots: Snake-Like Locomotors and Manipulators. Oxford University Press, Oxford (1993)

[3] Bayraktaroglu, Z.Y., Blazevic, P.: Snake-like locomotion with a minimal mechanism. In: Proceedings of the Third International Conference Climbing and Walking Robots CLAWAR, pp. 201–207. Madrid (2000)
[4] Transeth, A.A., Pettersen, K.V., Liljeback, P.: A survey on snake robot modelling and locomotion. Robotica **27**(7), 999–1015 (2009)
[5] Chernousko, F.L.: The motion of a multilink system along a horizontal plane. J. Appl. Math. Mech. **64**(1), 5–15 (2000)
[6] Chernousko, F.L.: The wavelike motion of a multilink system on a horizontal plane. J. Appl. Math. Mech. **64**(4), 497–508 (2000)
[7] Chernousko, F.L.: On the motion of a three-member linkage along a plane. J. Appl. Math. Mech. **65**(1), 3–18 (2001)
[8] Chernousko, F.L.: Controllable motions of a two-link mechanism along a horizontal plane. J. Appl. Math. Mech. **65**(4), 565–577 (2001)
[9] Smyshlyaev, A.S., Chernousko, F.L.: Optimization of the motion of multilink robots on a horizontal plane. J. Comput. Syst. Sci. Int. **40**(2), 340–348 (2001)
[10] Chernousko, F.L.: Snake-like locomotions of multilink mechanisms. J. Vib. Control **9**(1–2), 235–256 (2003)
[11] Chernousko, F.L.: Modelling of snake-like locomotion. J. Appl. Math. Comput. **164**(2), 415–434 (2005)
[12] Chernousko, F.L., Pfeiffer, F., Sobolev, N.A.: Experimental study of snake-like locomotion of a three-link mechanism. In: Proceedings of the IUTAM Symposium Vibration Control of Nonlinear Mechanisms and Structures, pp. 141–150. Springer, Dordrecht (2005)
[13] Chernousko, F.L., Shunderyuk, M.M.: The influence of friction forces on the dynamics of a two-link mobile robot. J. Appl. Math. Mech. **74**(1), 22–36 (2010)
[14] Darby, A.P., Pellegrino, S.: Inertial stick-slip actuators for active control of shape and vibration. J. Intell. Mater. Syst. Struct. **8**(12), 1001–1011 (1997)
[15] Breguet, J.-M., Clavel, R.: Stick and slip actuators: design, control, performances and applications. In: Proceedings of International Symposium Micromechatronics and Human Science (MHS), pp. 89–95. IEEE, New-York (1998)
[16] Schmoeckel, F., Worn, H.: Remotely controllable mobile mocrorobots acting as nano positioners and intelligent tweezers in scanning electron microscopes (SEMs). In: Proceedings of International Conference Robotics and Automation, pp. 3903–3913. IEEE, New York (2001)
[17] Fidlin, F., Thomsen, J.J.: Predicting vibration-induced displacement for a resonant friction slider. Eur. J. Mech. A/Solids **20**(1), 155–166 (2001)
[18] Lampert, P., Vakebtutu, A., Lagrange, B., De Lit, P., Delchambre, A.: Design and performances of a one-degree-of-freedom guided nano-actuator. Robot. Comput. Integr. Manuf. **19**(1–2), 89–98 (2003)
[19] Gradetsky, V., Solovtsov, V., Kniazkov, M., Rizzotto, G.G., Amato, P.: Modular design of electro-magnetic mechatronic microrobots. In: Proceedings of 6th International Conference Climbing and Walking Robots CLAWAR, pp. 651–658 (2003)
[20] Kim, B., Lee, S., Park, J.H., Park, J.O.: Design and fabrication of a locomotive mechanism for capsule-type endoscopes using shape memory alloys (SMAs). IEEE/ASME Trans. Mechatron. **10**(1), 77–86 (2005)
[21] Vartholomeos, P., Papadopoulos, E.: Dynamics, design and simulation of a novel microrobotic platform employing vibration microactuators. Trans. ASME J. Dyn. Syst. Meas. Control **128**(1), 122–133 (2006)
[22] Zimmermann, K., Zeidis, I., Behn, C.: Mechanics of Terrestrial Locomotion. Springer, Berlin (2009)
[23] Chernousko, F.L.: The optimum rectilinear motion of a two-mass system. J. Appl. Math. Mech. **66**(1), 1–7 (2002)
[24] Chernousko, F.L..: Analysis and optimization of the motion of a body controlled by a movable internal mass. J. Appl. Math. Mech. **70**(6), 915–941 (2006)

[25] Chernousko, F.L..: Dynamics of a body controlled by internal motions. In: Proceedings of IUTAM Symposium Dynamics and Control of Nonlinear Systems with Uncertainty, pp. 227–236. Springer, Dordrecht (2007)
[26] Chernousko, F.L..: The optimal periodic motions of a two-mass system in a resistant medium. J. Appl. Math. Mech. **72**(2), 116–125 (2008)
[27] Chernousko, F.L..: Analysis and optimization of the rectilinear motion of a two-body system. J. Appl. Math. Mech. **75**(5), 493–500 (2011)
[28] Figurina, T.Yu.: Quasi-static motion of a two-link system along a horizontal plane. Multibody Syst. Dyn. **11**(3), 251–272 (2004)
[29] Figurina, T.Yu.: Controlled slow motions of a three-link robot on a horizontal plane. J. Comput. Syst. Sci. Int. **44**(3), 473–480 (2005)
[30] Chernousko, F.L.: Equilibrum conditions for a solid on a rough plane. Mech. Solids **23**(6), 1–12 (1988)
[31] Figurina, T.Yu.: Optimal motion control for a system of two bodies on a straight line. J. Comput. Syst. Sci. Int. **46**(2), pp. 227–233 (2007)
[32] Bolotnik, N.N., Zeidis, I.M., Zimmermann, K., Yatsun, S.F.: Dynamics of controlled motion of vibration-driven system. J. Comput. Syst. Sci. Int. **45**(5), 831–840 (2006)
[33] Chernousko, F.L., Zimmermann, K., Bolotnik, N.N.,Yatsun, S.F., Zeidis, I.: Vibration-driven robots. In: Proceedings of Workshop on Adaptive and Intelligent Robots: Present and Future, vol. 1, pp. 26–31. Moscow (2005)
[34] Li, H., Furuta K., Chernousko, F.L.: A pendulum-driven cart via internal force and static friction. In: Proceedings of International Conference "Physics and Control", pp. 15–17. St. Petersburg, Russia (2005)
[35] Li, H., Furuta, K., Chernousko, F.L.: Motion generation of the Capsubot using internal force and static friction. In: Proceedings of 45th IEEE Conference Decision and Control, pp. 6575–6580. San Diego, USA (2006)
[36] Chernousko, F.L.: Optimal motion of a two-body system in a resistive medium. J. Optim. Theory Appl. **147**(2), 278–297 (2010)
[37] Chernousko, F.L.: Optimal control of multilink systems in a fluid. Cybern. Phys. **1**(1), 17–21 (2012)

# Method of Steepest Descent for Two-Dimensional Problems of Calculus of Variations

**M.V. Dolgopolik and G.Sh. Tamasyan**

**Abstract** In this paper we demonstrate an application of nonsmooth analysis and the theory of exact penalty functions to two-dimensional problems of the calculus of variations. We derive necessary conditions for an extremum in the problem under consideration and use them to construct a new direct numerical minimization method (the method of steepest descent). We prove the convergence of the method and give numerical examples that show the efficiency of the suggested method. The described method can be very useful for solving various practical problems of mechanics, mathematical physics and calculus of variations.

**Keywords** Method of steepest descent • Two-dimensional problem • Calculus of variations

## 1 Introduction

It is well known that different problems appearing in mechanics and mathematical physics can be formulated as problems of the calculus of variations [14–16]. In particular, the problem of finding the solution of the Poisson equation in a domain $D$

$$\frac{\partial^2 u(x,y)}{\partial x^2} + \frac{\partial^2 u(x,y)}{\partial y^2} = f(x,y) \quad \forall\, (x,y) \in D,$$

satisfying some boundary conditions, can be formulated as a problem of minimizing the functional

M.V. Dolgopolik • G.Sh. Tamasyan (✉)
Saint Petersburg State University, Universitetskii prospekt 35, Petergof
Saint-Petersburg, 198504, Russia
e-mail: maxim.dolgopolik@gmail.com; grigoriytamasjan@mail.ru

V.F. Demyanov et al. (eds.), *Constructive Nonsmooth Analysis and Related Topics*,
Springer Optimization and Its Applications 87, DOI 10.1007/978-1-4614-8615-2_7,

$$I(u) = \iint_D \left[ \left( \frac{\partial u(x,y)}{\partial x} \right)^2 + \left( \frac{\partial u(x,y)}{\partial y} \right)^2 + 2u(x,y)f(x,y) \right] dxdy,$$

over the set of functions satisfying the same boundary conditions.

There are many different numerical methods for solving problems of the calculus of variations [1, 3, 12, 14]. One of the most universal and commonly used direct numerical methods is the Galerkin method [14, 15]. Recently, there were suggested new direct numerical methods for solving one-dimensional problems of the calculus of variations, such as the method of steepest descent (MSD) and the method of hypodifferential descent [3, 10]. In [6] the efficiency of the new direct numerical methods in comparison with the Galerkin method was shown. These methods are based on the ideas of nonsmooth analysis [4, 5] and the theory of exact penalty functions [2, 3, 8, 11]. These ideas appeared to be very useful for studying various kinds of problems, such as problems of control theory [9] and mechanics [7, 17].

In this paper we study two-dimensional problems of the calculus of variations. We describe two different approaches to studying these problems. The first approach is based on the theory of exact penalty functions and unfortunately this approach is very complicated. The second approach is based on different ideas and uses some peculiarities of the problem under consideration. This approach helped us to construct an MSD for solving two-dimensional problems of the calculus of variations.

## 2 Problem Formulation

Consider the bounded domain

$$D = \left\{ (x,y) \in \mathbb{R}^2 \mid 0 < x < T_1,\, 0 < y < T_2 \right\}$$

and denote the boundary of $D$ by $\Gamma$. We denote by $C^2(\overline{D})$ a vector space consisting of all twice continuously differentiable functions $f\colon D \longrightarrow \mathbb{R}$, such that $f$ and all its first- and second-order partial derivatives are bounded and uniformly continuous in $D$ (then there exist unique, bounded, continuous extensions of the function $f$ and all its first- and second-order partial derivatives to the closure $\overline{D}$ of the set $D$).

Let $\psi\colon \Gamma \longrightarrow \mathbb{R}$, $p_0\colon [0,T_1] \longrightarrow \mathbb{R}$ and $q_0\colon [0,T_2] \longrightarrow \mathbb{R}$ be given functions. Introduce the set

$$\Omega = \Big\{ u \in C^2(\overline{D}) \;\Big|\; u(x,y) = \psi(x,y) \;\forall\, (x,y) \in \Gamma,$$
$$u'_x(x,0) = p_0(x) \;\forall\, x \in [0,T_1], \quad u'_y(0,y) = q_0(y) \;\forall\, y \in [0,T_2] \Big\}.$$

The set $\Omega$ consists of all functions $u \in C^2(\overline{D})$, $u = u(x,y)$, such that $u$ satisfies the boundary condition, defined by the functions $\psi$, $p_0$, $q_0$. We will assume that $\Omega$ is not empty.

Consider the functional

$$I(u) = \iint_D F(x,y,u(x,y),u'_x(x,y),u'_y(x,y))\,dxdy,$$

where $F\colon [0,T_1]\times[0,T_2]\times\mathbb{R}\times\mathbb{R}\times\mathbb{R}\longrightarrow\mathbb{R}$, $F=F(x,y,u,p,q)$ is a given twice continuously differentiable function, $u\in C^2(\overline{D})$.

**Problem 1.** Find a local minimizer of the functional $I$ on the set $\Omega$.

## 3 Problem Reformulation (First Version)

Consider the normed space $(C(\overline{D}),\|\cdot\|)$, where $C(\overline{D})$ is a vector space of continuous functions on $\overline{D}$, $\|z\|=\sqrt{\langle z,z\rangle}$ and

$$\langle z_1,z_2\rangle = \iint_D z_1(x,y)z_2(x,y)\,dxdy$$

is an inner product.

Let $u\in\Omega$ be an arbitrary function and denote $z=u''_{xy}$. Then the function $u$ can be represented in the form

$$u(x,y) = -\psi(0,0)+\psi(x,0)+\psi(0,y)+\int_0^x\int_0^y z(\tau,\gamma)\,d\gamma d\tau. \tag{1}$$

Also, one has that

$$u'_x(x,y) = p_0(x)+\int_0^y z(x,\gamma)\,d\gamma,$$

$$u'_y(x,y) = q_0(y)+\int_0^x z(\tau,y)\,d\tau.$$

Hereafter, we will write $u$ instead of representation (1), where it cannot cause misunderstanding.

Define the functional

$$f(z) = \iint_D F\left(x,\ y,\ u(x,y),\ p_0(x)+\int_0^y z(x,\gamma)\,d\gamma,\ q_0(y)+\int_0^x z(\tau,y)\,d\tau\right)dxdy,$$

and applying representation (1) introduce the set

$$Z=\Big\{z\in C(\overline{D}) \mid u(x,T_2)=\psi(x,T_2)\ \forall\,x\in[0,T_1],$$
$$u(T_1,y)=\psi(T_1,y)\ \forall\,y\in[0,T_2]\Big\}.$$

Note that Problem 1 is equivalent to the problem $f(z)\longrightarrow \inf\limits_{z\in Z}$.

## 4 Penalty Function

One can represent the set $Z$ as

$$Z = \{z \in C(\overline{D}) \mid \varphi(z) = 0\},$$

where

$$\varphi(z) = \sqrt{\int_0^{T_1} [u(x,T_2) - \psi(x,T_2)]^2\,dx + \int_0^{T_2} [u(T_1,y) - \psi(T_1,y)]^2\,dy}.$$

Let $\lambda \geqslant 0$ be fixed. Introduce the function

$$\Phi_\lambda(z) = f(z) + \lambda\varphi(z).$$

The function $\Phi_\lambda(z)$ is called a penalty function and $\lambda$ is referred to as a penalty parameter. The function $\Phi_\lambda$ is said to be an exact penalty function if every point of a local minimum of the function $f$ on the set $Z$ is a point of local minimum of the penalty function on $C(\overline{D})$. Exact penalty functions were introduced by Eremin (cf. [11]) and have a lot of different applications in the optimization theory.

It is easy to check that the function $\Phi_\lambda$ is directionally differentiable at every point $z \in C(\overline{D})$. Moreover, for any $z \notin Z$, the function $\Phi_\lambda$ is *Gâteaux* differentiable at the point $z$ and for any $z \in Z$ the function $\Phi_\lambda$ is subdifferentiable at the point $z$.

Different theorems that give conditions when the function $\Phi_\lambda$ is an exact penalty function were presented in [2–4]. Exact penalty functions allow one to study different constrained extremum problems and derive conditions for an extremum, both well-known and new ones. Also, exact penalty functions are a very useful tool for constructing effective numerical methods for solving optimization problems. The efficiency of the considered approach in studying one-dimensional problems of the calculus of variations was shown in [2, 3, 6].

The initial idea of this paper was to apply the theory of exact penalty functions to studying two-dimensional problems of the calculus of variations. However, we faced a lot of difficulties during the study of this problem and were not able to complete the research. Therefore, we formulate an open problem for the future study.

*Conjecture 1.* There exists $\lambda^* > 0$ such that for any $\lambda > \lambda^*$ the function $\Phi_\lambda$ is an exact penalty function.

*Remark 1.* As in the case of a functional, defined on the space of functions of one variable, one can construct the MSD for the functional $\Phi_\lambda$. However, in the general case, this method does not converge, since the subdifferential mapping of the functional $\Phi_\lambda$ is discontinuous in terms of the Hausdorff distance. Moreover, one can show that in contrast to the case of a functional, defined on the space of functions of one variable, for any $\lambda > 0$ the direction of steepest descent $g$ of the functional $\Phi_\lambda$ at the point $z_0 \in Z$ does not belong to the cone of feasible directions of the set $Z$ at the point $z_0$, i.e. for any $\alpha > 0$ one has that $z_0 + \alpha g \notin Z$.

## 5 Problem Reformulation (Second Version)

In this section we describe another approach to the problem under consideration. This approach uses the fact that the set $Z$ is a linear manifold in $C(\overline{D})$, i.e. a shifted linear subspace. It allows us to formulate new unconstrained extremum problem that is equivalent to the initial problem.

Introduce the set

$$X_0 = \Big\{ h \in C(\overline{D}) \;\Big|\; \int_0^{T_2} h(x,y)\,dy = 0 \;\; \forall x \in [0,T_1], \quad \int_0^{T_1} h(x,y)\,dx = 0 \;\; \forall y \in [0,T_2] \Big\} \tag{2}$$

and fix an arbitrary $z_0 \in Z$. It is clear that $X_0$ is a linear subspace of $C(\overline{D})$ and

$$Z = z_0 + X_0 = \{ z = z_0 + h \mid h \in X_0 \}.$$

Define the functional $f_0\colon X_0 \longrightarrow \mathbb{R}$, $f_0(h) = f(z_0 + h)$ for all $h \in X_0$. It is easy to show that the problem of minimizing the functional $f$ over the set $Z$ is equivalent to the problem of minimizing the functional $f_0$ over the space $X_0$.

In subsequent sections, we will solve the problem of minimizing the functional $f_0$ over the normed space $(X_0, \|\cdot\|)$ by the MSD.

## 6 Necessary Condition for an Extremum

Fix an arbitrary $h$, $\Delta h \in X_0$. Applying the theorem about differentiation under the integral sign and integrating by parts one gets

$$f_0(h + \alpha\Delta h) - f_0(h) = \alpha \iint_D \Big( \int_x^{T_1} \int_y^{T_2} F'_u(\tau,\gamma)\,d\gamma d\tau + \int_y^{T_2} F'_{u'_x}(x,\gamma)\,d\gamma + \int_x^{T_1} F'_{u'_y}(\tau,y)\,d\tau \Big) \Delta h(x,y)\,dxdy + o(\alpha\Delta h),$$

where $o(\alpha\Delta h)/\alpha \longrightarrow 0$ as $\alpha \longrightarrow 0$. Hence, the function $f_0$ is *Gâteaux* differentiable at every point $h \in X_0$. Hereafter, we will use the following abbreviation $F(x,y) = F\left(x,y,u(x,y),u'_x(x,y),u'_y(x,y)\right)$.

Denote

$$Q[h](x,y) = \int_x^{T_1}\int_y^{T_2} F'_u(\tau,\gamma)\,d\gamma d\tau + \int_y^{T_2} F'_{u'_x}(x,\gamma)\,d\gamma + \int_x^{T_1} F'_{u'_y}(\tau,y)\,d\tau.$$

The function $Q[h]\colon \overline{D} \longrightarrow \mathbb{R}$ is the "*Gâteaux* gradient" of the functional $f_0$ at the point $h$.

The following necessary optimality condition holds true.

**Theorem 1.** *Let $h^* \in X_0$ be a point of local extremum of the functional $f_0$. Then for any $x \in [0,T_1]$ and $y \in [0,T_2]$*

$$\frac{\partial^2 Q[h^*](x,y)}{\partial x \partial y} = F'_u(x,y) - \frac{\partial F'_{u'_x}(x,y)}{\partial x} - \frac{\partial F'_{u'_y}(x,y)}{\partial y} = 0, \tag{3}$$

*i.e. the Euler–Ostrogradsky equation is satisfied at the point $h^*$.*

*Proof.* Let $h^*$ be a point of local extremum of the functional $f_0$. Then taking into account the necessary condition for an extremum of a *Gâteaux* differentiable function one has

$$\langle Q[h^*],\Delta h\rangle = 0 \quad \forall\, \Delta h \in X_0.$$

Let $v\colon \overline{D} \longrightarrow \mathbb{R}$ be an arbitrary twice continuously differentiable function with compact support. Then it is easy to verify that $\partial^2 v/\partial x \partial y \in X_0$. Therefore, integrating by parts one gets

$$\left\langle Q[h^*], \frac{\partial^2 v}{\partial x \partial y} \right\rangle = \left\langle \frac{\partial^2 Q[h^*]}{\partial x \partial y}, v \right\rangle = 0.$$

Since the function $v$ is arbitrary, then applying the fundamental lemma of the calculus of variations one gets the required result. □

The point $h^*$ satisfying the necessary condition for an extremum (3) is called a stationary point of the functional $f_0$.

## 7 Method of Steepest Descent for Functional $f_0$

In this section we find the direction of steepest descent of the functional $f_0$ and describe the method of steepest descent for finding stationary points of the functional under consideration.

Fix an arbitrary $h \in X_0$ and suppose that the point $h$ is not a stationary point of the functional $f_0$. Let us find the direction of steepest descent of the functional $f_0$ at the point $h$, i.e. let us find a minimizer of the problem

$$\inf_{\Delta h \in X_0, \|\Delta h\|=1} f_0'[h](\Delta h) = \inf_{\Delta h \in X_0, \|\Delta h\|=1} \langle Q[h], \Delta h \rangle.$$

It is clear that the problem of finding the direction of steepest descent of the functional $f_0$ at the point $h$ is equivalent to the following convex programming problem:

$$g_0(\Delta h) = \langle Q[h], \Delta h \rangle \longrightarrow \inf_{\Delta h \in S}, \tag{4}$$

where $S = \{\Delta h \in X_0 \mid g_1(\Delta h) \leqslant 0\}$, $g_1(\Delta h) = \|\Delta h\|^2 - 1$. It is easy to check that the functionals $g_0$ and $g_1$ are *Gâteaux* differentiable at every point $\Delta h \neq 0$. Note that since $0 \in X_0$ and $g_1(0) < 0$, then the Slater condition in the problem (4) is satisfied. Therefore, by virtue of Theorem 1.1.2 from [12] one has that for a point $\Delta h^* \in X_0$ to be the direction of steepest descent of the functional $f_0$ at the point $h$, it is necessary and sufficient that there exists $\lambda \in \mathbb{R}$ such that

$$g_0'[\Delta h^*](v) + \lambda g_1'[\Delta h^*](v) = \langle Q[h] + \lambda \Delta h^*, v \rangle = 0 \quad \forall\, v \in X_0, \tag{5}$$

where $g_i'[\Delta h^*]$ is the *Gâteaux* gradient of the functional $g_i$ at the point $\Delta h^*$.

Suppose that $\Delta h^* \in C^2(\overline{D})$. Then arguing in the same way as in the proof of Theorem 1 one gets that the equality (5) is equivalent to the following partial differential equation:

$$\frac{\partial^2 Q[h](x,y)}{\partial x \partial y} + \lambda \frac{\partial^2 \Delta h^*(x,y)}{\partial x \partial y} = 0 \quad \forall\, x \in [0, T_1],\ y \in [0, T_2].$$

Let us show that $\lambda \neq 0$. Indeed, if $\lambda = 0$, then one gets that $h$ is a stationary point of the functional $f_0$, which contradicts our assumption. Thus $\lambda \neq 0$. Hence

$$\Delta h^*(x,y) = -\frac{1}{\lambda} Q[h](x,y) + r_1(x) + r_2(y),$$

where $r_1$ and $r_2$ are twice continuously differentiable functions that should be defined. Since $\Delta h^* \in X_0$, then taking into account (2) one gets

$$r_2(y) = \frac{1}{\lambda T_1} \int_0^{T_1} Q[h](x,y)\, dx - \frac{1}{T_1} \int_0^{T_1} r_1(x)\, dx,$$

consequently

$$r_1(x) - \frac{1}{T_1} \int_0^{T_1} r_1(x) = \frac{1}{T_2 \lambda} \int_0^{T_2} Q[h](x,y)\, dy - \frac{1}{\lambda T_1 T_2} \iint_D Q[h](x,y)\, dx dy.$$

Therefore

$$r_1(x) = \frac{1}{\lambda T_2}\int_0^{T_2} Q[h](x,y)\,dy,$$
$$r_2(y) = \frac{1}{\lambda T_1}\int_0^{T_1} Q[h](x,y)\,dx - \frac{1}{\lambda T_1 T_2}\iint_D Q[h](x,y)\,dxdy.$$

As a result one gets that the function

$$\Delta h^*(x,y) = \frac{1}{\lambda} G[h](x,y),$$

where

$$G[h](x,y) = -Q(x,y) + \frac{1}{T_2}\int_0^{T_2} Q[h](x,y)\,dy + \frac{1}{T_1}\int_0^{T_1} Q[h](x,y)\,dx - \frac{1}{T_1T_2}\iint_D Q[h](x,y)\,dxdy \tag{6}$$

and $\lambda = \|G[h]\|$ is the direction of steepest descent of the functional $f_0$ at the point $h$, since, as it is easy to check, the function $\Delta h^*$ satisfies the equality (5). Note that

$$f_0'[h](G[h]) = \langle Q[h], G[h]\rangle < 0, \tag{7}$$

since if $f_0'[h](G[h]) = 0$, then $f_0'[h](\Delta h) = 0$ for any $\Delta h \in X_0$, and therefore the point $h$ is a stationary point of the functional $f_0$, which contradicts our assumption.

**Theorem 2.** *Let $h \in X_0$ be arbitrary. Then $\|f_0'[h]\| = \|G[h]\|$.*

*Proof.* Let $h$ be a non-stationary point of the functional $f_0$. From (5) it follows that

$$\|f_0'[h]\| = \sup_{v\in X_0, \|v\|\leqslant 1} |\langle Q[h], v\rangle| = \lambda \sup_{v\in X_0, \|v\|\leqslant 1} |\langle \Delta h^*, v\rangle| = \lambda \|\Delta h^*\|.$$

Since $\|\Delta h^*\| = 1$, then $\|G[h]\| = \lambda = \|f_0'[h]\|$.

Suppose now that $h$ is a stationary point of the functional $f_0$. Then $\|f_0'[h]\| = 0$ and

$$\frac{\partial^2 Q[h](x,y)}{\partial x \partial y} = 0 \quad \forall\, x \in [0,T_1],\ y \in [0,T_2].$$

Hence, there exist twice continuously differentiable functions $w_1\colon [0,T_1] \longrightarrow \mathbb{R}$ and $w_2\colon [0,T_2] \longrightarrow \mathbb{R}$, such that $Q[h](x,y) = w_1(x) + w_2(y)$. Then taking into account (6) it is easy to check that $G[h](x,y) = 0$. Thus, $\|G[h]\| = 0 = \|f_0'[h]\|$ that completes the proof. □

Let us describe the MSD for the functional $f_0$. Choose an arbitrary $h_0 \in X_0$. Suppose that $h_k$ for some $k \in \mathbb{N}$ has already been found. Compute $G_k(x,y) = G[h_k](x,y)$. If $\|G_k\| = 0$, then the point $h_k$ is stationary and the process terminates. Otherwise, find

$$\inf_{\alpha \geqslant 0} f_0(h_k + \alpha G_k) = f_0(h_k + \alpha_k G_k)$$

and denote $h_{k+1} = h_k + \alpha_k G_k$. Taking into account (7) one has that

$$f_0(h_{k+1}) < f_0(h_k).$$

Continuing in the same manner one constructs a finite number of points $\{h_0, \ldots, h_{k+1}\}$, then the last of them is stationary, or a sequence $\{h_k\}$.

In the following section we will give a convergence theorem of the described method.

## 8 Convergence of the Method of Steepest Descent

In order to prove the convergence of the MSD for the functional $f_0$, we will use general convergence theorems of the MSD given in [13].

*Remark 2.* Note, that although the MSD is studied in [13] for the functional defined in a Banach space, the completeness of the space is not used in the proofs of the main convergence theorems. Therefore, despite the incompleteness of the space $(X_0, \|\cdot\|)$, we can use the results given in [13] to prove the convergence of the MSD for the functional $f_0$.

The following convergence theorem holds true.

**Theorem 3.** *Let for any $(x,y) \in \overline{D}$ the function*

$$(u,p,q) \longrightarrow F(x,y,u,p,q)$$

*be convex. Suppose also that the set $A(h_0) = \{h \in X_0 \mid f_0(h) \leqslant f_0(h_0)\}$ is bounded, $\inf_{h \in X_0} f_0(h) > -\infty$ and the Gâteaux derivative of the functional $f_0$ is Lipschitz continuous in $\{h \in X_0 \mid \|h\| \leqslant R\}$ for some $R > \sup_{h \in A(h_0)} \|h\|$. Then*

$$\|f_0'[h_k]\| \longrightarrow 0, \quad f_0(h_k) - \inf_{h \in X_0} f_0(h) \longrightarrow 0 \text{ as } k \longrightarrow \infty.$$

*Moreover, if the Gâteaux derivative of the functional $f_0$ is Lipschitz continuous in $X_0$, then*

$$f_0(h_k) - \inf_{h \in X_0} f_0(h) = O\left(\frac{1}{k}\right).$$

*Proof.* Under our assumptions on the function $F$, the functional $f_0$ is convex. Then it remains to use the general convergence theorems of the MSD given in [13]. □

## 9 Numerical Experiments

*Example 1.* Let $T_1 = T_2 = 1$, the functional

$$I(u) = \iint_D \left[ \left(\frac{\partial u}{\partial x}\right)^2 + \left(\frac{\partial u}{\partial y}\right)^2 \right] dxdy,$$

and the boundary conditions

$$u(x,0) = \sin(\pi x), \quad u(x,1) = \mathrm{e}^{-\pi}\sin(\pi x), \quad u'_x(x,0) = \pi\cos(\pi x) \qquad \forall\, x \in [0,1],$$
$$u(0,y) = u(1,y) = 0, \quad u'_y(0,y) = 0 \qquad \forall\, y \in [0,1].$$

The exact solution of the problem under consideration is

$$u_*(x,y) = \mathrm{e}^{-\pi y}\sin(\pi x),$$

where $I(u_*) = \pi(1-\mathrm{e}^{-2\pi})/2 \approx 1.5677863$. Hereafter we will use the following abbreviation $z_k = z_0 + h_k$ and $z_* = \partial^2 u_*/\partial x \partial y$.

In Table 1 we give some computational results of the MSD, where $z_0(x,y) = -\pi^2\mathrm{e}^{-\pi y}\cos(\pi x)$ and the starting point is $h_0(x,y) = 100(2x-1)(2y-1)$.

All computations were carried out in Mathcad.

*Example 2.* Let $T_1 = T_2 = 1$. We consider the problem of finding a solution of the Laplace equation

$$\Delta u = 0,$$

**Table 1** Some computational results of the MSD

| $k$ | $f_0(h_k)$ | $\|G_k\|$ | $\|z_k - z_*\|$ | $\|u_k - u_*\|$ |
|---|---|---|---|---|
| 0 | 223.790085 | 13.3729 | 33.3333 | 3.3333 |
| 1 | 2.283958 | 0.564294 | 2.566338 | 0.080757 |
| 2 | 1.573173 | 0.04765 | 0.380643 | 0.011319 |
| 3 | 1.5699509 | 0.025408 | 0.320768 | 0.003482 |

satisfying the following boundary conditions:

$$\begin{array}{llll} u(x,0)=0, & u(x,1)=0, & u'_x(x,0)=0 & \forall\, x\in[0,1], \\ u(0,y)=0, & u(1,y)=\sin(\pi y), & u'_y(0,y)=0 & \forall\, y\in[0,1]. \end{array}$$

The exact solution of the problem is

$$u_*(x,y)=\frac{\sin(\pi y)}{\sin(\pi)}\sinh(\pi x),$$

where $I(u_*)\approx 1.576674047$.

In Table 2 we give some computational results of the MSD, where $z_0(x,y)=\pi\cos(\pi y)$ and the starting point is $h_0(x,y)=0$.

*Example 3.* Let $T_1=T_2=1$, the functional

$$I(u)=\iint_D\left[\left(\frac{\partial u(x,y)}{\partial x}\right)^2+\left(\frac{\partial u(x,y)}{\partial y}\right)^2+2u(x,y)x\sin(\pi y)\right]dxdy,$$

and the boundary conditions

$$\begin{array}{lll} & u(x,0)=u(x,1)=u'_x(x,0)=0 & \forall\, x\in[0,1], \\ u(0,y)=0, & u(1,y)=\sin(\pi y),\quad u'_y(0,y)=0 & \forall\, y\in[0,1]. \end{array}$$

The exact solution of the problem is

$$u_*(x,y)=\frac{\sin(\pi y)}{\pi^2\sinh(\pi)}\left((\pi^2+1)\sinh(\pi x)-x\sinh(\pi)\right),$$

where $I(u_*)\approx 1.789020079$.

In Table 3 we give some computational results of the MSD, where $z_0(x,y)=\pi\cos(\pi y)$ and the starting point is $h_0(x,y)=0$.

*Example 4.* Let $T_1=T_2=1$. We consider the problem of finding a solution of the Poisson equation

$$\Delta u=-1,$$

**Table 2** Some computational results of the MSD

| $k$ | $f_0(h_k)$ | $\|G_k\|$ | $\|z_k-z_*\|$ | $\|u_k-u_*\|$ |
|---|---|---|---|---|
| 0 | 2.1449341 | 0.662306 | 1.740218 | 0.16275 |
| 1 | 1.587024 | 0.071756 | 0.290971 | 0.013401 |
| 2 | 1.576912 | 0.012906 | 0.038473 | 0.002992 |
| 3 | 1.576687 | 0.002541 | 0.010983 | 0.000385 |
| 4 | 1.5766748 | 0.000745 | 0.002253 | 0.000171 |

**Table 3** Some computational results of the MSD

| $k$ | $f_0(h_k)$ | $\|G_k\|$ | $\|z_k - z_*\|$ | $\|u_k - u_*\|$ |
|---|---|---|---|---|
| 0 | 2.4782674 | 0.729412 | 1.916539 | 0.17924 |
| 1 | 1.801574 | 0.079027 | 0.320453 | 0.014758 |
| 2 | 1.789309 | 0.014213 | 0.042371 | 0.003295 |
| 3 | 1.789036 | 0.002799 | 0.012096 | 0.000424 |
| 4 | 1.789021 | 0.000821 | 0.002481 | 0.000188 |

**Table 4** Some computational results of the MSD

| $k$ | $f_0(h_k)$ | $\|G_k\|$ | $\|u_k - u_*\|$ |
|---|---|---|---|
| 0 | 0 | 0.166667 | 0.04126 |
| 1 | −0.035 | 0.012859 | 0.001904 |
| 2 | −0.035942 | 0.002552 | 0.00071 |
| 3 | −0.0351089 | 0.002088 | 0.000527 |

vanishing in $\Gamma$ and

$$u'_x(x,0) = 0 \quad \forall\, x \in [0,1], \qquad u'_y(0,y) = 0 \quad \forall\, y \in [0,1].$$

Recall that $\Gamma$ is the boundary of the domain $D$.

The exact solution of the problem is

$$u_*(x,y) = \sum_{k,m=1,3,5,\ldots} \frac{16\, \sin(k\pi x)\, \sin(m\pi y)}{\pi^4 km(k^2+m^2)}.$$

In Table 4 we give some computational results of the MSD, where $z_0(x,y) = h_0(x,y) = 0$.

**Acknowledgements** The work is supported by the Russian Foundation for Basic Research (project No. 12-01- 00752).

## References

[1] Bliss, G.A.: Lectures on the Calculus of Variations. University of Chicago Press, Chicago (1946)

[2] Demyanov, V.F.: Exact penalty functions in problems of nonsmooth optimization [in Russian]. Vestn. St-Peterbg. Univ. **4**, Ser. I, 21–27 (1994); [English transl.: Vestn. St. Petersbg. Univ. Math. **27**, 16–22 (1994)]

[3] Demyanov, V.F.: Extremality Conditions and Variational Problems. Vyssh. Shkola, Moscow (2005) [in Russian]

[4] Demyanov, V.F.: Nonlinear optimization. In: Di Pillo, G., Schoen, F. (eds.) Lecture Notes in Mathematics, vol. 1989, pp. 55–163. Springer (2010)

[5] Demyanov, V.F., Rubinov, A.M.: Constructive Nonsmooth Analysis. Peter Lang Verlag, Frankfurt a/M (1995)
[6] Demyanov, V.F., Tamasyan, G.Sh.: Exact penalty functions in isoperimetric problems. Optimization **60**, 153–177 (2011)
[7] Demyanov, V.F., Stavroulakis, G.E., Polyakova, L.N., Panagiotopoulos, P.D.: Quasidifferentiability and Nonsmooth Modelling in Mechanics, Engineering and Economics. Kluwer Academic, Doordrecht (1996)
[8] Demyanov, V.F., Di Pillo, G., Facchinei, F.: Exact penalization via Dini and Hadamard conditional derivatives. Optim. Methods Softw. **9**, 19–36 (1998)
[9] Demyanov, V.F., Giannessi, F., Karelin, V.V.: Optimal control problems via exact penalty functions. J. Glob. Optim. **12**, 215–223 (1998)
[10] Demyanov, V.F., Giannessi, F., Tamasyan, G.Sh.: Variational control problems with constraints via exact penalization. In: Giannessi, F., Maugeru, A. (eds.) Nonconvex Optimization and Its Applications, vol. 79. Variational Analysis and Applications, pp. 301–342. Springer, New York (2005)
[11] Eremin, I.I.: A method of "penalties" in convex programming. Sov. Math. Dokl. **143**, 748–751 (1967)
[12] Ioffe, A.D., Tikhomirov, V.M.: Theory of Extremal Problems. North-Holland, Amsterdam etc. (1979)
[13] Kantorovich, L.V., Akilov, G.P.: Functional Analysis, 2nd edn. Pergamon Press, Oxford, XIV (1982)
[14] Kantorovich, L.V., Krylov, V.I.: Approximate Methods of Higher Analysis. Wiley, NewYork (1964)
[15] Mihlin, S.G.: Calculus of Variational and Its Applications (Proceedings of the Eighth Symposium in Applied Mathematics). McGraw–Hill B. Comp, New York (1958)
[16] Mihlin, S.G.: Numerical Implementation of Variational Methods. Nauka Publishers, Moscow (1966) [In Russian]
[17] Moreau, J.J., Panagiotopoulos, P.D.: Nonsmooth Mechanics and Applications. Springer, Berlin (1988)

# On a Quantitative Semicontinuity Property of Variational Systems with Applications to Perturbed Quasidifferentiable Optimization

**A. Uderzo**

**Abstract** Lipschitz lower semicontinuity is a quantitative stability property for set-valued maps with relevant applications to perturbation analysis of optimization problems. The present paper reports on an attempt of studying such property, by starting with a related result valid for variational systems in metric spaces. Elements of nonsmooth analysis are subsequently employed to express and apply such result and its consequences in more structured settings. This approach leads to obtain a solvability, stability, and sensitivity condition for perturbed optimization problems with quasidifferentiable data.

**Keywords** Lipschitz lower semicontinuity • Perturbed quasidifferentiable optimization • Solvability condition • Stability condition • Sensitivity condition

## 1 Introduction

A theme pervading at various levels modern continuous optimization is perturbation analysis of extremum problems. This is not only due to concrete needs arising when a real-world phenomenon is to be modelized, so effects of errors and/or inaccuracies entering problem data must be taken into account. As illustrated by popular monographs devoted to this and related topics (among them, see [2,4,15,17–19,21,28]), perturbation analysis can also afford useful theoretical insights into the very nature of the problem. In fact, such analysis approach happened to yield as by-products results concerning optimality conditions.

The perturbation analysis of continuous optimization problems is performed as a rule by considering various kinds of parameterizations of problem data. Throughout

A. Uderzo (✉)
Dipartimento di Matematica e Applicazioni, Università di Milano-Bicocca,
Via Cozzi, 53 - 20125 Milano, Italy
e-mail: amos.uderzo@unimib.it

V.F. Demyanov et al. (eds.), *Constructive Nonsmooth Analysis and Related Topics*, Springer Optimization and Its Applications 87, DOI 10.1007/978-1-4614-8615-2_8,

this paper, the basic format of parametric optimization problems will be according to [4]. Let $\varphi : P \times X \longrightarrow \mathbb{R} \cup \{\mp\infty\}$ and $h : P \times X \longrightarrow Y$ be given functions and let $C$ be a nonempty closed subset of $Y$. Here $P$, $X$, and $Y$ stand for (at least) metric spaces. For any given $p \in P$, the above data define the constrained extremum problem

$$\min_{x \in X} \varphi(p,x) \quad \text{subject to} \quad h(p,x) \in C.$$

This class of parametric optimization problems will be denoted by $(\mathcal{P}_p)$. The feasible region $\mathcal{R} : P \longrightarrow 2^X$ of $(\mathcal{P}_p)$ is therefore the set-valued map

$$\mathcal{R}(p) = \{x \in X : \ h(p,x) \in C\}, \tag{1}$$

whereas the optimal value function $\mathrm{val}_{\mathcal{P}} : P \longrightarrow \mathbb{R} \cup \{\pm\infty\}$ is given by

$$\mathrm{val}_{\mathcal{P}}(p) = \inf_{x \in \mathcal{R}(p)} \varphi(p,x). \tag{2}$$

The convention $\inf \varnothing = +\infty$ is adopted throughout the paper. The solution map $\mathrm{Argmin}_{\mathcal{P}} : P \longrightarrow 2^X$ associated with $(\mathcal{P}_p)$ is

$$\mathrm{Argmin}_{\mathcal{P}}(p) = \{x \in \mathcal{R}(p) : \ \varphi(p,x) = \mathrm{val}_{\mathcal{P}}(p)\}. \tag{3}$$

Notice that under the format $(\mathcal{P}_p)$ many exemplary classes of extremum problems fall. Let us mention, among them, problems of nonlinear programming, semi-definite programming, semi-infinite programming, optimal control, and min-max optimization.

As clearly appears from the above recalled elements connected with $(\mathcal{P}_p)$, an aspect, which seems to be unavoidable, characterizing perturbation analysis is the need to deal with set-valued maps and nonsmooth functions. Dealing with set-valued maps naturally leads to handle stability concepts, that often rely on various forms of semicontinuity. After the seminal papers [1, 26], Lipschitzian properties were soon understood to play a crucial role in such context. According to [17], the below definition fixes a quantitative notion of semicontinuity for set-valued maps, which, postulating more than a mere topological condition, can be classified under the broad umbrella known as Lipschitz behavior. In what follows, given $x \in X$ and $S \subseteq X$, $\mathrm{B}(x,r)$ and $\mathrm{B}(S,r)$ denote the closed ball with center $x$ and the closed enlargement of $S$ with radius $r > 0$, respectively. The distance of an element $x$ from a set $S$ will be indicated by $d_S(x)$ or $\mathrm{dist}(x,S)$. Given a set-valued map $\Phi : P \longrightarrow 2^X$, its domain is denoted by $\mathrm{dom}\,\Phi = \{p \in P : \ \Phi(p) \neq \varnothing\}$.

**Definition 1.** A set-valued map $\Phi : P \longrightarrow 2^X$ between metric spaces is said to be *Lipschitz lower semicontinuous* (for short, *Lipschitz l.s.c.*) at $(\bar{p},\bar{x}) \in \mathrm{gph}\,\Phi$ if there exist positive real constants $\zeta$ and $l$ such that

$$\Phi(p) \cap \mathrm{B}(\bar{x}, l d(p,\bar{p})) \neq \varnothing, \quad \forall p \in \mathrm{B}(\bar{p},\zeta). \tag{4}$$

The value

$$\operatorname{Liplsc} \Phi(\bar{p},\bar{x}) = \inf\{l > 0 : \ \exists \zeta \text{ for which (4) holds } \}$$

is called *modulus of Lipschitz lower semicontinuity* of $\Phi$ at $(\bar{p},\bar{x})$.

*Remark 1.* From Definition 1 it is clear that, whenever a map $\Phi$ is Lipschitz l.s.c. at each point $(\bar{p},x) \in P \times X$, with $x$ varying in $\Phi(\bar{p})$, then it is also l.s.c. at $\bar{p}$. Recall that $\Phi$ is defined to be l.s.c. at $\bar{p}$ if for every open set $V \subseteq X$ such that $V \cap \Phi(\bar{p}) \neq \varnothing$ there exists $\delta_V > 0$ such that $V \cap \Phi(p) \neq \varnothing$, for every $p \in \mathrm{B}(\bar{p}, \delta_V)$. Thus, the notion under study on one hand localizes the lower semicontinuity property of a map at a given point of its graph, on the other hand it quantifies by a Lipschitzian estimate nonemptiness of intersections. In Example 1 below a map is presented, which is l.s.c. at $\bar{p} = 0$, but is not Lipschitz l.s.c. at $(\bar{p},0)$.

The present paper concentrates on the analysis of the Lipschitz behavior of maps in connection with problems $(\mathcal{P}_p)$. In particular, this study is concerned with conditions for the Lipschitz lower semicontinuity of the feasible region map $\mathcal{R}$ and the solution map $\mathrm{Argmin}_{\mathcal{P}}$, which are generally set-valued. The reader should notice that the Lipschitz behavior captured by (4) not only ensures local nonemptiness of $\Phi$ around a reference pair (solvability), but also provides useful quantitative information about the sensitivity of $\Phi$. Both these features seem to be appealing in consideration of the bizarre geometry of solution maps to constraint systems or to optimization problems, whose graph may happen to switch from multivaluedness to single-valuedness or even to emptiness.

Another property relating to the Lipschitz behavior of set-valued maps involved in this paper is calmness (see [17,28]).

**Definition 2.** A set-valued map $\Phi : P \longrightarrow 2^X$ between metric spaces is said to be *calm* at $(\bar{p},\bar{x}) \in \operatorname{gph} \Phi$ if there exist positive real constants $\delta$ and $\ell$ such that

$$\Phi(p) \cap \mathrm{B}(\bar{x},\delta) \subseteq \mathrm{B}(\Phi(\bar{p}), \ell d(p,\bar{p})) \quad \forall p \in \mathrm{B}(\bar{p},\delta). \tag{5}$$

The value

$$\operatorname{clm} \Phi(\bar{p},\bar{x}) = \inf\{\ell > 0 : \ \exists \delta \text{ for which (5) holds}\}$$

is called *modulus of calmness* of $\Phi$ at $(\bar{p},\bar{x})$.

*Remark 2.* Clearly, whenever a map $\Phi$ is single-valued, Lipschitz lower semicontinuity and calmness, along with their respective moduli, coincide, provided that $\Phi(p) \cap \mathrm{B}(\bar{x},\delta) \neq \varnothing$. In this event, the term prevailing in the literature is "calm", so it will be adopted throughout this paper. In the case of set-valued maps the two properties may yield substantially different behaviors, as illustrated by the following examples.

*Example 1.* Let $X = P = \mathbb{R}$ be endowed with its usual (Euclidean) metric structure. Let $\bar{p} = \bar{x} = 0$. The set-valued map $\Phi : \mathbb{R} \longrightarrow 2^{\mathbb{R}}$, given by

$$\Phi(p) = \{x \in \mathbb{R} : x \geq \sqrt{|p|}\},$$

is calm at $(0,0)$ but not Lipschitz l.s.c. at the same point.

*Example 2.* Again, let $X = P = \mathbb{R}$ be endowed with its usual metric structure. Let $\bar{p} = \bar{x} = 0$. The set-valued map $\Phi : \mathbb{R} \longrightarrow 2^{\mathbb{R}}$, defined by

$$\Phi(p) = \{x \in \mathbb{R} : x \geq -\sqrt{|p|}\},$$

is Lipschitz l.s.c. at $(0,0)$, but it fails to be calm at that point.

Besides being interesting in itself, Lipschitz lower semicontinuity may be exploited to characterize other stability property of set-valued maps (especially, the Aubin property). Studies on its relationships with different specializations of Lipschitz behavior do already exist (see, for instance, [17]). As defining a quite general form of stability, it may be invoked when other stronger stability properties fail.

Much work has been done in the last decades to establish conditions or criteria for detecting Lipschitz behavior of general multifunctions. In the resulting theory nonsmooth analysis tools and methods occupy an important part. Nevertheless, when passing to consider set-valued maps representing feasible regions and, overall, solution sets of parametric optimization problems, the situation becomes more complicated. Indeed, according to deep-rooted trends of research (see, for instance, [2, 4, 15]), the perturbation analysis of extremum problems is usually performed by assuming the data to be (at least) $C^2$ smooth. Lipschitz stability results for parametric programs involving $C^{1,1}$ data can be found in Chap. 8 of [17]. This is due to specific technical reasons, having to do with the methodology of study. In the present paper the approach adopted avoids the direct employment of Lagrange optimality conditions and the related Kojima function. It relies instead on an analysis of Lipschitz lower semicontinuity of variational systems, which is conducted in metric spaces with tools adequate to such setting. The findings of these investigations are subsequently interpreted in more structured settings (Banach and Hilbert spaces) by means of nonsmooth analysis constructions. The extensive development of this area of mathematics offers a broad choice of approaches and machineries to accomplish such task. In the current paper, the reader will find an attempt to tackle the question by employing elements of quasidifferential calculus, which is a well-known subject in nonsmooth analysis. Of course, there are other more comprehensive and refined generalized differentiation concepts that could be considered to the same purpose. Nonetheless, the author believes that quasidifferential calculus can provide an already broad and suggestive view on what can be done if extremum problems under examination exhibit nondifferentiable data. It is worth noting that, even if with reference to smooth parametric nonlinear optimization problems, quasidifferentiability appeared in the study of the optimal solution map also in [9], while in the book [19] relevant sensitivity results on the optimal value function associated to perturbed programs with quasidifferentiable data are presented.

The contents of the paper are arranged in four main sections, included the current one. In Sect. 2 a general result for the Lipschitz lower semicontinuity of variational systems in metric spaces is recalled along with the generalized differentiation tools needed to state it. Moreover, a first consequence on the Lipschitz behavior of the solution set to problems $(\mathcal{P}_p)$ is derived. Section 3 furnishes more involved constructions of nonsmooth analysis, mainly taken from quasidifferential calculus, in view of the perturbation analysis of optimization problems to be conducted in Banach and Hilbert spaces. The results of such analysis are exposed in Sect. 4. Here, a first subsection contains a condition for the Lipschitz lower semicontinuity of the constraint system appearing in $(\mathcal{P}_p)$. In the subsequent subsection stability and sensitivity results in perturbed quasidifferentiable optimization are discussed.

## 2 Lipschitz Behavior in Metric Spaces

Since all the aforementioned stability properties make sense in a metric space setting, the starting point of the proposed analysis will be a sufficient condition for Lipschitz lower semicontinuity of the solution map associated with a parameterized generalized equation in metric spaces. In such context, a parameterized generalized equation is a problem which can be formalized as follows: Let $(P,d)$, $(X,d)$ and $(Y,d)$ be metric spaces; given $f : P \times X \longrightarrow Y$ and $F : P \times X \longrightarrow 2^Y$, for any given value of $p \in P$, find $x \in X$ such that

$$f(p,x) \in F(p,x). \tag{6}$$

Here $P$ plays the role of parameter space, whereas $X$ denotes the space where solutions, if any, have to be sought. The solution map $G : P \longrightarrow 2^X$ associated with (6), namely

$$G(p) = \{x \in X : \ f(p,x) \in F(p,x)\},$$

is often referred to as a *variational system*, according to a widespread terminology. Such an abstract formalism can cover a variety of problems arising in optimization and control theory. In this paper, it will be essentially utilized to represent constraint systems and optimality "situations." In this way, it will come possible to study feasible region as well as $\mathrm{Argmin}_{\mathcal{P}}$ map of $(\mathcal{P}_p)$.

Even in such a poor environment one can conduct a fruitful analysis of the property in consideration by employing tools, which have been devised for variational analysis in metric spaces. Among them, those needed in accordance with the current approach are briefly recalled.

Given a function $\psi : X \longrightarrow \mathbb{R} \cup \{\mp\infty\}$, a basic well-known tool for locally measuring the decrease rate of $\psi$ near a reference point $\bar{x} \in X$ is the *strong slope* of $\psi$ at $\bar{x}$, usually denoted by $|\nabla\psi|(\bar{x})$ (see [8]). Dealing with parameterized functions, a partial version of this notion is needed. Given $\psi : P \times X \longrightarrow \mathbb{R} \cup \{\mp\infty\}$ and $(\bar{p},\bar{x}) \in P \times X$, the value

$$|\nabla_x \psi|(\bar{p},\bar{x}) = \begin{cases} 0, & \text{if } \bar{x} \text{ is a local minimizer for } \psi(\bar{p},\cdot) \\ \limsup\limits_{x \longrightarrow \bar{x}} \dfrac{\psi(\bar{p},\bar{x}) - \psi(\bar{p},x)}{d(x,\bar{x})}, & \text{otherwise} \end{cases}$$

is called *partial strong slope* of $\psi$ at $(\bar{p},\bar{x})$. Based on the partial strong slope, the following regularization, introduced in [14] to the aim of expressing error bound criteria, has revealed to work properly in the current investigations. The value

$$\overline{|\nabla\psi|}^{>}(\bar{p},\bar{x}) = \liminf_{\varepsilon \downarrow 0} \{|\nabla_x \psi|(p,x) : \ x \in \mathrm{B}(\bar{x},\varepsilon), \ p \in \mathrm{B}(\bar{p},\varepsilon),$$
$$\psi(\bar{p},\bar{x}) < \psi(p,x) \leq \psi(\bar{p},\bar{x}) + \varepsilon\}$$

is called *partial strict outer slope* (with respect to $x$) of $\psi$ at $(\bar{p},\bar{x})$.

*Remark 3.* Let $\psi_i : X \longrightarrow \mathbb{R} \cup \{\mp\infty\}$ be given functions, with $i = 1, 2$. Whenever one of them, say $\psi_2$, is locally Lipschitz near a point $x \in X$, having Lipschitz constant $\kappa$, then the below estimates can be easily proved to hold

$$|\nabla(\psi_1 + \psi_2)|(x) \geq |\nabla\psi_1|(x) - \kappa, \qquad \overline{|\nabla(\psi_1 + \psi_2)|}^{>}(x) \geq \overline{|\nabla\psi_1|}^{>}(x) - \kappa.$$

Of course, both the above inequalities continue to hold if the strong slope and the strict outer slope are replaced by their partial counterparts, respectively, provided that function $\psi_2(\cdot, p)$ is locally Lipschitz near $x$, with constant $\kappa$, for every $p$ near $\bar{p}$.

A sufficient condition for the Lipschitz lower semicontinuity of the solution map to parameterized generalized equations (6) has been recently achieved in [33]. It is based on a nondegeneracy requirement on the strict outer slope of the function $\mathrm{dist}(f,F) : P \times X \longrightarrow \mathbb{R}$

$$\mathrm{dist}(f,F)(p,x) = \mathrm{dist}(f(p,x),F(p,x)).$$

**Theorem 1.** *Let $f : P \times X \longrightarrow Y$ and $F : P \times X \longrightarrow Y$ be given maps defining a parameterized generalized equation (6), with solution map $G : P \longrightarrow 2^X$. Let $\bar{p} \in P$, $\bar{x} \in G(\bar{p})$ and $\bar{y} = f(\bar{p},\bar{x})$. Suppose that:*

*(i) $(X,d)$ is metrically complete;*
*(ii) There is $\delta_1 > 0$ such that for every $p \in \mathrm{B}(\bar{p},\delta_1)$ map $F(p,\cdot) : X \longrightarrow 2^Y$ is u.s.c. at each point of $\mathrm{B}(\bar{x},\delta_1)$;*
*(iii) The set-valued map $F(\cdot,\bar{x})$ is Lipschitz l.s.c. at $(\bar{p},\bar{y})$;*

*(iv)* *There is* $\delta_2 > 0$ *such that for every* $p \in \mathrm{B}(\bar{p}, \delta_2)$ *map* $f(p,\cdot) : X \longrightarrow Y$ *is continuous at each point of* $\mathrm{B}(\bar{x}, \delta_2)$*;*
*(v)* *Map* $f(\cdot, \bar{x}) : P \longrightarrow Y$ *is calm at* $\bar{p}$*;*
*(vi)* *It holds*

$$\overline{|\nabla \mathrm{dist}(f,F)|}^{>}(\bar{p},\bar{x}) > 0. \tag{7}$$

*Then, G is Lipschitz l.s.c. at* $(\bar{p},\bar{x})$.

*Remark 4.* By a perusal of the proof of Theorem 1, one readily sees that the thesis of such theorem can be complemented with the following estimate of the Lipschitz lower semicontinuity modulus of $G$:

$$\mathrm{Liplsc}\, G(\bar{p},\bar{x}) \leq \frac{\mathrm{Liplsc}\, F(\cdot,\bar{x})(\bar{p},\bar{y}) + \mathrm{clm}\, f(\bar{x},\cdot)(\bar{p})}{\overline{|\nabla \mathrm{dist}(f,F)|}^{>}(\bar{p},\bar{x})}. \tag{8}$$

As a consequence of the above general theorem, the following result, which focusses on the Lipschitz behavior of the solution map associated with the class of problems $(\mathcal{P}_p)$, can be established already in a metric space setting.

**Proposition 1.** *With reference to a class of parametric optimization problems* $(\mathcal{P}_p)$*, let* $\bar{p} \in P$ *and* $\bar{x} \in \mathrm{Argmin}_{\mathcal{P}}(\bar{p})$*. Suppose that:*

*(i')* $(X,d)$ *is metrically complete;*
*(ii')* *The set-valued map* $\mathcal{R} : P \longrightarrow 2^X$ *is Lipschitz l.s.c. at* $(\bar{p},\bar{x})$*;*
*(iii')* *There exists* $\delta > 0$ *such that, for every* $p \in \mathrm{B}(\bar{x}, \delta)$*, function* $\varphi(p,\cdot)$ *is continuous in* $\mathrm{B}(\bar{x}, \delta)$*;*
*(iv')* *There exist positive* $\tilde{\delta}$ *and* $l$ *such that*

$$|\varphi(p,\bar{x}) - \mathrm{val}_{\mathcal{P}}(p)| \leq l d(p,\bar{p}) \quad \forall p \in \mathrm{B}(\bar{p}, \tilde{\delta});$$

*(v')* *It holds*

$$\overline{|\nabla|\varphi - \mathrm{val}_{\mathcal{P}}||}^{>}(\bar{p},\bar{x}) > 1.$$

*Then,* $\bar{p} \in \mathrm{int}\,(\mathrm{dom}\,(\mathrm{Argmin}_{\mathcal{P}}))$*, map* $\mathrm{Argmin}_{\mathcal{P}}$ *is Lipschitz l.s.c. at* $(\bar{p},\bar{x})$ *and the following estimate holds:*

$$\mathrm{Liplsc}\, \mathrm{Argmin}_{\mathcal{P}}(\bar{p},\bar{x}) \leq \frac{\mathrm{Liplsc}\, \mathcal{R}(\bar{p},\bar{x}) + l}{\overline{|\nabla|\varphi - \mathrm{val}_{\mathcal{P}}||}^{>}(\bar{p},\bar{x}) - 1}. \tag{9}$$

*Proof.* The thesis can be achieved by applying Theorem 1 to the parameterized generalized equation (6), which is defined by $f : P \times X \longrightarrow (\mathbb{R} \cup \{\mp\infty\}) \times X$ and $F : P \times X \longrightarrow 2^{(\mathbb{R} \cup \{\mp\infty\}) \times X}$ as follows

$$f(p,x) = (\varphi(p,x) - \mathrm{val}_{\mathcal{P}}(p), x) \quad \text{and} \quad F(p,x) = \{0\} \times \mathcal{R}(p).$$

With such positions, it is clear that

$$G(p) = \mathrm{Argmin}_{\mathcal{P}}(p), \quad \forall p \in P.$$

Let us check that all hypotheses of Theorem 1 are currently in force. Since $F$ does not depend on $x$, $(ii)$ is automatically fulfilled. It is readily seen that the assumed Lipschitz lower semicontinuity of $\mathcal{R}$ at $(\bar{p},\bar{x})$ implies the same property for $F(\cdot,\bar{x})$ at $(\bar{p},(0,\bar{x}))$, with $\mathrm{Liplsc}\, F(\cdot,\bar{x})(\bar{p},(0,\bar{x})) = \mathrm{Liplsc}\, \mathcal{R}(\bar{p},\bar{x})$. As to the hypotheses concerning $f$, observe that, for every $p \in \mathrm{B}(\bar{p},\delta)$, the continuity of function $x \mapsto \varphi(p,x)$ entails the continuity of the function $x \mapsto (\varphi(p,x) - \mathrm{val}_{\mathcal{P}}(p),x)$, for every $p$ near $\bar{p}$. In particular, notice that $\mathrm{val}_{\mathcal{P}}$ takes values in $\mathbb{R}$ for every $p$ near $\bar{p}$, by virtue of $(iv')$. Assumption $(iv')$ is also equivalent to the calmness of $f(\cdot,\bar{x})$ at $\bar{p}$. Finally, it is to be shown that assumption $(v')$ leads to the fulfilment of condition (7). To this aim, it is convenient to recall what has been noticed in Remark 3. Since function $x \mapsto \mathrm{dist}(x,\mathcal{R}(p))$ is Lipschitz continuous on $X$ with constant 1, uniformly in $p \in P$, if equipping the product space $(\mathbb{R} \cup \{\mp\infty\}) \times X$ with the sum distance, one finds

$$\mathrm{dist}((\varphi(p,x) - \mathrm{val}_{\mathcal{P}}(p),x),\{0\} \times \mathcal{R}(p)) = |\varphi(p,x) - \mathrm{val}_{\mathcal{P}}(p)| + \mathrm{dist}(x,\mathcal{R}(p)),$$

for every $(p,x) \in P \times X$, whence

$$\overline{|\nabla \mathrm{dist}((\varphi(\cdot,\cdot) - \mathrm{val}_{\mathcal{P}}(\cdot),\cdot),\{0\} \times \mathcal{R}(p))|}^{>}(\bar{p},\bar{x}) \geq \overline{|\nabla|\varphi - \mathrm{val}_{\mathcal{P}}||}^{>}(\bar{p},\bar{x}) - 1.$$

It is now clear that assumption $(v')$ implies condition (7). The estimate of the modulus of Lipschitz lower semicontinuity in the thesis is a straightforward consequence of (8). The proof is complete. □

*Remark 5.* The assumptions of Proposition 1 not only refer to problem data ($\varphi$, $h$, and $C$), but they invoke more involved terms associated with $(\mathcal{P}_p)$, such as the set-valued map $\mathcal{R}$ and the functional $\mathrm{val}_{\mathcal{P}}$. In this regard, a condition in order for $\mathcal{R}$ to satisfy assumption $(ii')$ will be provided later in Sect. 4, in terms of problem data. Notice that assumption $(iv')$ turns out to be satisfied whenever $\varphi(\cdot,\bar{x})$ and $\mathrm{val}_{\mathcal{P}}$ are both calm at $\bar{p}$. Several conditions in order for $\mathrm{val}_{\mathcal{P}}$ to be calm already exist in literature (see, for instance, [33]). They may be established by imposing appropriate requirements on the behavior of $\varphi$ and $\mathcal{R}$. On the other hand, calmness (even in a weaker form, called *calmness from below*) of $\mathrm{val}_{\mathcal{P}}$ is known to imply problem calmness in the sense of [5].

## 3 Nonsmooth Analysis Constructions

Throughout the current and the subsequent sections, $(\mathbb{X},\|\cdot\|)$ and $(\mathbb{Y},\|\cdot\|)$ will be real Banach spaces. The (topological) dual space of $\mathbb{X}$ is marked by $\mathbb{X}^*$, with $\mathbb{X}^*$ and $\mathbb{X}$ being paired in duality by the bilinear form $\langle \cdot,\cdot \rangle : \mathbb{X}^* \times \mathbb{X} \longrightarrow \mathbb{R}$. The null vector in a linear space is indicated by $\mathbf{0}$, while the null functional by $\mathbf{0}^*$. The unit sphere centered at the null vector of a Banach space is denoted by $\mathbb{S}$, whereas the unit ball and the unit sphere centered at $\mathbf{0}^*$ are denoted by $\mathbb{B}^*$ and $\mathbb{S}^*$, respectively. Given a subset $A$ of a Banach space, its support function is denoted by $\varsigma(\cdot, A)$. The semigroup (with respect to the Minkowski sum of sets) of all nonempty convex and weak* compact subsets of $\mathbb{X}^*$ is denoted by $\mathcal{K}(\mathbb{X}^*)$.

## 3.1 *Quasidifferentiable Functionals*

Given a functional $\vartheta : \mathbb{X} \longrightarrow \mathbb{R} \cup \{\mp\infty\}$ and $\bar{x} \in \operatorname{dom} \vartheta$, let $\vartheta'(\bar{x}; v)$ denote the directional derivative of $\vartheta$ at $\bar{x}$ in the direction $v \in \mathbb{X}$. According to [10], functional $\vartheta$ is said to be *quasidifferentiable* at $\bar{x}$ if it admits directional derivative at $\bar{x}$ in all directions and there exist two (Lipschitz) continuous sublinear functionals $\underline{\vartheta}, \overline{\vartheta} : \mathbb{X} \longrightarrow \mathbb{R}$ such that

$$\vartheta'(\bar{x}; v) = \underline{\vartheta}(v) - \overline{\vartheta}(v) \quad \forall v \in \mathbb{X}.$$

In the light of the Minkowski–Hörmander semigroup duality between sublinear continuous functionals on $\mathbb{X}$ and elements of $\mathcal{K}(\mathbb{X}^*)$ (see, for instance, [22]), this means that

$$\vartheta'(\bar{x}; v) = \varsigma(v, \partial \underline{\vartheta}(\mathbf{0})) - \varsigma(v, \partial \overline{\vartheta}(\mathbf{0})) \quad \forall v \in \mathbb{X}, \tag{10}$$

where $\partial$ denotes the subdifferential operator in the sense of convex analysis. Clearly, the dual representation (10) of $\vartheta'(\bar{x}; \cdot)$, as well as the previous one, is by no means unique. Nevertheless, every pair of elements of $\mathcal{K}(\mathbb{X}^*)$ representing $\vartheta'(\bar{x}; \cdot)$ belongs to the same class with respect to an equivalence relation $\sim$ defined on $\mathcal{K}(\mathbb{X}^*) \times \mathcal{K}(\mathbb{X}^*)$, according to which $(A, B) \sim (C, D)$ if $A + D = B + C$. The $\sim$-equivalence class containing the pair $(\partial \underline{\vartheta}(\mathbf{0}), -\partial \overline{\vartheta}(\mathbf{0}))$ is called *quasidifferential* of $\vartheta$ at $\bar{x}$ and will be denoted in the below constructions by $\mathcal{D}\vartheta(\bar{x})$. Any pair in the class $\mathcal{D}\vartheta(\bar{x})$ will be henceforth indicated by $(\underline{\partial}\vartheta(\bar{x}), -\overline{\partial}\vartheta(\bar{x}))$, so

$$\vartheta'(\bar{x}; v) = \varsigma(v, \underline{\partial}\vartheta(\bar{x})) - \varsigma(v, -\overline{\partial}\vartheta(\bar{x})) \quad \forall v \in \mathbb{X},$$

with $\mathcal{D}\vartheta(\bar{x}) = [\underline{\partial}\vartheta(\bar{x}), -\overline{\partial}\vartheta(\bar{x})]$.

In the early 1980s a complete calculus for quasidifferentiable functionals has been developed, which finds a geometric counterpart in the calculus for $\sim$-equivalence classes of pairs in $\mathcal{K}(\mathbb{X}^*) \times \mathcal{K}(\mathbb{X}^*)$ (see [12, 13, 29]). This, along with a notable computational tractability of the resulting constructions, made such approach a recognized and successful subject within nonsmooth analysis.

In view of the employment of quasidifferential calculus, an estimate of the strong slope in terms of directional derivative is provided by the following lemma, whose elementary proof is given for the sake of completeness.

**Lemma 1.** *Given a functional $\vartheta : \mathbb{X} \longrightarrow \mathbb{R} \cup \{\mp\infty\}$ and an element $\bar{x} \in \operatorname{dom} \vartheta$, suppose that $\vartheta$ is directionally differentiable at $\bar{x}$. Then*

$$|\nabla\vartheta|(\bar{x}) \geq \sup_{u \in \mathbb{S}} [-\vartheta'(\bar{x}; u)]. \tag{11}$$

*Proof.* If $\bar{x}$ is a local minimizer for $\vartheta$, then one trivially has

$$-\vartheta'(\bar{x}; u) \leq 0 = |\nabla\vartheta|(\bar{x}) \quad \forall u \in \mathbb{S},$$

whence the inequality (11) follows. Otherwise, one obtains

$$\begin{aligned} |\nabla\vartheta|(\bar{x}) &= \limsup_{v \longrightarrow \mathbf{0}} \frac{\vartheta(\bar{x}) - \vartheta(\bar{x} + v)}{\|v\|} = -\liminf_{v \longrightarrow \mathbf{0}} \frac{\vartheta(\bar{x} + v) - \vartheta(\bar{x})}{\|v\|} \\ &\geq -\liminf_{t \downarrow 0} \frac{\vartheta(\bar{x} + tu) - \vartheta(\bar{x})}{t} = -\vartheta'(\bar{x}; u) \quad \forall u \in \mathbb{S}. \end{aligned}$$

Again, inequality (11) follows. □

The next example shows that, even when $\mathbb{X}$ is a finite-dimensional space, so that the supremum in (11) is attained, such inequality may be strict.

*Example 3.* Let $\mathbb{X} = \mathbb{R}^2$, $\bar{x} = \mathbf{0} = (0, 0)$, let $\vartheta : \mathbb{R}^2 \longrightarrow \mathbb{R}$ be defined by

$$\vartheta(x_1, x_2) := \begin{cases} -\sqrt[4]{(3x_1^2 - x_2)(x_2 - x_1^2)} & \text{if } x_1^2 < x_2 < 3x_1^2, \\ 0 & \text{otherwise.} \end{cases}$$

By direct calculations it is readily seen that

$$|\nabla\vartheta|(\bar{x}) = 1 > 0 = \max_{u \in \mathbb{S}} [-\vartheta'(\bar{x}; u)].$$

## 3.2 Scalarly Quasidifferentiable Mappings

Since, according to the current analysis, among the problem data to deal with are vector-valued functions, quasidifferential calculus must be extended to such a more general context. This has been done in several fashions. Unlike in [11], here a scalarization approach is considered.

**Definition 3.** A map $f : U \longrightarrow \mathbb{Y}$, with $U$ open subset of $\mathbb{X}$, is said to be *scalarly quasidifferentiable* (for short, *scalarly q.d.*) at $\bar{x} \in U$ if $f$ is directionally differentiable at $\bar{x}$ and for every $y^* \in \mathbb{Y}^*$ the scalar functional $x \mapsto \langle y^*, f(x) \rangle$ is q.d. at $\bar{x}$, that is, there is $(A_{y^*}, B_{y^*}) \in \mathcal{K}(\mathbb{X}^*) \times \mathcal{K}(\mathbb{X}^*)$ such that

$$(y^* \circ f)'(\bar{x}; v) = \varsigma(v, A_{y^*}) - \varsigma(v, -B_{y^*}) = \max_{x^* \in A_{y^*}} \langle x^*, v \rangle + \min_{x^* \in B_{y^*}} \langle x^*, v \rangle \quad \forall v \in \mathbb{X}.$$

The $\sim$-equivalence class (depending on $y^* \in \mathbb{Y}^*$) containing the pair $(A_{y^*}, -B_{y^*})$ is called the *scalarized quasidifferential* of $f$ at $\bar{x}$, with respect to $y^*$, and it will be denoted by

$$\mathcal{D}_{y^*} f(\bar{x}) = [\underline{\partial}_{y^*} f(\bar{x}), -\overline{\partial}_{y^*} f(\bar{x})].$$

The map $\mathcal{D}f(\bar{x}) : \mathbb{Y}^* \longrightarrow \mathcal{K}(\mathbb{X}^*) \times \mathcal{K}(\mathbb{X}^*)/\sim$, defined by

$$\mathcal{D}f(\bar{x})(y^*) = \mathcal{D}_{y^*} f(\bar{x}),$$

is called the *scalarized coquasiderivative* of $f$ at $\bar{x}$.

*Example 4.* Let $\mathbb{Y} = \mathbb{R}^m$. As one checks immediately, a map $f : \mathbb{X} \longrightarrow \mathbb{Y}$, with $f(x) = (f_1(x), \ldots, f_m(x))$, turns out to be scalarly q.d. at $\bar{x}$ iff each component $f_i : \mathbb{X} \longrightarrow \mathbb{R}$, with $i = 1, \ldots, m$, is q.d. at $\bar{x}$. For any $y^* = (y_1^*, \ldots, y_m^*) \in \mathbb{R}^m$, one has the following representation of $\mathcal{D}f(\bar{x})(y^*)$ in terms of the quasidifferentials of each component of $f$

$$\underline{\partial}_{y^*} f(\bar{x}) = \sum_{i=1}^{m} [y_i^*]_+ \underline{\partial} f_i(\bar{x}) - \sum_{i=1}^{m} [y_i^*]_- \overline{\partial} f_i(\bar{x}),$$

and

$$\overline{\partial}_{y^*} f(\bar{x}) = \sum_{i=1}^{m} [y_i^*]_+ \overline{\partial} f_i(\bar{x}) - \sum_{i=1}^{m} [y_i^*]_- \underline{\partial} f_i(\bar{x}),$$

where $[r]_+ = \max\{r, 0\}$ and $[r]_- = \max\{-r, 0\}$. It is worth noting that this class of maps includes in particular those having each component $f_i$ which is convex or concave or DC, that is expressible as difference of two convex functionals.

*Example 5.* Assume now the Banach space $(\mathbb{Y}, \|\cdot\|)$ to be partially ordered by a convex cone $K$ (with apex at the null vector of $\mathbb{Y}$). Let us recall that its positive dual cone $K^+ = \{y^* \in \mathbb{Y}^* : \langle y^*, y \rangle \geq 0 \quad \forall y \in K\}$ is said to be *reproducing*, if the equality $\mathbb{Y}^* = K^+ - K^+$ holds true. If $\mathbb{Y}$ has finite dimension, the above equality happens to hold, whenever the cone $K$ is pointed. In general, a sufficient condition for $K^+$ to be reproducing is that it admits nonempty algebraic interior. Some characterizations for a positive dual cone to be reproducing are known, which are formulated in terms of normality of $K$ (see, for instance, [23]). Suppose further $(\mathbb{Y}, \|\cdot\|)$ to be a Banach–Kantorovich space, that is a Banach space, where the partial order $\leq_K$ induced by a convex cone $K$ fulfils the following two properties: any subset $\leq_K$-bounded from above admits least upper bound; the norm $\|\cdot\|$ is monotone, i.e.,

$$\max\{z, -z\} \leq_K \max\{y, -y\} \quad \text{implies} \quad \|z\| \leq \|y\| \quad \forall z, y \in \mathbb{Y}.$$

The notion of quasidifferentiability has been extended to maps taking values in such class of spaces in [11, 13]. More precisely, a map $f : \mathbb{X} \longrightarrow \mathbb{Y}$ admitting directional derivative at $\bar{x} \in \mathbb{X}$ is called *K-quasidifferentiable* at $\bar{x}$, if there are a $K$-sublinear map $\underline{f} : \mathbb{X} \longrightarrow \mathbb{Y}$ (in the sense that $\underline{f}(v_1 + v_2) \leq_K \underline{f}(v_1) + \underline{f}(v_2)$ for every $v_1, v_2 \in \mathbb{X}$) and a $K$-superlinear map $\overline{f} : \mathbb{X} \longrightarrow \mathbb{Y}$ allowing for the representation

$$f'(\bar{x}; v) = \underline{f}(v) + \overline{f}(v) \quad \forall v \in \mathbb{X}.$$

If $f$ is $K$-quasidifferentiable at $\bar{x}$, with $K^+$ being reproducing, then $f$ turns out to be scalarly quasidifferentiable at the same point. Indeed, taken any $y^* \in \mathbb{Y}^*$, by the above assumption, one has $y^* = y^*_+ + y^*_-$, where $y^*_+$ and $y^*_-$ are proper elements of $K^+$ and $-K^+$, respectively. It suffices therefore to observe that $y^*_+ \circ \underline{f} + y^*_- \circ \overline{f}$ is a sublinear functional, whereas $y^*_- \circ \underline{f} + y^*_+ \circ \overline{f}$ is a superlinear functional.

**Definition 4.** The inner operation $\underline{\triangle} : \mathcal{K}(\mathbb{X}^*) \times \mathcal{K}(\mathbb{X}^*) \longrightarrow \mathcal{K}(\mathbb{X}^*)$, defined by

$$A \,\underline{\triangle}\, B = \partial^\circ(\varsigma(\cdot, A) - \varsigma(\cdot, B))(\mathbf{0}),$$

where $\partial^\circ$ denotes the Clarke subdifferential operator (see [6, 13]), is called the *Demyanov difference* of $A$ and $B$.

*Remark 6.* When $A$ and $B$ are compact convex subsets of a finite-dimensional space, their Demyanov difference $A \,\underline{\triangle}\, B$ can be constructively computed as convex closure of all differences of exposed points in $A$ and $B$. Indeed, by virtue of the Rademacher theorem function $\varsigma(\cdot, A) - \varsigma(\cdot, B)$ is (Fréchet) differentiable at each point of a full measure set, say $\tilde{\mathbb{X}}$. Take into account that the gradient $\mathrm{D}\varsigma(\cdot, A)(x)$ coincides with the exposed point of set $A$ by the linear functional $\langle \cdot, x \rangle$, which is unique whenever $\varsigma(\cdot, A)$ happens to be differentiable at $x$. Hence, the aforementioned representation of $\underline{\triangle}$ is a consequence of the following equality due to J.-B. Hiriart-Urruty (see [29, 31])

$$A \,\underline{\triangle}\, B = \mathrm{cl\,co}\,\{\mathrm{D}\varsigma(\cdot, A)(x) - \mathrm{D}\varsigma(\cdot, B)(x) : \; x \in \tilde{\mathbb{X}}\},$$

where $\mathrm{cl\,co}$ denotes the convex closure of a given set.

Some properties of the Demyanov difference that will be exploited in the sequel are collected in the below lemma. Their proofs as well as further material on this topic can be found in [13, 29–31].

**Lemma 2.**

*(1) For every $A, B, C, D \in \mathcal{K}(\mathbb{X}^*)$, if $(A, B) \sim (C, D)$, then $A \,\underline{\triangle}\, B = C \,\underline{\triangle}\, D$;*
*(2) For every $A \in \mathcal{K}(\mathbb{X}^*)$, it holds $A \,\underline{\triangle}\, A = \{\mathbf{0}^*\}$;*
*(3) For every $A, B \in \mathcal{K}(\mathbb{X}^*)$, if $B \subseteq A$, then $\mathbf{0}^* \in A \,\underline{\triangle}\, B$;*
*(4) For every $A, B, C, D \in \mathcal{K}(\mathbb{X}^*)$, it results in $(A + B) \,\underline{\triangle}\, (C + D) \subseteq (A \,\underline{\triangle}\, C) + (B \,\underline{\triangle}\, D)$.*

By combining the scalarized approach to quasidifferentiation of maps with the Demyanov difference, one gets a generalized differentiation tool, which is appropriate for the purposes of the present analysis. Note that such combination of nonsmooth analysis elements was already employed for investigating metric regularity of parameterized generalized systems in [32].

Given a map $f : U \longrightarrow \mathbb{Y}$ scalarly q.d. at $\bar{x} \in U$, the following generalized derivative construction $\mathcal{D}^{\triangle} f(\bar{x}) : \mathbb{Y}^* \longrightarrow 2^{\mathbb{X}^*}$ defined by

$$\mathcal{D}^{\triangle} f(\bar{x})(y^*) = \underline{\partial}_{y^*} f(\bar{x}) \,\underline{\triangle}\, (-\overline{\partial}_{y^*} f(\bar{x})) \quad y^* \in \mathbb{Y}^*,$$

will be referred to as the *Demyanov coquasiderivative* of $f$ at $\bar{x}$. Note that $\mathcal{D}^{\triangle} f(\bar{x})$ is a positively homogeneous set-valued map, taking nonempty, weak$^*$ compact and convex values. With each of its values it is possible to associate the quantity

$$|||\mathcal{D}^{\triangle} f(\bar{x})(y^*)||| = \mathrm{dist}(\mathbf{0}^*, \underline{\partial}_{y^*} f(\bar{x}) \,\underline{\triangle}\, (-\overline{\partial}_{y^*} f(\bar{x})) \quad y^* \in \mathbb{Y}^*.$$

This will be used to quantify a nondegeneracy condition expressed in terms of scalarized quasidifferential through the following lemma.

**Lemma 3.** *Let $\vartheta : U \longrightarrow \mathbb{R}$ be a function q.d. at $x \in U \subseteq \mathbb{X}$ and let $\gamma > 0$. Then $|||\mathcal{D}^{\triangle} \vartheta(x)||| > \gamma$ implies the existence of $u_0 \in \mathbb{S}$ such that $\vartheta'(x; u_0) < -\gamma$.*

*Proof.* Recall that, according to well-known facts in q.d. calculus, function $\gamma\|\cdot - x\|$ is q.d. at $x$, and it holds

$$\mathcal{D}(\gamma\|\cdot - x\|)(x) = [\gamma\mathbb{B}^*, \{\mathbf{0}^*\}].$$

By virtue of the sum rule for quasidifferentials and of Lemma 2(4), one has

$$\begin{aligned}
\left(\underline{\partial}\vartheta(x) \,\underline{\triangle}\, (-\overline{\partial}\vartheta(x))\right) + \gamma\mathbb{B}^* &= \left(\underline{\partial}\vartheta(x) \,\underline{\triangle}\, (-\overline{\partial}\vartheta(x))\right) \\
&\quad + \left(\underline{\partial}\gamma\|\cdot - x\|(x) \,\underline{\triangle}\, (-\overline{\partial}\gamma\|\cdot - x\|(x))\right) \\
&\supseteq (\underline{\partial}\vartheta(x) + \underline{\partial}\gamma\|\cdot - x\|(x)) \,\underline{\triangle} \left(-\overline{\partial}\vartheta(x) - \overline{\partial}\gamma\|\cdot - x\|(x)\right) \\
&= \underline{\partial}(\vartheta + \gamma\|\cdot - x\|)(x) \,\underline{\triangle}\, -\overline{\partial}(\vartheta + \gamma\|\cdot - x\|)(x).
\end{aligned}$$

The hypothesis $|||\mathcal{D}^{\triangle} \vartheta(x)||| > \gamma$ implies $\mathbf{0}^* \notin [\mathcal{D}^{\triangle} \vartheta(x)] + \gamma\mathbb{B}^*$. This fact, in the light of the above inclusion along with Lemma 2(3), gives

$$-\overline{\partial}(\vartheta + \gamma\|\cdot - x\|)(x) \not\subseteq \underline{\partial}(\vartheta + \gamma\|\cdot - x\|)(x). \tag{12}$$

Since, given $A, B \in \mathcal{K}(\mathbb{X}^*)$, the inclusion $B \subseteq A$ holds true iff $\varsigma(v, B) \leq \varsigma(v, A)$ for every $v \in \mathbb{X}$, from (12), one has that there exists $u_0 \in \mathbb{S}$ such that

$$\varsigma(u_0, \underline{\partial}(\vartheta + \gamma\|\cdot - x\|)(x)) < \varsigma(u_0, -\overline{\partial}(\vartheta + \gamma\|\cdot - x\|)(x)),$$

Thus, one obtains

$$\underline{\vartheta}(u_0) + \gamma \| \cdot - x\|'(x;u_0) < \overline{\vartheta}(u_0),$$

whence

$$\vartheta'(x;u_0) = \underline{\vartheta}(u_0) - \overline{\vartheta}(u_0) < -\gamma \| \cdot - x\|'(x;u_0) = -\gamma.$$

This completes the proof. □

*Remark 7.* Since $\underline{\triangle}$ is a $\sim$-invariant operation [remember Lemma 2(1)], the Demyanov coquasiderivative of a map is uniquely defined, even if the representative pair $(\underline{\partial}_{y^*} f(\bar{x}), -\overline{\partial}_{y^*} f(\bar{x}))$ is not. To connect such concept to classical calculus, observe that, whenever map $f$ is Gâteaux differentiable at $\bar{x}$, one obtains

$$\begin{aligned} \mathcal{D}^{\triangle} f(\bar{x})(y^*) &= \underline{\partial}(y^* \circ f)(\bar{x}) \,\underline{\triangle}\, (-\overline{\partial}(y^* \circ f)(\bar{x})) = \{\mathrm{D}(y^* \circ f)(\bar{x})\} \,\underline{\triangle}\, (-\{\mathbf{0}^*\}) \\ &= \{\mathrm{D}f(\bar{x})^* y^*\}, \end{aligned}$$

where $\mathrm{D}f(\bar{x})^* : \mathbb{Y}^* \longrightarrow \mathbb{X}^*$ indicates the adjoint operator to the Gâteaux derivative of $f$ at $\bar{x}$. Therefore, in such special case, it is $|||\mathcal{D}^{\triangle} f(\bar{x})(y^*)||| = \|\mathrm{D}f(\bar{x})^* y^*\|$.

### 3.3 Prox-Regular Sets

The next element of nonsmooth analysis to be introduced can be viewed as a sort of variational generalization of convexity and will enable to extend the perturbation analysis at the issue also to problems $(\mathcal{P}_p)$ with datum $C$ being nonconvex. The price to be paid for that is the requirement on the space $\mathbb{Y}$ to be a Hilbert one. So, assume now $(\mathbb{Y}, (\cdot,\cdot))$ to be a Hilbert space. Let us denote by $\mathrm{N}(y,C)$ the cone of all *normals* to $C$ at $y$ in the *general sense*, i.e.,

$$\mathrm{N}(y,C) = \{y^* \in \mathbb{Y}^* : \exists (y_n)_{n\in\mathbb{N}},\, y_n \in C,\, y_n \longrightarrow y,\, \exists (y_n^*)_{n\in\mathbb{N}},\, y_n^* \in \hat{\mathrm{N}}(y_n,C),\, y_n^* \rightharpoonup y^*\},$$

where $\hat{\mathrm{N}}(y_n,C)$ is the normal cone to $C$ at $y_n$ in the regular (alias Fréchet) sense and $\rightharpoonup$ indicates weak convergence. According to [25, 28], a closed subset $C \subset \mathbb{Y}$ is said to be *prox-regular* at $\bar{y}$, with $\bar{y} \in C$, if there exist positive $\varepsilon$ and $\rho$ such that for every $y \in C \cap \operatorname{int} \mathrm{B}(\bar{y},\varepsilon)$ and for every $y^* \in \mathrm{N}(y,C) \cap \operatorname{int} \mathrm{B}(\mathbf{0}^*,\varepsilon)$ it holds

$$(y^*, z - y) \le \frac{\rho}{2}\|z - y\|^2 \quad \forall z \in C \cap \operatorname{int} \mathrm{B}(\bar{y},\varepsilon).$$

Here int$A$ indicates the topological interior of a subset $A$. A set is called *prox-regular* if it is prox-regular at each of its points.[1] Of course, every closed convex set is prox-regular. Moreover, a good deal of examples of sets exhibiting prox-regularity in finite dimensions has been provided in [24], in particular sets admitting a smooth constraint representation.

The key property making prox-regularity of $C$ so appealing within the proposed approach is the deep connection with the local differentiability theory of the distance function $d_C : \mathbb{Y} \longrightarrow \mathbb{R}$, what historically motivated its introduction. In this vein, for the purposes of the present analysis, some facts concerning prox-regular sets collected in the next lemma will be relevant (for its proof and further equivalences see Theorem 1.3 in [25]). In what follows, as usually, $C^{1,1}(\Omega)$ denotes the class of all Fréchet differentiable functions $\phi$ on $\Omega$, such that the Fréchet derivative map $\hat{\mathrm{D}}\phi : \Omega \longrightarrow \mathbb{Y}^*$ is Lipschitz on $\Omega$.

**Lemma 4.** *For a closed set $C \subset \mathbb{Y}$ and any point $\bar{y} \in C$, the following assertions are equivalent:*

$(a_1)$ *$C$ is prox-regular at $\bar{y}$;*
$(a_2)$ *Function $d_C$ is continuously differentiable on $[\mathrm{int}\,\mathrm{B}(\bar{y},r)] \backslash C$, for some $r > 0$;*
$(a_3)$ *Function $d_C^2 \in C^{1,1}([\mathrm{int}\,\mathrm{B}(\bar{y},r)] \backslash C)$, for some $r > 0$.*

# 4 Conditions for Lipschitz Lower Semicontinuity in Structured Spaces

Since the notion of prox-regularity is going to be employed, throughout the current section, the space $\mathbb{Y}$, where the constraining map $h$ takes values, will be supposed to be a Hilbert space.

## 4.1 Scalarly q.d. Constraint Systems

Let us start with considering parameterized constraint systems as in $(\mathcal{P}_p)$, that is

$$\mathcal{R}(p) = \{x \in \mathbb{X} : \ h(p,x) \in C\} = h^{-1}(p,\cdot)(C). \tag{13}$$

The latter equality in (13) makes clear that, whenever map $h(p,\cdot)$ is continuous on $\mathbb{X}$, $\mathcal{R}$ takes closed (possibly empty) values in $2^{\mathbb{X}}$. The stability properties of these

[1] Actually, this is not the original definition, but an alternative one, as resulting from Proposition 1.2 in [25]. The notion of prox-regularity was introduced in [24] in a finite-dimensional setting. It is worth noting that, quite recently, the notion of prox-regular set has been extended to the uniformly convex Banach space setting in [3]. Since such an extension requires a certain amount of technicalities and the case of Hilbert spaces is already significant, consequent generalizations of results here presented to such a broader setting will be disregarded here.

important solution maps have been the subject of intensive investigations within nonlinear programming and variational analysis (see, for instance, [16, 17, 20, 27, 34]). Here, according to the main theme of the paper, the focus is on Lipschitz lower semicontinuity. If assuming $h$ to be scalarly q.d. with respect to $x$ at each point of a neighborhood of a reference pair $(\bar{p},\bar{x}) \in P \times \mathbb{X}$ and $C$ to be prox-regular, then it is possible to define the following constant:

$$c^{\vartriangle}[h,C](\bar{p},\bar{x}) = \liminf_{\varepsilon \downarrow 0} \{|||\mathcal{D}^{\vartriangle} h(p,\cdot)(x)(\hat{\mathrm{D}}d_C(h(p,x)))||| :$$
$$x \in \mathrm{B}(\bar{x},\varepsilon),\, p \in \mathrm{B}(\bar{p},\varepsilon) :\ 0 < d_C(h(p,x)) \le \varepsilon\},$$

where $\hat{\mathrm{D}}d_C(y)$ denotes the Fréchet derivative of function $d_C$ at $y \in \mathbb{Y}$; under the above assumptions, it is shown to exist, provided that $y$ is chosen in a proper way near $h(\bar{p},\bar{x})$ (see the proof of the next theorem). For a better understanding of constant $c^{\vartriangle}[h,C](\bar{p},\bar{x})$, take into account that, being $d_C$ Lipschitz continuous with constant 1, wherever $\hat{\mathrm{D}}d_C(y)$ exists, it is $\|\hat{\mathrm{D}}d_C(y)\| \le 1$. More precisely, by virtue of a known characterization of the Fréchet subdifferential of $d_C$ at out-of-set points (see Theorem 1.99 in [21]), whenever $\hat{\mathrm{D}}d_C(y)$ exists, one has

$$\hat{\mathrm{D}}d_C(y) \in \hat{\mathrm{N}}(y, \mathrm{B}(C, d_C(y))) \cap \mathbb{S}^* \quad y \notin C.$$

The above constant enables to formulate the following condition for the Lipschitz lower semicontinuity of $\mathcal{R}$.

**Theorem 2.** *With reference to a parameterized constraint system (13), let $\bar{p} \in P$ and $\bar{x} \in \mathcal{R}(\bar{p})$. Suppose that:*

*(i) $C$ is prox-regular at $h(\bar{p},\bar{x})$;*
*(ii) There exists $\delta_* > 0$ such that for every, $p \in \mathrm{B}(\bar{p},\delta_*)$, each map $h(p,\cdot)$ is continuous on $\mathrm{B}(\bar{x},\delta_*)$;*
*(iii) $h(\cdot,\bar{x})$ is calm at $\bar{p}$;*
*(iv) $h$ is scalarly q.d. with respect to $x$ at each point of $B(\bar{p},\delta_2) \times \mathrm{B}(\bar{x},\delta_2)$ and it is*

$$c^{\vartriangle}[h,C](\bar{p},\bar{x}) > 0. \tag{14}$$

*Then, $\bar{p} \in \mathrm{int}(\mathrm{dom}\,\mathcal{R})$, $\mathcal{R}$ is Lipschitz l.s.c. at $(\bar{p},\bar{x})$, and it holds*

$$\mathrm{Liplsc}\,\mathcal{R}(\bar{p},\bar{x}) \le \frac{\mathrm{clm}\,h(\cdot,\bar{x})(\bar{p})}{c^{\vartriangle}[h,C](\bar{p},\bar{x})}. \tag{15}$$

*Proof.* The thesis can be proved by applying Theorem 1 to the more specific context of the parameterized constraint systems. To do so, observe that, regarding system (13) as a parameterized generalized equation (6), the field $F$ is the set-valued map taking constantly the value $C$. Thus, $F(p,\cdot)$ is u.s.c. on $\mathbb{X}$, whereas $F(\cdot,\bar{x})$ is Lipschitz l.s.c. at each point of its graph, that is $P \times C$. Therefore, it remains to show that

$$\overline{|\nabla(d_C \circ h)|^{>}}(\bar{p},\bar{x}) > 0. \tag{16}$$

According to assumption $(iv)$, corresponding to $c \triangleq [h,C](\bar{p},\bar{x})/2$, there exists $\delta_c > 0$ such that

$$\inf\Big\{ |||\mathfrak{D}^{\triangle} h(p,\cdot)(x)(\hat{D}d_C(h(p,x)))||| : x \in B(\bar{x},\delta_c),\, p \in B(\bar{p},\delta_c) : \tag{17}$$
$$0 < d_C(h(p,x)) \leq \delta_c \Big\} > \frac{c \triangleq [h,C](\bar{p},\bar{x})}{2}.$$

Consider an arbitrary point $(p,x) \in B(\bar{p},\delta_c) \times B(\bar{x},\delta_c)$, with the property that $0 < d_C(h(p,x)) \leq \delta_c$. Since by hypothesis $C$ is prox-regular at $h(\bar{p},\bar{x}) \in C$, without loss of generality, it is possible to assume that function $d_C$ is continuously differentiable in $[\operatorname{int} B(h(\bar{p},\bar{x}),r)] \setminus C$, with $\delta_c < r$, where $r > 0$ is as in Lemma 4$(a_2)$. By virtue of hypothesis $(ii)$, there exists $\delta_* > 0$ such that

$$\|h(p,x) - h(p,\bar{x})\| < r/2 \quad \forall p \in B(\bar{p},\delta_*) \;\; \forall x \in B(\bar{x},\delta_*).$$

Hypothesis $(iii)$ ensures the existence of positive $\delta_\kappa$ and $\kappa$ such that

$$\|h(p,x) - h(\bar{p},\bar{x})\| \leq \kappa d(p,\bar{p}) \quad \forall p \in B(\bar{p},\delta_\kappa).$$

Thus, by taking $0 < \delta_c < \min\{\delta_*,\, \delta_\kappa,\, r\delta_\kappa/2\kappa\}$, one obtains

$$\begin{aligned}\|h(p,x) - h(\bar{p},\bar{x})\| &\leq \|h(p,x) - h(p,\bar{x})\| + \|h(p,\bar{x}) - h(\bar{p},\bar{x})\| \\ &< \frac{r}{2} + \kappa\frac{r}{2\kappa} = r \quad \forall x \in B(\bar{x},\delta_c) \;\; \forall p \in B(\bar{p},\delta_c).\end{aligned}$$

In other words, if $(p,x) \in B(\bar{p},\delta_c) \times B(\bar{x},\delta_c)$ and $0 < d_C(h(p,x)) \leq \delta_c$, it is

$$h(p,x) \in [\operatorname{int} B(h(\bar{p},\bar{x}),r)] \setminus C,$$

with the consequence that $d_C$ is Fréchet differentiable at any such point. Now, fix $(p_0,x_0) \in B(\bar{x},\delta_c) \times B(\bar{p},\delta_c)$, with $0 < d_C(h(p_0,x_0)) \leq \delta_c$. The above exposed argument, along with hypothesis $(iv)$, guarantees that the functional $d_C \circ h(p_0,\cdot)$ is directionally differentiable with respect to $x$ at $x_0$. So, according to the chain rule, it results in

$$(d_C \circ h(p_0,\cdot))'(x_0;v) = \langle \hat{D}d_C(h(p_0,x_0)), h(p_0,\cdot)'(x_0;v)\rangle \quad \forall v \in \mathbb{X}.$$

Since $h$ is scalarly q.d. with respect to $x$ at $(p_0,x_0)$, from the last equality, the following representation of the Demyanov coquasiderivative holds:

$$\mathfrak{D}^{\triangle}(d_C \circ h(p_0,\cdot))(x_0) = \mathfrak{D}^{\triangle} h(p_0,\cdot)(x_0)(\hat{D}d_C(h(p_0,x_0))).$$

In the light of inequality (17), one has

$$|||\mathcal{D}^{\triangle}(d_C \circ h(p_0,\cdot))(x_0)||| > c/2.$$

By virtue of Lemma 3, the last inequality entails the existence of $u_0 \in \mathbb{S}$ such that

$$(d_C \circ h(p_0,\cdot))'(x_0;u_0) < -c/2.$$

This fact, according to Lemma 1, in turn allows to obtain

$$|\nabla(d_C \circ h(p_0,\cdot))|(x_0) \geq \sup_{u \in \mathbb{S}}[-(d_C \circ h(p_0,\cdot))'(x_0;u)] > c/2.$$

Thus, by arbitrariness of $(p_0,x_0)$, the above inequality leads to prove the existence of $\delta_c > 0$ such that

$$\inf\{|\nabla_x(d_C \circ h)|(p,x) : \ x \in \mathrm{B}(\bar{x},\delta_c), \ p \in \mathrm{B}(\bar{p},\delta_c), \ 0 < d_C(h(p,x)) \leq \delta_c\} \geq c/2.$$

According to the definition of partial strict outer slope of a functional at $(\bar{p},\bar{x})$, it follows

$$\overline{|\nabla_x(d_C \circ h)|}^{>}(\bar{p},\bar{x}) \geq c/2 > 0.$$

Therefore, condition (7) is fulfilled and this completes the proof. □

*Remark 8.*

(1) Note that in Theorem 2, while spaces $\mathbb{X}$ and $\mathbb{Y}$ are equipped with a linear structure, the parameter space $P$ remains a merely metric space, thereby allowing for very general problem perturbations. In the case in which $P$ also should be a vector space, hypothesis $(iii)$ might be replaced by specific conditions ensuring calmness, which are expressed in terms of scalarized coquasidifferentials of $h$, with respect to $p$.
(2) The prox-regularity assumption on $C$ appearing in hypothesis $(ii)$ can be evidently replaced by proximal smoothness in the sense of [7]. Recall that a closed subset $C$ of a Hilbert space is defined to be *proximally smooth* if function $d_C$ is (norm-to-norm) continuously differentiable on a whole (tube) set $(\mathrm{int}\, B(C,r))\backslash C$, for some $r > 0$.
(3) When, in particular, $h$ is Gâteaux differentiable, with respect to $x$, at each point of $B(\bar{p},\delta_2) \times \mathrm{B}(\bar{x},\delta_2)$, to verify hypothesis $(iv)$ of Theorem 2, the following estimate can be exploited in the light of Remark 7:

$$c^{\triangle}[h,C](\bar{p},\bar{x}) \geq \liminf_{\varepsilon \downarrow 0} \ \{\inf_{y^* \in \mathbb{S}^*} \|\mathrm{D}h(p,\cdot)(x)^* y^*\| : x \in \mathrm{B}(\bar{x},\varepsilon), p \in \mathrm{B}(\bar{p},\varepsilon) :$$
$$0 < d_C(h(p,x)) \leq \varepsilon\}.$$

Recall that in linear functional analysis the value $\inf_{y^* \in \mathbb{S}^*} \|\Lambda^* y^*\|$ is often referred to as the *dual Banach constant* of a linear operator $\Lambda : \mathbb{X} \longrightarrow \mathbb{Y}$ between Banach spaces. According to a modern reformulation of the Banach–Schauder theorem, the positivity of the dual Banach constant is known to characterize openness of bounded linear operators. Thus, the nondegeneracy condition expressed via the constant $c^{\triangle}[h,C](\bar{p},\bar{x})$ can be regarded as a nonsmooth/set-valued generalization of the regularity condition appearing in the celebrated open mapping theorem.

## 4.2 A Stability and Sensitivity Result in Perturbed Nondifferentiable Optimization

Now, let us suppose that, for a parameter $\bar{p}$ fixed in $P$, problem $(\mathcal{P}_{\bar{p}})$ admits a (global) solution at $\bar{x} \in \mathcal{R}(\bar{p})$. The next result shows that, under proper assumptions, all problems $(\mathcal{P}_p)$ corresponding to small perturbations of the value of $\bar{p}$ are still (globally) solvable. Moreover, the solution map turns out to be Lipschitz l.s.c. at $(\bar{p},\bar{x})$, and its Lipschitz lower semicontinuity modulus can be estimated in terms of problem data and of $\mathrm{val}_{\mathcal{P}}$. Such result can be regarded as a nonsmooth implicit multifunction theorem.

**Theorem 3.** *With reference to problems $(\mathcal{P}_p)$, let $\bar{p} \in P$ and $\bar{x} \in \mathrm{Argmin}_{\mathcal{P}}(\bar{p})$. Suppose that:*

$(i)$ *$C$ is prox-regular at $h(\bar{p},\bar{x})$;*
$(ii_1)$ *There exists $\delta_1 > 0$ such that for every $p \in \mathrm{B}(\bar{p},\delta_1)$, map $h(p,\cdot)$ is continuous in $\mathrm{B}(\bar{x},\delta_1)$;*
$(ii_2)$ *Map $h(\cdot,\bar{x})$ is calm at $\bar{p}$;*
$(ii_3)$ *There exists $\delta_2 > 0$ such that $h$ is scalarly q.d. with respect to $x$ at each point of $\mathrm{B}(\bar{p},\delta_2) \times \mathrm{B}(\bar{x},\delta_2)$ and $c^{\triangle}[h,C](\bar{p},\bar{x}) > 0$;*
$(iii_1)$ *There exists $\delta_3 > 0$ such that, for every $p \in \mathrm{B}(\bar{p},\delta_3)$, $\varphi(p,\cdot)$ is continuous on $\mathrm{B}(\bar{x},\delta_3)$;*
$(iii_2)$ *The functional $\varphi(\cdot,\bar{x}) - \mathrm{val}_{\mathcal{P}}$ is calm at $\bar{p}$;*
$(iii_3)$ *It holds $\overline{|\nabla|\varphi - \mathrm{val}_{\mathcal{P}}||}^{>}(\bar{p},\bar{x}) > 1$.*

*Then, $\bar{p} \in \mathrm{int}\,(\mathrm{dom}\,\mathrm{Argmin}_{\mathcal{P}})$ and $\mathrm{Argmin}_{\mathcal{P}}$ is Lipschitz l.s.c. at $(\bar{p},\bar{x})$. Moreover, the following estimate holds:*

$$\mathrm{Liplsc}\,\mathrm{Argmin}_{\mathcal{P}}(\bar{p},\bar{x}) \leq \frac{\mathrm{clm}\,h(\cdot,\bar{x}) + c^{\triangle}[h,C](\bar{p},\bar{x}) \cdot \mathrm{clm}\,(\varphi(\cdot,\bar{x}) - \mathrm{val}_{\mathcal{P}})(\bar{p})}{c^{\triangle}[h,C](\bar{p},\bar{x}) \left[\overline{|\nabla|\varphi - \mathrm{val}_{\mathcal{P}}||}^{>}(\bar{p},\bar{x}) - 1\right]}.$$

*Proof.* According to Theorem 2, by assumptions $(i)$, $(ii_1)$–$(ii_3)$, the feasible set map $\mathcal{R}$ is Lipschitz l.s.c. at $(\bar{p},\bar{x})$, and inequality (15) holds true. Since function $\varphi(\cdot,\bar{x}) - \mathrm{val}_{\mathcal{P}}$ is calm at $\bar{p}$, for every $\varepsilon > 0$, there exists $\delta_\varepsilon > 0$ such that

$$|\varphi(p,\bar{x}) - \mathrm{val}_{\mathcal{P}}(p)| \leq (\mathrm{clm}\,(\varphi(\cdot,\bar{x}) - \mathrm{val}_{\mathcal{P}})(\bar{p}) + \varepsilon)d(p,\bar{p}), \quad \forall p \in \mathrm{B}(\bar{p},\delta_{\varepsilon}).$$

Then, under the current hypotheses, it is possible to apply Proposition 1. Concerning the estimation of the Lipschitz lower semicontinuity modulus of $\mathrm{Argmin}_{\mathcal{P}}$, from (9) one obtains

$$\mathrm{Liplsc}\,\mathrm{Argmin}_{\mathcal{P}}(\bar{p},\bar{x}) \leq \frac{\mathrm{clm}\,h(\cdot,\bar{x}) + c \triangleq [h,C](\bar{p},\bar{x}) \cdot [\mathrm{clm}\,(\varphi(\cdot,\bar{x}) - \mathrm{val}_{\mathcal{P}})(\bar{p}) + \varepsilon]}{c \triangleq [h,C](\bar{p},\bar{x}) \left[\overline{|\nabla|\varphi - \mathrm{val}_{\mathcal{P}}||}^{>}(\bar{p},\bar{x}) - 1\right]}.$$

By arbitrariness of $\varepsilon > 0$, the above inequality gives the estimate in the thesis. □

*Remark 9.* Notice that, whenever $\bar{x}$ is the unique solution to problem $(\mathcal{P}_{\bar{p}})$, Theorem 3 entails in particular the lower semicontinuity of $\mathrm{Argmin}_{\mathcal{P}}$ at $\bar{p}$. Furthermore, if map $\mathrm{Argmin}_{\mathcal{P}}$ is locally single-valued near $\bar{p}$, then one obtains that $\mathrm{Argmin}_{\mathcal{P}}$ is calm at $\bar{p}$. A similar situation can be reproduced even when map $\mathrm{Argmin}_{\mathcal{P}}$ is not single-valued, by passing to a selector of it. Such useful possibility is emphasized by the next corollary.

**Corollary 1.** *Under the hypotheses of Theorem 3, for every $l$ such that*

$$l > \frac{\mathrm{clm}\,h(\cdot,\bar{x}) + c \triangleq [h,C](\bar{p},\bar{x}) \cdot \mathrm{clm}\,(\varphi(\cdot,\bar{x}) - \mathrm{val}_{\mathcal{P}})(\bar{p})}{c \triangleq [h,C](\bar{p},\bar{x}) \left[\overline{|\nabla|\varphi - \mathrm{val}_{\mathcal{P}}||}^{>}(\bar{p},\bar{x}) - 1\right]},$$

*there exists $\zeta_l > 0$ and a selector* argmin : $\mathrm{B}(\bar{p},\zeta_l) \longrightarrow \mathbb{X}$ *of* $\mathrm{Argmin}_{\mathcal{P}}$, *such that* $\mathrm{argmin}(\bar{p}) = \bar{x}$, *the function* argmin *is calm at $\bar{p}$ and it holds*

$$\mathrm{clm}\,\mathrm{argmin}(\bar{p}) \leq \frac{\mathrm{clm}\,h(\cdot,\bar{x}) + c \triangleq [h,C](\bar{p},\bar{x}) \cdot \mathrm{clm}\,(\varphi(\cdot,\bar{x}) - \mathrm{val}_{\mathcal{P}})(\bar{p})}{c \triangleq [h,C](\bar{p},\bar{x}) \left[\overline{|\nabla|\varphi - \mathrm{val}_{\mathcal{P}}||}^{>}(\bar{p},\bar{x}) - 1\right]}. \tag{18}$$

*Proof.* In the light of Theorem 3 the set-valued map $\mathrm{Argmin}_{\mathcal{P}}$ is Lipschitz l.s.c. at $(\bar{p},\bar{x})$. Thus for every $l > \mathrm{Liplsc}\,\mathrm{Argmin}_{\mathcal{P}}(\bar{p},\bar{x})$, there exists $\zeta_l > 0$ such that

$$\mathrm{Argmin}_{\mathcal{P}}(p) \cap \mathrm{B}(\bar{x}, l d(p,\bar{p})) \neq \varnothing \quad \forall p \in \mathrm{B}(\bar{p},\zeta_l).$$

This means that, for every $p \in \mathrm{B}(\bar{p},\zeta_l)$, it is possible to choose an element $x_p \in \mathrm{Argmin}_{\mathcal{P}}(p)$ such that

$$\|x_p - \bar{x}\| \leq l d(p,\bar{p}).$$

Then the proof is accomplished by setting $\mathrm{argmin}(p) = x_p$. □

**Acknowledgements** The revised version of this paper took benefit from the careful reading and the useful remarks by two anonymous referees. The author thanks both of them.

## References


[1] Aubin, J.-P.: Lipschitz behavior of solutions to convex minimization problems. Math. Oper. Res. **9**, 87–111 (1984)

[2] Bank, B., Guddat, J., Klatte, D., Kummer, B., Tammer, K.: Non-Linear Parametric Optimization. Akademie-Verlag, Berlin (1982)

[3] Bernard, F., Thibault, L., Zlateva, N.: Characterizations of prox-regular sets in uniformly convex Banach spaces. J. Convex Anal. **13**, 525–559 (2006)

[4] Bonnans, J.F., Shapiro, A.: Perturbation Analysis of Optimization Problems. Springer, New York (2000)

[5] Clarke, F.H.: A new approach to Lagrange multipliers. Math. Oper. Res. **1**, 165–174 (1976)

[6] Clarke, F.H.: Optimization and Nonsmooth Analysis. Wiley Interscience, New York (1983)

[7] Clarke, F.H., Stern R.J., Wolenski P.R.: Proximal smoothness and lower-$C^2$ property. J. Convex Anal. **2**, 117–144 (1995)

[8] De Giorgi, E., Marino, A., Tosques, M.: Problems of evolution in metric spaces and maximal decreasing curves. Atti Accad. Naz. Lincei Rend. Cl. Sci. Fis. Mat. Natur. **68**(8), 180–187 (1980) [in Italian]

[9] Dempe, S., Pallaschke, D.: Quasidifferentiability of optimal solutions in parametric nonlinear optimization. Optimization **40**, 1–24 (1997)

[10] Demyanov, V.F., Rubinov, A.M.: Quasidifferentiable functionals. Dokl. Akad. Nauk SSSR **250**, 21–25 (1980) [in Russian]

[11] Demyanov, V.F., Rubinov, A.M.: On quasidifferentiable mappings. Math. Oper. forsch. Stat. Ser. Optim. **14**, 3–21 (1983)

[12] Demyanov, V.F., Rubinov, A.M.: Quasidifferential Calculus. Optimization Software, New York (1986)

[13] Demyanov, V.F., Rubinov, A.M.: Constructive Nonsmooth Analysis. Peter Lang, Frankfurt am Main (1995)

[14] Fabian, M.J., Henrion, R., Kruger, A.Y., Outrata, J.: Error bounds: necessary and sufficient conditions. Set-Valued Var. Anal. **18**, 121–149 (2010)

[15] Fiacco, A.V.: Introduction to Sensitivity and Stability Analysis. Academic, New York (1983)

[16] Henrion, R., Outrata, J.V.: Calmness of constraint systems with applications. Math. Program. **104**, Ser. B, 437–464 (2005)

[17] Klatte, D., Kummer, B.: Nonsmooth Equations in Optimizations. Regularity, Calculus, Methods and Applications. Kluwer, Dordrecht (2002)

[18] Levitin, E.S.: Perturbation Theory in Mathematical Programming and Its Applications. Wiley, Ltd., Chichester (1994)

[19] Luderer, B., Minchenko, L., Satsura, T.: Multivalued Analysis and Nonlinear Programming Problems with Perturbations. Kluwer, Dordrecht (2002)

[20] Muselli, E.: Upper and lower semicontinuity for set-valued mappings involving constraints. J. Optim. Theory Appl. **106**, 527–550 (2000)

[21] Mordukhovich, B.S.: Variational Analysis and Generalized Differentiation, I: Basic Theory. Springer, Berlin (2006)

[22] Pallaschke, D., Urbański, R.: Pairs of Compact Convex Sets. Fractional Arithmetic with Convex Sets. Kluwer, Dordrecht (2002)

[23] Peressini, A.L.: Ordered Topological Vector Spaces. Harper and Row, New York-London (1967)

[24] Poliquin, R.A., Rockafellar, R.T.: Prox-regular functions in variational analysis. Trans. Am. Math. Soc. **348**, 1805–1838 (1996)

[25] Poliquin, R.A., Rockafellar R.T., Thibault, L.: Local differentiability of distance functions. Trans. Am. Math. Soc. **352**, 5231–5249 (2000)

[26] Robinson, S.M.: Generalized equations and their solutions, Part I: Basic theory. Math. Program. Study **10**, 128–141 (1979)

[27] Robinson, S.M.: Local structure of feasible sets in nonlinear programming, Part III: Stability and sensitivity. Math. Program. Study **30**, 45–66 (1987)
[28] Rockafellar, R.T., Wets, R.J.-B.: Variational Analysis. Springer, Berlin (1998)
[29] Rubinov, A.M.: Differences of convex compact sets and their applications in nonsmooth analysis. In: Giannessi, F. (ed.) Nonsmooth Optimization. Methods and Applications, pp. 366–378. Gordon and Breach, Amsterdam (1992)
[30] Rubinov, A.M., Akhundov, I.S.: Differences of compact sets in the sense of Demyanov and its applications to nonsmooth analysis. Optimization **23**, 179–188 (1992)
[31] Rubinov, A.M.,Vladimirov, A.A.: Differences of convex compacta and metric spaces of convex compacta with applications: a survey. In: Demyanov, V.F., Rubinov, A.M. (eds.) Quasidifferentiability and Related Topics, pp. 263–296. Kluwer, Dordrecht (2000)
[32] Uderzo A.: Convex difference criteria for the quantitative stability of parametric quasidifferentiable systems. Set-Valued Anal. **15**, 81–104 (2007)
[33] Uderzo A.: On Lipschitz semicontinuity properties of variational systems with application to parametric optimization. Submitted paper, 1–26 preprint (2012), arXiv:1305.3459
[34] Yen, N.D.: Stability of the solution set of perturbed nonsmooth inequality systems and application. J. Optim. Theory Appl. **93**, 199–225 (1997)

# Some Remarks on Bi-level Vector Extremum Problems

**Carla Antoni and Franco Giannessi**

**Abstract** The present note aims at introducing a new approach for handling bi-level vector extremum problems. After having defined a class of nonconvex functions on which it seems promising to carry on the research for such problems, the note is concentrated on the convex-linear bi-level problems; in this case, the results are compared with the existing literature. Suggestions for further research are given.

**Keywords** Bi-level vector optimization • Vector optimization • Multiobjective optimization • Cone functions • Scalarization

## 1 Introduction

Many real world problems can be formulated mathematically as extremum problems, where there are several objective functions. Rarely, such functions achieve the extremum at a same point. This has led, in the last decades, to a rapid mathematical development of this field, whose origin goes back to more than one century ago. Almost independently of this, in some fields of engineering dealing with the design, the need has gradually emerged of taking into account the competition of some variables, which were previously condensed in just one. Roughly speaking, the researches, carried on in the mathematical optimization area, can be split into those which aim at detecting properties of the set of solutions and those which aim at providing us with methods for finding such a set by using, in general, scalarizing

C. Antoni (✉)
Naval Academy, 72 Viale Italia, Leghorn, Italy
e-mail: carla.antoni5@gmail.com

F. Giannessi
Department of Mathematics, University of Pisa, Pisa, Italy
e-mail: gianness@dm.unipi.it

V.F. Demyanov et al. (eds.), *Constructive Nonsmooth Analysis and Related Topics*, Springer Optimization and Its Applications 87, DOI 10.1007/978-1-4614-8615-2_9,

techniques; the former are extremely important as a base for any other research; the latter should take into consideration the fact that, in most of the applications, the designer has a bi-level problem: an extremum problem (upper level), whose feasible region is the set of (vector) extremum points of a vector constrained extremum problem (lower level); consequently, the methods of solution of the bi-level problem should require to run on such a feasible set, namely the set of vector extremum points, as less as possible (unlike what some existing methods try to do). Here, based on previous results [6], a method for solving the bi-level problem is described. Our main scope consists in outlining the method and letting it be easily understood to a wide audience more than deliver a detailed, rigorous exposition of the method; this will be done in a forthcoming paper [1]. Consequently, to make the text plain, we make some assumptions, which are somewhat strong, and which can be easily weakened; moreover, again for the sake of simplicity, we take for granted the existence of the extrema we meet. In Sect. 2, we consider a class of nonconvex functions, which enjoy the nice property to have convex level sets, and for which constructive sufficient condition can be established in order to state whether or not a given function belongs to such a class. In Sect. 3, after having proved some properties of such a class, we define a new type of scalarization for a multiobjective problem (lower level), and then we outline an approach to the bi-level problem. While this will be the subject of a forthcoming paper, in this note (Sect. 4), we will develop the case where the multiobjective problem (lower level) is linear and the upper level is convex, and we consider strong solutions with Pareto-cone. It will be shown that, in this case, we improve the existing literature, in as much as the lower level requires to handle a linear problem (while the existing literature is faced with a nonlinear one and, in general, for weak solutions). Section 5 contains some numerical examples. In the final section, we discuss shortly some further developments.

Many real world problems lead to the minimization (or maximization) of a scalar function over the set of minimum points of a vector problem. Hence, we are faced with a bi-level problem.

Let $l$, $m$, and $n$ be positive integer, $X \subseteq \mathbb{R}^n$, $C \subseteq \mathbb{R}^l$ be a convex, closed, and pointed cone with apex at the origin; the functions $\Phi : \mathbb{R}^n \longrightarrow \mathbb{R}$, $f : \mathbb{R}^n \longrightarrow \mathbb{R}^l$, $g : \mathbb{R}^n \longrightarrow \mathbb{R}^m$ are given. In the sequel, $\operatorname{int} S$ and $\operatorname{ri} S$ will denote the topological interior and relative interior of the set $S$, respectively.

Consider the problem (lower level):

$$\min_{C_0} f(x) \;\; \text{s.t. } x \in K := \{x \in \mathbb{R}^n : g(x) \geq 0\}, \tag{1}$$

where $\min_{C_0}$ marks vector minimum with respect to the cone $C_0 := C \setminus \{0\}$: $y$ is a (global) vector minimum point (in short, VMP) of (1) if and only if

$$f(y) \not\geq_{C_0} f(x), \quad \forall x \in K, \tag{2}$$

where the inequality in (2) means $f(y) - f(x) \notin C_0$. At $C = \mathbb{R}^l_+$, (1) becomes the classic Pareto vector problem.

Finally, consider the problem (upper level):

$$\min \Phi(x) \text{ s.t. } x \in K^0, \tag{3}$$

where $K^0$ is the set of VMPs of (1).

## 2 C-Functions

The definitions of $A$ and $D$ in this section are independent of those of the other sections.

**Definition 1.** The (positive) polar of the cone $C$ is given by:

$$C^* := \{x \in \mathbb{R}^n : \langle y, x \rangle \geq 0, \ \forall y \in C\}. \tag{4}$$

The following set of functions will be the base of the present approach. More precisely, in the present paper, we establish the theory, based on $C-$functions, which is the background of the approach to the bi-level problems we want to carry on. In the present paper, we begin with the class of problems, say convex-linear problems, which have $\Phi$ convex, $f$ linear, and $K$ polyhedral; other classes will be studied in further coming papers.

**Definition 2.** Let $X$ be convex; $f$ is a $C$-function [4] if and only if $\forall x^1, x^2 \in X$, $\forall \alpha \in [0,1]$:

$$(1-\alpha)f(x^1) + \alpha f(x^2) - f((1-\alpha)x^1 + \alpha x^2) \in C. \tag{5}$$

When $C \subseteq \mathbb{R}^l$ or $C \supseteq \mathbb{R}^l$, then f is called $C$-convex. At $l = 1$ and $C = \mathbb{R}_+$, $f$ is the classic convex function. In most of the literature, regardless of the occurrence of such inclusions, a $C$-function is often called $C$-convex; this is not suitable. For instance, the $(\mathbb{R}_-)$-function, which turns out to be a concave function, should be called $(\mathbb{R}_-)$-convex; this, even if formally correct, is unnecessarily far from the intuitive sense and the common language. The definition is a cornerstone of mathematics; consequently, it should be handled very cautiously, without distorting the already possessed concepts. The following property of $C$-functions will be fundamental in the sequel.

**Proposition 1.** *If $f$ is a $C$-function on $X$ and $c^* \in C^*$, then $\langle c^*, f \rangle$ is convex on $X$.*

*Proof.* Since $f$ is a $C$-function, $\forall c^* \in C^*$, $\forall x^1, x^2 \in X$, $\forall \alpha \in [0,1]$,

$$\langle c^*, (1-\alpha)f(x^1) + \alpha f(x^2) - f((1-\alpha)x^1 + \alpha x^2) \rangle \geq 0$$

or, equivalently,

$$(1-\alpha)\langle c^*, f(x^1)\rangle + \alpha\langle c^*, f(x^2)\rangle - \langle c^*, f((1-\alpha)x^1 + \alpha x^2)\rangle \geq 0.$$

Hence, the convexity of $\langle c^*, f\rangle$ follows. □

As it is well known, the drawback of most of the extensions of convex functions is the lack of conditions which allow one to detect, through viable numerical calculus, whether or not a given function fulfils the definition of such an extension. The $C$-functions are among the few extensions for which some viable conditions can be established. Suppose that the cone $C$ be polyhedral, so that there exists a matrix $A \in \mathbb{R}^{r \times l}$, whose generic entry is denoted by $a_{ij}$, such that:

$$C = \{u \in \mathbb{R}^l : Au \geq 0\}. \tag{6}$$

In this case, $f$ is a $C$-function if and only if, $\forall x^1, x^2 \in X,\ \forall \alpha \in [0,1]$,

$$A[(1-\alpha)f(x^1) + \alpha f(x^2) - f((1-\alpha)x^1 + \alpha x^2)] \geq 0 \tag{7}$$

or

$$(1-\alpha)\phi_i(x^1) + \alpha\phi_i(x^2) - \phi_i((1-\alpha)x^1 + \alpha x^2) \geq 0, \quad i = 1,\dots,r,$$

where

$$\phi_i(x) := \sum_{j=1}^{l} a_{ij} f_j(x), \quad i = 1,\dots,r. \tag{8}$$

Thus, the following result holds.

**Proposition 2.** *$f$ is a $C$-function on $X$ with respect to the polyhedral cone (6) if and only if the functions $\phi_i$, $i = 1,\dots,r$ of (8) are convex on $X$.*

Observe that the functions $\phi_1,\dots,\phi_r$ can be convex, even if some (all) the functions $f_1,\dots,f_l$ are not, as the following example shows.

*Example 1.* Let $X = \mathbb{R}^2$, $C = \{u \in \mathbb{R}^2 : 2u_1 + u_2 \geq 0,\ u_1 + 2u_2 \geq 0\}$. Let $f_1(x) = -x_1^2/2 + 3x_2^2$, $f_2(x) = 3x_1^2 - x_2^2/2$. $f_1$ and $f_2$ are not convex, but $\phi_1 = 2f_1 + f_2$ and $\phi_2 = f_1 + 2f_2$ are convex and then $f = (f_1, f_2)$ is a $C$-function.

The preceding example suggests a condition for $f$ to be a $C$-function when $C$ is like in (6). Set:

$$f = (f_1,\dots,f_l), \quad \text{where } f_i(x) = \langle x, D_i x\rangle,\ D_i \in \mathbb{R}^{n \times n}, \quad i = 1,\dots l, \tag{9}$$

and put $x(\alpha) := (1-\alpha)x^1 + \alpha x^2$, $\alpha \in [0,1]$. Condition (7) is fulfilled if and only if, $\forall i = 1,\dots,l$, $\forall x^1, x^2$, $\forall \alpha \in [0,1]$,

$$(1-\alpha)\langle x^1, \sum_{j=1}^{l} a_{ij} D_j x^1 \rangle + \alpha \langle x^2, \sum_{j=1}^{l} a_{ij} D_j x^2 \rangle - \langle x(\alpha), \sum_{j=1}^{l} a_{ij} D_j x(\alpha) \rangle \geq 0.$$

Thus, the following result holds.

**Proposition 3.** *The (vector-valued) quadratic function (9) is a $C$-function on $X$ with respect to the cone (6), if and only if each of the matrices*

$$Q_i := \sum_{j=1}^{l} a_{ij} D_j, \quad i = 1, \dots, l,$$

*has nonnegative eigenvalues.*

*Remark 1.* The cone of Example 1 contains $\mathbb{R}^2_+$, but it does not differ much from $\mathbb{R}^2_+$. In several applications, like the design of aircrafts, the cone is the Pareto one; however, the designers may desire to explore what happens, if such a cone is relaxed a little bit.

*Remark 2.* Example 2 suggests a condition for $f$ to be a $C$-function on $X$ with respect to a not necessarily polyhedral cone $C$.

Now, let us consider the case, where C is not necessarily polyhedral; let it be defined by its supporting half-spaces, or

$$C := \bigcap_{t \in T} \{x \in \mathbb{R}^l : \langle a_t, x \rangle \geq 0\}, \tag{10}$$

where $T$ is an interval of $\mathbb{R}$, and, $\forall t \in T,\ a_t \in \mathbb{R}^l$.

When the cone $C$ is given by (10), a function $f$ is a $C$-function if and only if, $\forall x^1, x^2,\ \forall \alpha \in [0,1]$,

$$\langle a_t, (1-\alpha) f(x^1) + \alpha f(x^2) - f(x(\alpha)) \rangle \geq 0, \quad \forall t \in T,$$

that is,

$$(1-\alpha)\varphi_t(x^1) + \alpha \varphi_t(x^2) - \varphi_t(x(\alpha)) \geq 0, \quad \forall t \in T,$$

where

$$\varphi_t = \langle a_t, f \rangle. \tag{11}$$

We have thus obtained:

**Proposition 4.** *The function $f$ is a $C$-function on $X$ with respect to the cone $C$ defined in (10) if and only if, $\forall t \in T$, the function $\varphi_t$ defined in (11) is convex on $X$.*

The functions $\varphi_t, t \in T$, can be convex on $X$ even if some (all) the functions $f_j$, $j = 1,\dots,l$ are not, as the following example shows.

*Example 2.* Let $X = \mathbb{R}^2$ and $C = \{u \in \mathbb{R}^3 : u_3 \geq \sqrt{u_1^2 + u_2^2}\}$. The family of all the supporting half-spaces of $C$, namely (10), is easily found to be:

$$\bigcap_{t\in[-\sqrt{2},\sqrt{2}]} \{u \in \mathbb{R}^3 : -tu_1 \pm \sqrt{2-t^2}u_2 + \sqrt{2}u_3 \geq 0\}.$$

Consider the vector function $f = (f_1, f_2, f_3)$ with:

$$f_i(x) = \langle x, D_i x\rangle,$$

being

$$D_1 = \begin{pmatrix} 1 & 0 \\ 0 & -1 \end{pmatrix}, \; D_2 = \begin{pmatrix} -1 & 0 \\ 0 & 1 \end{pmatrix}, \; D_3 = \begin{pmatrix} 2 & 0 \\ 0 & 2 \end{pmatrix}.$$

We have now:

$$a_t = (-t, \pm\sqrt{2-t^2}, \sqrt{2}), \quad t \in [-\sqrt{2}, \sqrt{2}],$$

so that

$$\phi_t(x) = \left\langle x, \begin{pmatrix} -t \pm \sqrt{2-t^2} + \sqrt{2} & 0 \\ 0 & -t \pm \sqrt{2-t^2} + \sqrt{2} \end{pmatrix} x \right\rangle.$$

It is easy to see that, for each $t \in [-\sqrt{2}, \sqrt{2}]$, $\phi_t$ is convex on $X$, while $f_1$ and $f_2$ are not convex.

## 3 Scalarization of the Lower Level

Now, let us consider the scalarization of (1) by exploiting the method, which was introduced in [5] ; see also Sect. 8 of [6].

For each $y \in X$ and $p \in C^*$, consider the sets:

$$S(y) := \{x \in X : f(x) \in f(y) - C\},$$
$$S_p(y) := \{x \in X : \langle p, f(x)\rangle \leq \langle p, f(y)\rangle\}.$$

$S(y)$ is evidently a level set of $f$ with respect to $C$. Indeed, when $X = \mathbb{R}^n$ and $C = \mathbb{R}^l_+$, then it is precisely the lower set of $f$; in the affine case, $S(y)$ is a cone with apex at $y$ , and $S_p(y)$ becomes a supporting half-space at its apex.

Note that, apart from $S(y)$, $S_p(y)$, and Proposition 5 (where $p$ is a parameter), $p$ will be considered fixed, in particular in the case of the algorithm. Instead, $y$ will now play the role of a parameter and, later, that of the unknowns; this will be reported. The concept of level set, in strict or extended sense, plays a fundamental role in the present scalarization; in order to describe it, we need to establish some properties.

**Proposition 5.**

*(i) If $f$ is a C-function on $X$, then, $\forall y \in X$, $S(y)$ is convex.*
*(ii) If $p \in C^*$, then, $\forall y \in X$,*

$$S(y) \subseteq S_p(y), \quad y \in S(y) \cap S_p(y).$$

*Proof.*

(i) For $x^i \in S(y)$ there exist $c^i$ such that $f(x^i) = f(y) - c^i$, $i = 1,2$. Then, $\forall \alpha \in [0,1]$,

$$(1-\alpha)f(x^1) + \alpha f(x^2) = f(y) - ((1-\alpha)c^1 + \alpha c^2).$$

Since $C$ is convex, then $(1-\alpha)c^1 + \alpha c^2) \in C$. Moreover, if $f$ is a $C$ function, then $\forall \alpha \in [0,1]$ there exists $c'(\alpha)$ such that

$$f(x(\alpha)) = (1-\alpha)f(x^1) + \alpha f(x^2) - c'(\alpha).$$

It follows that

$$f(x(\alpha)) = f(y) - (1-\alpha)c^1 - \alpha c^2 - c'(\alpha)$$

and then $x(\alpha) \in S(y)$, $\forall \alpha \in [0,1]$, $\forall y \in X$.
(ii) $0 \in C \iff y \in S(y)$; $y \in S_p(y)$ is trivial. The thesis follows.

□

Now, let $p \in C^*$; as announced, *unlike what in general happens in the field of scalarization, p will remain fixed* in the rest of this section. Let us introduce the following problem (in the unknown $x$, depending on the parameter $y$):

$$\min \langle p, f(x) \rangle, \quad x \in K(y) := K \cap S(y). \tag{12}$$

Borrowing the terminology of variational inequalities, we call (12) *quasi-minimum problem*. Its feasible region depends (parametrically) on $y$; we will see that, for our scalarization method, it will be important to consider the case $y = x$, where the feasible region depends on the unknown; i.e., the feasible points are "fixed points" of the point-to-set map $y \mapsto K(y)$.

*Remark 3.* Under suitable assumptions, the first-order necessary condition of (12) is

$$\langle p^T \nabla f(x), y - x \rangle \geq 0, \quad x \in K(y), \tag{13}$$

which is a particular case of a quasi-variational inequality.

In general, problem (13) looks difficult. The following proposition identifies a class of (12), which can be handled easily.

**Proposition 6.** *Let $X$ be convex, $f$ be a $C$-function, on $X$ and $g$ be concave on $X$ and $p \in C^*$. Then (12) is convex.*

*Proof.* We have to show that the restriction of $\langle p, f(\cdot) \rangle$ to $X$ and $K(y)$ are convex. Since $p \in C^*$ and $f$ is a $C$-function, then $\forall x^1, x^2 \in X$, it holds

$$\langle p, (1-\alpha) f(x^1) + \alpha f(x^2) - f(x(\alpha)) \rangle \geq 0, \quad \forall \alpha \in [0,1],$$

and, equivalently,

$$(1-\alpha)\langle p, f(x^1) \rangle + \alpha \langle p, f(x^2) \rangle - \langle p, f(x(\alpha)) \rangle \geq 0, \quad \forall \alpha \in [0,1],$$

that is, the convexity of $x \mapsto \langle p, f(x) \rangle$. The convexity of $X$ and the concavity of $g$ imply the convexity of $K$; then, the convexity of $S(y)$ (Proposition 5) implies that of $K(y)$. □

As announced, we want to run on the set of VMPs of (1) in such a way to make the resolution of (1) as easy as possible. In other words, by exploiting the properties of (12), it will be possible to define a method which solves (1) without being obliged to find in advance $K^0$.

**Proposition 7.** $y \in X,\ x \in S(y) \implies S(x) \subseteq S(y)$.

*Proof.* $x \in S(y)$ if and only if there is $c \in C$ such that $f(x) = f(y) - c$, and $\hat{x} \in S(x)$ if and only if there is $\hat{c} \in C$ such that $f(\hat{x}) = f(x) - \hat{c}$. It follows that $f(\hat{x}) = f(y) - (c + \hat{c})$, that is, $\hat{x}$ belongs to $S(y)$. Then, the inclusion $S(x) \subseteq S(y)$ is proved. □

The above proposition shows that if the "apex"of the level set $S(y)$ is shifted to a point belonging to it, then the translated level set is contained in it. This property will allow us to find a VMP of (1).

**Proposition 8.** *If $x^0$ is a (global) minimum point of (12) at $y = y^0$, then $x^0$ is a (global) minimum point of (12) at $y = x^0$.*

*Proof.* Proposition 5 guarantees that $x^0 \in S(x^0)$. Ab absurdo, suppose that $x^0$ be not a (global) minimum point of (12) at $y = x^0$. Then,

$$\exists \hat{x} \in K \cap S(x^0),\ \langle p, f(\hat{x}) \rangle < \langle p, f(x^0) \rangle. \tag{14}$$

Proposition 7 implies $\hat{x} \in S(y_0)$, and the conditions

$$\hat{x} \in K \cap S(y^0),\ \langle p, f(\hat{x}) \rangle < \langle p, f(x^0) \rangle. \tag{15}$$

contradict the assumptions. Necessarily $x^0$ is a global minimum point of (12) at $y = x^0$. □

*Remark 4.* Taking into account Propositions 7 and 8, Problem (12) can be formulated as:

$$\text{find } x^0 \in K \text{ s.t. } \min_{x \in K(x^0)} \langle p, f(x) \rangle = \langle p, f(x^0) \rangle, \tag{16}$$

which justifies, once more, the quasi-minimum problem.

The following proposition connects the optimality of (1), its image, and the optimality of (16). It is trivial to note that (2) is satisfied if and only if the system (in the unknown $x$):

$$f(y) - f(x) \geq_{C_0} 0, \quad g(x) \geq 0, \quad x \in X \tag{17}$$

is impossible.

**Proposition 9.** *Let* $p \in \operatorname{int} C^*$.

*(i) y is a VMP of (1) if and only if the system (in the unknown x):*

$$\langle p, f(y) - f(x) \rangle > 0, \quad f(y) - f(x) \in C, \quad g(x) \geq 0, \quad x \in X, \tag{18}$$

*is impossible.*

*(ii) The impossibility of (18) is a necessary and sufficient condition for y to be a (scalar) minimum point of (12) or (16).*

*Proof.*

(i) If $\hat{x}$ satisfies (18), then $f(y) - f(\hat{x}) \neq 0$; consequently $\hat{x}$ satisfies (17). Vice versa, suppose there is $\hat{x} \in X$ such that $f(y) - f(\hat{x}) \geq_{C_0} 0$, $g(\hat{x}) \geq 0$. This implies $\langle p, f(y) - f(\hat{x}) \rangle > 0$. In fact, since $p \in \operatorname{int} C^*$, there is $r > 0$ such that $p + N_r \subseteq C^*$, where $N_r = \{\delta \in \mathbb{R}^l : \| \delta \| < r\}$, then

$$\langle p + \delta, f(y) - f(\hat{x}) \rangle \geq 0, \quad \forall\, \delta \in N_r. \tag{19}$$

Ab absurdo, suppose

$$\langle p, f(y) - f(\hat{x}) \rangle = 0. \tag{20}$$

Since there exists $\varepsilon > 0$ such that $\varepsilon \| f(y) - f(\hat{x}) \| < r$, and $f(y) - f(\hat{x}) \neq 0$, then, from (19) and (20), it follows that

$$0 \leq \langle p - \varepsilon(f(y) - f(\hat{x})), f(y) - f(\hat{x}) \rangle = -\varepsilon \| f(y) - f(\hat{x}) \| < 0. \tag{21}$$

This is absurd.

(ii) Follows from the definition of scalar minimum point. □

*Remark 5.* System (18) allows one to associate (1) with its image space (IS) and perform an useful analysis. To this end pose:

$$u = f(y) - f(x), \quad v = g(x), \quad x \in X; \tag{22}$$

the image of $X$ through the function $x \mapsto (f(y) - f(x), g(x))$ is the IS associated with (1). For details, see [4, 6].

We are now able to define the steps of an approach for finding (all) the VMPs of (1).

(A) Choose any $p \in \text{int}\, C^*$ ($p$ will remain fixed in the sequel).
(B) Choose any $y^0 \in K$ and solve the (scalar) problem (12) at $y = y^0$; let $x^0$ be a solution; according to Proposition 8, $x^0$ is a VMP of (1).
(C) Consider (12) as a parametric problem in the parameter y: start at $y = x^0$ and find its solutions. According to Propositions 8 and 9, all the solutions of (1) will be found. This approach, which will be developed in [1], seems promising independently of the bi-level problem. If we apply it to the bi-level problem, then it becomes the following set of steps.

As said in Sect. 1, in general, in the real applications, the problem to solve is just (3) and not that of finding all the solutions of (1); hence, it is desired to meet, among the solutions of (1), only those, which allow one to solve (3). The approach described above serves to satisfy such a need. To this end, the method described at the end of the previous section can be integrated this way:

(A) Choose any $p \in \text{int}\, C^*$ ($p$ will remain fixed in the sequel).
(B) Choose any $y^1 \in K$ and solve the (scalar) problem (12) at $y = y^1$; let $x^1$ be a minimum point; call $K_1^0$ the set of solutions of (12) obtained by varying $y$ from $y = y^1$ ; of course $x^1 \in K_1^0$; according to Propositions 8 and 9, all the elements of $K_1^0$ are VMPs of (1).
(C) Solve the problem:

$$\min \Phi(y) \text{ s.t. } y \in K_1^0; \tag{23}$$

if we can conclude that the solutions of this problem are such also on $K^0$, then (3) is solved; otherwise, we must continue.
(D) Jump to a subset of $K^0$, adjacent to $K_1^0$; let it be $K_2^0$; repeat (C) on it; and so on.

As is easily seen, thanks to the method of the above section, in solving (3) we do not meet all the solutions of (1). The above method is a general scheme, which requires to be implemented; the implementation takes advantage, if it is done within a certain class of functions; an instance of this is shown below.

## 4 The Upper Level. The Convex/Linear Case

The symbols of this section are independent of those of the previous sections. Let us now consider problem (3), where $\Phi$ is convex; even if it is not necessary, for the sake of simplicity, $\Phi$ will be assumed to be differentiable.

Now, suppose that the lower level be linear, and consider the case where $X = \mathbb{R}^n_+$ and $C = \mathbb{R}^l_+$, which, although a particular one, is among the most important formats in the applications. Thus, without any loss of generality, we can set:

$$f_i(x) = \langle d^i, x\rangle,\ d^i \in \mathbb{R}^n,\ i = 1,\dots,l,\ D = \begin{pmatrix} d^1 \\ \vdots \\ d^l \end{pmatrix},\ f(x) = Dx$$

$$\langle p, f(x)\rangle = \langle pD, x\rangle$$

where $d^i$ and $p$ are considered as row vectors. The set $K(y)$ can be cast in the form:

$$K(y) = \{x \in \mathbb{R}^n_+ : -Dx \geq -Dy,\ \Gamma x \geq \gamma\},$$

with $\Gamma \in \mathbb{R}^{m\times n}$, $\gamma \in \mathbb{R}^m$. By setting

$$A = \begin{pmatrix} -D \\ \Gamma \end{pmatrix},\ b = \begin{pmatrix} 0 \\ \gamma \end{pmatrix},\ E = \begin{pmatrix} -D \\ 0 \end{pmatrix},\ c = pD$$

problem (12) takes the form:

$$\min \langle c, x\rangle,\ Ax \geq b + Ey,\ x \geq 0, \tag{24}$$

where, without any loss of generality, we assume that the rank of $A$ be $n$ and that

$$A = \begin{pmatrix} B \\ N \end{pmatrix},\ b = \begin{pmatrix} b_B \\ b_N \end{pmatrix},\ E = \begin{pmatrix} E_B \\ E_N \end{pmatrix},$$

where $B$ is a feasible basis, that is,

$$B^{-1}(b_B + E_B y) \geq 0,\ N(B^{-1}(b_B + E_B y)) \geq b_N + E_N y. \tag{25}$$

The vector

$$x = B^{-1}(b_B + E_B y) \tag{26}$$

is a solution of (24) if and only if

$$cB^{-1} \geq 0, \tag{27}$$

and remains optimal until $y$ fulfils (25). Note that the performance of step $(B)$ of the method leads at a point, namely $x^0$, where the inequalities, which defines $S(y)$, are verified as equalities; consequently, in general, $x^0$ will be overdetermined. Therefore, we assume that an anti-cycling ordering is adopted. Moreover, it is not restrictive to suppose that $B$ contains at least one row of $D$; this will understood in the sequel.

Now, with a small abuse of notation, problem (23) becomes:

$$\min \Phi(y), \text{ s.t. } y \in K^0_{B_1}, \tag{28}$$

where

$$K^0_{B_1} = \{y \in \mathbb{R}^n : (I_n - B_1^{-1}E_{B_1})y = B_1^{-1}b_{B_1},\ (N_1B_1^{-1}E_{B_1} - E_{N_1})y \geq b_{N_1} - N_1B_1^{-1}b_{B_1}\},$$

and $B_1$ is a base which identifies $K^1_0$. We can now specify the method (A)–(D) to the present case; it finds a local minimum point of (3).

(a) Choose any $p$ such that $p \in \operatorname{int} C^*$; $p$ will remain fixed in the sequel.
(b) Choose any $y^1 \in K$. We have to solve (24), which is assumed to have minimum. By a standard use of simplex method, we find an optimal basis, say $B_1$ and the minimum point given by:

$$x^1 = B_1^{-1}(b_{B_1} + E_{B_1}y^1).$$

Now, replace $y^1$ with the parameter $y$, but keep $B_1$ as basis. According to the Propositions 8 and 9, all the VMPs of (1), corresponding to $B_1$, are obtained as those solutions of (24) which equal the very parameter $y$; this is equivalent to say that $y$ must be such that $B_1$ is both primal and dual feasible and $y$ must be a fixed point of the map: $y \mapsto B_1^{-1}(b_{B_1} + E_{B_1}y)$. Since $B_1$ is dual feasible if and only if $cB_1^{-1} \geq 0$ (which is a by-product of the construction of $x^1$ and does not depend on $y$), in conclusion, such VMPs are the solutions of the system:

$$\begin{cases} B_1(N_1B_1{}^{-1}E_{B_1} - E_{N_1})y \geq b_{N_1} - N_1B_1{}^{-1}b_{B_1} & \text{primal feasibility} \\ y = B_1^{-1}(b_{B_1} + E_{B_1}y) & \text{fixed point} \end{cases} \tag{29}$$

Call $K^0_{B_1}$ the set of solutions of (29).

(c) Consider the problem

$$\min \Phi(y), \text{ s.t. } y \in K^0_{B_1}, \tag{30}$$

and let $y^2$ be a minimum point of it. If $y^2 \in \operatorname{ri} K^0_{B_1}$, then stop; otherwise, perform next step.
(d) If $\Phi$ does not decrease, when we try, through a pivot, to exchange $B_1$ with one of its adjacent bases (such an exchange will be performed under an anti-cycling

rule), then again with $y^2$, we have reached a solution of (30) and we stop. Otherwise, we replace $B_1$ with an adjacent basis, which allows $\Phi$ to decrease, and repeat the step (c).

*Remark 6.* It is worthy to stress the fact that the previous method may reduce the bi-level problem to a finite sequence of scalar extremum problems. For instance, if both (1) and (3) are linear, then, performing the steps (a)–(d) of this section amounts to execute a finite steps of simplex method; if $\Phi$ is convex and (1) is linear (as assumed in this section), then performing (a)–(d) in this section amounts to execute a finite number of steps of the gradient method.

Now, we will give a justification of the above method. Let $K$ be a polyhedron of $\mathbb{R}^n$ and $Q$ a convex cone having, as apex, the origin, which does not belong to it. Given a vector $x$, $Q_x$ denotes the translation of $Q$, which has $x$ as apex or

$$Q_x = \{y \in \mathbb{R}^n : y = x + q,\ q \in Q\}.$$

Moreover, in the sequel, $H^0$ denotes any hyperplane of $\mathbb{R}^n$, defined by

$$\{x \in \mathbb{R}^n : \langle a, x\rangle = b\}$$

and $H^-$ and $H^+$ denote, respectively, the half-spaces

$$\{x \in \mathbb{R}^n : \langle a, x\rangle \leq b\}, \quad \{x \in \mathbb{R}^n : \langle a, x\rangle \geq b\}.$$

**Definition 3.** Let $\mathcal{F}$ be a set of proper faces of a polyhedron $K$. $\mathcal{F}$ is said to be connected, if and only if, for each pair of element of $\mathcal{F}$, say $F'$ and $F''$, there exists a set of proper faces of $K$, say $F_0, F_1, \ldots, F_r$, contained in $\mathcal{F}$, such that $F_0 = F'$, $F_r = F''$, $F_{i-1} \cap F_i \neq \emptyset$, and $F_i$ is not a subface of $F_{i-1}$, for $i = 1, \ldots, r$.

**Lemma 1.** *Let $F$ be a face of $K$ and $x^0 \in \operatorname{int} F$. If*

$$Q_{x^0} \cap K = \emptyset, \tag{31}$$

*then, for all $x \in F$,*

$$Q_x \cap K = \emptyset.$$

*Proof.* Let the polyhedron $K$ be the set

$$\{x \in \mathbb{R}^n : \Gamma x \geq \gamma\},$$

where $\Gamma \in \mathbb{R}^{m\times n}$, $\gamma \in \mathbb{R}^m$. Without any loss of generality, as face $F$, we can consider

$$F = \{x \in \mathbb{R}^n : \Gamma_1 x = \gamma_1,\ \Gamma_2 x \geq \gamma_2\},$$

where

$$\Gamma = \begin{pmatrix} \Gamma_1 \\ \Gamma_2 \end{pmatrix}, \quad \gamma = \begin{pmatrix} \gamma_1 \\ \gamma_2 \end{pmatrix}.$$

The hypothesis $x^0 \in \mathrm{ri}\, F$ means

$$\Gamma_1 x^0 = \gamma_1, \; \Gamma_2 x^0 > \gamma_2. \tag{32}$$

From (31) we draw that not only one of the inequalities which define $K$ is violated, but, account taken of (32), also that such inequality corresponds to $\Gamma_1$: with obvious notation, let us denote it by

$$\langle (\Gamma_1)_i, x^0 + q \rangle < (\gamma_1)_i. \tag{33}$$

In fact, if ab absurdo such a violated inequality corresponded to $\Gamma_2$, then, by letting $q \longrightarrow 0$, we would obtain

$$\langle (\Gamma_2)_i, x^0 \rangle \leq (\gamma_2)_i$$

which contradicts (32). Now, let $x \in F$. The equalities

$$\langle (\Gamma_1)_i, x + q, a >= \langle (\Gamma_1)_i, x^0 + q \rangle,$$

and (33) lead to $Q_x \cap K = \emptyset$. □

**Lemma 2.** *$K^0$ is connected.*

*Proof.* First, observe that $K^0 = \{x \in K : Q_x \cap K = \emptyset\}$. If $F_1$ and $F_2$ are proper faces of $K$, which belong also to $K^0$, then there exist two hyperplanes they are not parallel (to deny this leads to contradict that one of the two faces does not belong to $K^0$):

$$H_1^0 = \{x : \langle a_1, x \rangle = b_1\}, \quad H_2^0 = \{x : \langle a_2, x \rangle = b_2\}$$

which support $K$ and, respectively, contain $F_1$ and $F_2$, and such that

$$Q_x \subseteq H_i^-, \; \forall x \in H_i^0, \; i = 1, 2.$$

Put

$$V^- := H_1^- \cap H_2^-, \quad V^+ := H_1^+ \cap H_2^+,$$

and set, $\forall t \in [0,1]$,

$$H_t^0 := \{x \in \mathbb{R}^n : \langle (1-t)a_1 + ta_2, x \rangle = (1-t)b_1 + tb_2\}.$$

Since, $\forall t \in [0,1]$,

$$K \subseteq V^+ \subseteq H_t^+, \quad Q_x \subseteq V^-, \forall x \in H_1^0 \cap H_2^0,$$

then $\forall t \in [0,1]$, there is a hyperplane, say $\mathcal{H}_t^0$, parallel to $H_t^0$, supporting $K$ and such that

$$K \subseteq \mathcal{H}_t^+, \quad Q_x \subseteq \mathcal{H}_t^-, \quad \forall x \in \mathcal{H}_t^0.$$

This means that the face $\mathcal{H}_t^0 \cap K$ is a subset of $K^0$. Finally, the interval $[0,1]$ is partitioned into a finite number of subintervals, and, this way, two consecutive intervals correspond to adjacent faces of $K^0$. □

**Proposition 10.** *Suppose that the function $\Phi : \mathbb{R}^n \longrightarrow \mathbb{R}$ be convex and differentiable, and suppose that its infimum (minimum) occurs on $\mathbb{R}^n \setminus K$. Assume that $f$ and $g$ be as above. Then, the algorithm (a)–(d) finds a local minimum point of (3) in a finite number of steps.*

*Proof.* First of all observe that $K^0$ is a connected set of faces of $K$ (Lemma 2). Observe also that, in going from basis $B$ to an adjacent one, maintaining a solution of (24) (note that $B$ contains at least one row of $D$; such an assumption is not restrictive because of Proposition 8 and allows us to parametrize the faces of $K^0$), we pass from a face of $K^0$ to an adjacent face of $K$ still included in $K^0$. Then, by adopting any (but fixed) ordering of the combinations of class $n$ extracted from $\{1,2,\ldots,l+m\}$, and an anti-cycling order, the algorithm $(a)$–$(d)$ can visit all the faces of $K$ and then of $K^0$ if it is necessary for the minimization of (3).

The stationary point at which the algorithm stops is a local minimum point of (3). In fact, it holds that

$$\nabla \Phi(x^0) \in \operatorname{conv}\{a_i,\ i \in \mathcal{I}(F^0)\}, \tag{34}$$

where $\mathcal{I}(F^0)$ is the set of indexes of the constraints of (24), which are also of $K$ and are binding at $x^0$. From (34) we draw that zero belongs to a convex combination, at $x^0$, of the gradients of $\Phi$ and of the constraints of (24), which are binding at $x^0$, identified by $\mathcal{I}(F^0)$. Due to the convexity of all the implicated functions, such a condition is sufficient besides necessary. □

## 5 Examples

*Example 3.* In (1) and (3), set $n = m = l = 2,\ X = C = \mathbb{R}^2_+$ and

$$f(x) = \begin{pmatrix} 2x_1 - x_2 \\ -x_1 + 2x_2 \end{pmatrix},\ g(x) = \begin{pmatrix} 2x_1 + x_2 - 1 \\ -x_1 + 2x_2 - 1 \end{pmatrix},\ \Phi(x) = x_1^2 + \left(x_2 - 1/2\right)^2.$$

Let us perform (a)–(d) of Sect. 4.

(a) Since $C^* = C$, we can choose $p = (2,3)$; it will remain fixed.
(b) Now we have:

$$D = \begin{pmatrix} 2 & -1 \\ -1 & 2 \end{pmatrix}, \; \Gamma = \begin{pmatrix} 2 & 1 \\ 1 & 2 \end{pmatrix}, \; \gamma = \begin{pmatrix} 1 \\ 1 \end{pmatrix}, \; c = \begin{pmatrix} 1 & 4 \end{pmatrix},$$

so that (28) becomes

$$\min(x_1 + 4x_2)$$
$$\text{s.t.} \tag{35}$$
$$\begin{pmatrix} -2 & 1 \\ 1 & -2 \\ 2 & 1 \\ 1 & 2 \end{pmatrix} \begin{pmatrix} x_1 \\ x_2 \end{pmatrix} \geq \begin{pmatrix} 0 \\ 0 \\ 1 \\ 1 \end{pmatrix} + \begin{pmatrix} -2 & 1 \\ 1 & -2 \\ 0 & 0 \\ 0 & 0 \end{pmatrix} \begin{pmatrix} y_1 \\ y_2 \end{pmatrix}, \; x \in X$$

We choose $y^1 = \begin{pmatrix} 2/3 & 4/9 \end{pmatrix}$. By means of a straightforward use of simplex method, we find that the basis of A, formed with the first and fourth rows, or

$$B_1 = \begin{pmatrix} -2 & 1 \\ 1 & 2 \end{pmatrix}$$

gives the unique solution of (35) (with $y = y^1$), i.e., $x^1 = \begin{pmatrix} 5/9 & 2/9 \end{pmatrix}$. The set $K^0_{B_1}$, given by the system (29), becomes

$$K^0_{B_1} = \{y \in \mathbb{R}^2_+ : y_1 + 2y_2 = 1, \; 3y_1 \geq 1\}.$$

(c) Problem (30) becomes

$$\min \left( y_1^2 + (y_2 - 1/2)^2 \right), \; \text{s.t.} y \in K^0_{B_1}.$$

Now replace $B_1$ with

$$B_2 = \begin{pmatrix} -2 & 1 \\ 2 & 1 \end{pmatrix}.$$

By means of the gradient method, we easily find its unique minimum point and minimum:

$$y^2 = \begin{pmatrix} 1/3 & 1/3 \end{pmatrix}, \quad \Phi(y^2) = 5/36.$$

Since $y^2 \in \partial K^0_{B_1}$, we perform (d).

(d) From the equalities

$$\nabla\Phi(y_2) = \left(\,-2/3\;-1/3\,\right)$$

we draw that $y^2$ is not either a global or a local minimum point of $\Phi$ on $K^0$ and that (the first constraint of $K$ being binding and the second being redundant), by replacing $B_1$ with $B_2$, $\Phi$ decreases with respect to 5/36. Then perform again the step (b) with $B_2$ and $y^2$ in place of $B_1$ and $y^1$, respectively. (b)′ When $B_2^B$ system (29) gives the set

$$K^0_{B_2} = \{y \in \mathbb{R}^2_+ : 2y_1 + y_2 = 1,\ 3y_1 \leq 1\}.$$

(c)′ Problem (30) becomes

$$\min\left(y_1^2 + (y_2 - 1/2)^2\right),\ \text{s.t.} y \in K^0_{B_2}.$$

By means of the gradient method, we easily find its unique minimum point and minimum:

$$y^3 = \left(\,1/5\;3/5\,\right),\quad \Phi(y^3) = 1/20 < \Phi(y^2).$$

Since $y^3 \in \partial K^0_{B_2}$, $y^3$ is a global besides local minimum point of (3).

*Example 4.* Consider the previous example, replacing $\Phi$ with the following one:

$$\Phi(x) = (x_1 - 2)^2 + (x_2 - 5/6)^2.$$

Perform the steps (a), (b), and (c), but with the present $\Phi$. Solving

$$\min\left((x_1 - 2)^2 + (x_2 - 5/6)^2\right),\ \text{s.t.} y \in K^0_{B_1}, \tag{36}$$

we find

$$y^2 = \left(\,2/5\;3/10\,\right) \in \text{ri}K^0_{B_1},\quad \Phi(y^2) = 16/45.$$

Despite of this, if we consider in (36) $K^0_{B_2}$ instead of $K^0_{B_1}$, we find

$$\tilde{y} = \left(\,1/5\;3/5\,\right),\quad \Phi(\tilde{y}) = 49/180 < 16/45,$$

which shows that, notwithstanding the fact that $y^2$ be a global minimum point of $\Phi$ on $K^0_{B_1}$, it is not a global minimum point on $K^0$.

*Example 5.* Let us now briefly discuss a classic scalarization method, namely that introduced in [3]; see also [2, 8]. It aims a finding the weak VMPs of (1), and thus the comparison with the method described in the previous sections is not perfectly fitting; however, we disregard this aspect since one might think of extending it; hence, we want to see what would happen if it were extended to the case of a cone $C$ and not int$C$. Consider again Example 3. To find a (weak) VMP, we must consider the problem:

$$\forall(u_1,u_2) \in f(K), \text{ find } F(u) = \min_{x\in K} \max\left(f_1(x)-u_1, f_2(x)-u_2\right); \tag{37}$$

the result is a (weak) VMP. Note that the minimization in (37) is a nonsmooth problem. For instance, when $f$ is the function of Example 3, and $u = \begin{pmatrix} 2 & 1 \end{pmatrix}$, such a minimization becomes

$$\min_{x\in K} \max\left(2x_1 - x_2 - 2, -x_1 + 2x_2 - 1\right), \tag{38}$$

which leads to $x^1 = \begin{pmatrix} 5/9 & 2/9 \end{pmatrix}$ of Example 3. The practically impossible problem is to express $\Phi$ as function of a vector running on the set of solutions of (37), even if only those of a subset of $K^0$ like $K^0_{B_1}$.

*Example 6.* Let us set $n = 1$, $l = m = 2$, $X\mathbb{R}$, $C = \mathbb{R}^2_+$,

$$f_1(x) = x,\ f_2(x) = x^2,\ g_1(x)0x + 1,\ g_2(x) = -x.$$

Obviously, $K = [-1,0]$, and all the elements of $K$ are VMPs of (1). Set $y = 0$. For $p_1, p_2 > 0$, consider the classic scalarized problem:

$$\min(p_1x + p_2x),\ x \in K.$$

Note that $x = 0$ is not a (global) minimum point of the classic scalarized problem whatever $p_1,\ p_2 > 0$ may be.

*Example 7.* Let us set $n = 1$, $l = m = 2$, $X = \mathbb{R}$, $C = \mathbb{R}^2$, and

$$f = \begin{pmatrix} 2x - x^2 & 1 - x^2 \end{pmatrix},\ g = \begin{pmatrix} x & 1 - x \end{pmatrix}.$$

we find $S(y) = \{y\}$, $\forall y \in [0,1]$. Hence, the unique solution of (3) is $y$ itself. By varying $y$, (3) gives, with its solutions, the interval $\sigma = [0,1]$, which is the set of VMP of (1), as it is trivial to check. Now, let us use the classic scalarization [7] outside the classic assumption of convexity, i.e., the scalar parametric problem which, here, becomes:

$$\min(c_1 f_1(x) + c_2 f_2(x)), \quad x \in \sigma$$

that is,

$$\min -(c_1 + c_2)x^2 + 2c_1 x + c_2, \quad x \in \sigma \tag{39}$$

where $\left(c_1\ c_2\right) \in \text{int}\, C^* = \text{int}\, \mathbb{R}^2$ is a pair of parameters. Every minimum point of (39) is a VMP of (1). In the present example it is easy to see that the only solutions of (39) are $x = 0$, or $x = 0$ and $x = 1$, or $x = 1$ according to, respectively, $c_2 < c_1$ or $c_2 = c_1$ or $c_2 > c_1$. Hence, the scalarized problem (39) does not detect all the solutions of (1) (the same happens obviously to (3), if $S(y)$ is deleted).

# 6 Further Developments

The development carried out in the previous sections is, deliberately, much simplified. In fact, the scope of this paper is to stress the importance, for the applications, of addressing some research efforts to the study of the bi-level vector problem. Some possible extensions are outlined below.

(i) A first effort will be devoted to let the previous method be able to find global minima. Some of the assumptions, made to simplifying the exposition, are too restrictive; it should be useful to remove them.

(ii) In order to stress the importance of the bi-level approach, let us bring an example. An extremely important application of vector optimization is to aerospace design. In this field, the first fundamental quantities are lift, drug, and cost (of course, in reality, besides them, we have many other quantities or their splitting). To formulate (1) with such three objectives ($l = 3$) should me meaningless; a competition between the cost and the lift or the drug should be a nonsense. A meaningful approach is to formulate (1) with 2 objectives, the lift and the drug ($l = 2$), and (3) with $\Phi$ to represent the cost.

(3i) In Sect. 4, the general method of Sect. 3 has been applied to a particular (even if particularly important) class of problems, and it has been shown how the bi-level problem can be reduced to a (finite) sequence of single problems. It should be interesting to obtain a similar result for other classes of problems; for instance, exploiting Sect. 2, the class of C-functions. Extensions to infinite dimensional spaces are also of great importance. We note that the method of the previous sections may reduce the bi-level problem to a finite sequence of scalar extremum problems. For instance, if both (1) and (3) are linear, then, performing (a)–(d) of Sect. 4 amounts to execute a finite steps of simplex method; if $\Phi$ is convex and (1) is linear (as assumed in Sect. 4) then performing (a)–(d) of Sect. 4 amounts to execute a finite steps of the gradient method. It should interesting to identify other classes of bi-level problems for which such a reduction holds.

(4i) As it is well known, not always an equilibrium can be expressed as the extremum of any functional; this led to formulate the theory of variational inequalities (VI). Furthermore, some equilibria are characterized by more than one operator and a blending of the involved operator may be not sufficient; this led to formulate the theory of vector variational inequalities (VVI). As shown for VOP in the previous sections, also in this case a bi-level approach is suitable for the applications. Consequently, it should be useful to extend the method of Sect. 4 to the case of VVI. In other words, (1) must be replaced by a VVI: let $F : \mathbb{R}^n \longrightarrow \mathbb{R}^{l \times n}$ be a matrix-valued function and consider the VVI, which consists in finding $y \in K$ such that:

$$F(y)(x-y) \not\leq_{C_0} 0, \quad \forall x \in K, \tag{40}$$

where $C_0$ and $K$ are as in Sect. 1. Denote by $K^0$ the set of solutions to (40). Now, consider the (scalar) VI, which consists in finding $y \in K^0$ such that:

$$\langle \Psi(y), x-y \rangle \geq 0, \quad \forall x \in K^0, \tag{41}$$

where $\Phi : \mathbb{R}^n \longrightarrow \mathbb{R}^n$. When both (40) and (41) admit the primitives (see the so-called symmetry principle), then they can be cast in the formats (1) and (3), respectively. The scalarization method for (40) described in Sect. 9 of [6] should allow one to define, for (40) and (41), a method like that of Sect. 4, avoiding to be obliged to find necessarily all the solutions of (40), namely $K^0$. The above VVI and VI are of Stampacchia type; same question about scalarization can be posed for the Minty type VVI and VI.

(5i) Another development may deal with the perturbation function of (3). There exists a wide literature as it concerns with scalar optimization and a few with (40), but they are independent each other. In as much as the important problem is (3), the study of the perturbation function of (1) should be *auxiliary* to (3) and not autonomous. Let the constraints of (1) be $g \geq \xi$, where $\xi$ plays the role of a parameter. Then $K$ and $K^0$ depend on $\xi$; denote them by $K(\xi)$ and $K^0(\xi)$, respectively. Hence, the minimum in (3) will depend on $\xi$, say $\Phi^{\downarrow}(\xi)$. The study of the properties of $K^0(\xi)$ is extremely important, while to find it is, in general, very difficult, but also useless, if (3) is the main scope.

(6i) Another subject, strictly connected with the previous one, is that of duality. The literature on duality for (1) is wide. Here too, in as much as the important problem is (3), the study of duality of (1) should be dependent on that of (3). Let us restrict to the Lagrangian duality, whose study is naturally located in the image space associated with the given problem. In fact the dual space is that of the functionals, whose zero level sets are considered to separate two suitable sets of the IS [4]. Hence, we have an IS associated with (1) and an IS associated with (3). In general, (3) has a positive duality gap. Sensitivity is a further topic, which is fundamental for the applications.

(7i) An extension of the present approach to set-valued, in particular interval-valued, extremum problems is conceivable. The infinite dimensional vector extremum problems, especially those of isoperimetric type, and the stochastic version of the previously mentioned problems, are surely interesting fields of research.

# References


[1] Antoni, C., Al-Shahrani, M.: On Bi-Level Vector Optimization (in preparation)

[2] Eichfelder, G.: Adaptive Scalarization Methods in Multiobjective Optimization. Springer, Heidelberg (2008)

[3] Gerstewitz (Tammer), C.: Nichtkonvexe Dualität in der Vektoroptimierung, Wissensch. Zeitschr. TH Leuna-Merseburg **25**, 357–364 (1983)

[4] Giannessi, F.: Constrained Optimization and Image Space Analysis, vol. I: Separation of Sets and Optimality Conditions. Springer, New York (2005)

[5] Giannessi, F., Pellegrini, L.: Image space analysis for vector optimization and variational inequalities. Scalarization. In: Pardalos, P.M., Migdalas, A., Burkard, R.E. (eds.) Combinatorial and Global Optimization, pp. 97–110. Kluwer, Dordrecht (2001)

[6] Giannessi, F., Mastroeni, G., Pellegrini, L.: On the theory of vector optimization and variational inequalities. Image space analysis and separation. In: Vector Variational Inequalities and Vector Equilibria. Mathematical Theories, Series Nonconvex Optimization and Its Applications, pp. 141–215. Kluwer, Dordrecht (2000)

[7] Luc, D.T.: Theory of Vector Optimization. Springer, Berlin (1989)

[8] Mastroeni, G.: Optimality conditions in image space analysis for vector optimization problems. In: Ansari, Q.H., Yao, J.-C. (eds.) Recent Developments in Vector Optimization, Series Vector Optimization, pp. 169–220. Springer (2012)

# Well-Posedness for Lexicographic Vector Equilibrium Problems

**L.Q. Anh, T.Q. Duy, A.Y. Kruger, and N.H. Thao**

**Abstract** We consider lexicographic vector equilibrium problems in metric spaces. Sufficient conditions for a family of such problems to be (uniquely) well posed at the reference point are established. As an application, we derive several results on well-posedness for a class of variational inequalities.

**Keywords** Lexicographic order • Equilibrium problem • Well-posedness

## 1 Introduction

Equilibrium problems first considered by Blum and Oettli [20] have been playing an important role in optimization theory with many striking applications particularly in transportation, mechanics, economics, etc. Equilibrium models incorporate many other important problems such as optimization problems, variational inequalities, complementarity problems, saddle point/minimax problems, and fixed points. Equilibrium problems with scalar and vector objective functions have been widely studied. The crucial issue of solvability (the existence of solutions) has attracted the most considerable attention of researchers; see, e.g., [17, 24, 27, 29, 42]. A relatively new but rapidly growing topic is the stability of solutions, including semicontinuity

L.Q. Anh
Department of Mathematics, Teacher College, Cantho University, Cantho, Vietnam
e-mail: quocanh@ctu.edu.vn

T.Q. Duy
Department of Mathematics, Cantho Technical and Economic College, Cantho, Vietnam
e-mail: tqduy@ctec.edu.vn

A.Y. Kruger (✉) • N.H. Thao
Centre for Informatics and Applied Optimization, School of Science, Information Technology and Engineering, University of Ballarat, Ballarat, Victoria, Australia
e-mail: a.kruger@ballarat.edu.au; hieuthaonguyen@students.ballarat.edu.au; nhthao@ctu.edu.vn

V.F. Demyanov et al. (eds.), *Constructive Nonsmooth Analysis and Related Topics*, Springer Optimization and Its Applications 87, DOI 10.1007/978-1-4614-8615-2_10, 

properties in the sense of Berge and Hausdorff (see, e.g., [2, 4, 5, 7, 16]) and the Hölder/Lipschitz continuity of solution mappings (see, e.g., [1,3,6,10,12,15,34,35]) and the (unique) well-posedness of approximate solutions in the sense of Hadamard and Tikhonov (see, e.g., [8,9,11,12,26,39,41]). The ultimate issue of computational methods for solving equilibrium problems has also been considered in the literature; see, e.g., [21, 30, 40].

With regard to vector equilibrium problems, most of existing results correspond to the case when the order is induced by a closed convex cone in a vector space. Thus, they cannot be applied to lexicographic cones, which are neither closed nor open. These cones have been extensively investigated in the framework of vector optimization; see, e.g., [18, 19, 22, 25, 28, 32, 33, 37]. However, for equilibrium problems, the main emphasis has been on the issue of solvability/existence. To the best of our knowledge, there have not been any works on well-posedness for lexicographic vector equilibrium problems.

In this article, we establish necessary and/or sufficient conditions for such problems to be (uniquely) well posed. As an application, we consider the special case of variational inequalities.

## 2 Preliminaries

We first recall the concept of lexicographic cone in finite-dimensional spaces and models of equilibrium problems with the order induced by such a cone.

The lexicographic cone of $\mathbb{R}^n$, denoted $C_l$, is the collection of zero and all vectors in $\mathbb{R}^n$ with the first nonzero coordinate being positive, i.e.,

$$C_l := \{0\} \cup \{x \in \mathbb{R}^n \mid \exists i \in \{1,2,\dots,n\} : x_i > 0 \text{ and } x_j = 0 \quad \forall j < i\}.$$

This cone is convex and pointed and induces the total order as follows:

$$x \geq_l y \Longleftrightarrow x - y \in C_l.$$

We also observe that it is neither closed nor open. Indeed, when comparing with the cone $C_1 := \{x \in \mathbb{R}^n \mid x_1 \geq 0\}$, we see that $\operatorname{int} C_1 \subsetneq C_l \subsetneq C_1$, while

$$\operatorname{int} C_l = \operatorname{int} C_1 \quad \text{and} \quad \operatorname{cl} C_l = C_1.$$

In what follows, $K : \Lambda \rightrightarrows X$ is a set-valued mapping between metric spaces and $f = (f_1, f_2, \dots, f_n) : K(\Lambda) \times K(\Lambda) \times \Lambda \to \mathbb{R}^n$ is a vector-valued function. For each $\lambda \in \Lambda$, the lexicographic vector equilibrium problem is

(LEP$_\lambda$) find $\bar{x} \in K(\lambda)$ such that

$$f(\bar{x}, y, \lambda) \geq_l 0 \quad \forall y \in K(\lambda).$$

*Remark 1.* This model covers parameterized bilevel optimization problems: minimize $g_2(\cdot,\lambda)$ over the solution set of the problem of minimizing $g_1(\cdot,\lambda)$ over $K(\lambda)$, where $g_1$ and $g_2$ are real-valued functions on $\operatorname{gph} K$. Recall that the graph of a (set-valued) mapping $Q : X \rightrightarrows Y$ is defined by $\operatorname{gph} Q := \{(x,y) \in X \times Y \mid y \in Q(x)\}$.

We denote $(\mathbf{LEP}) := \{(\mathrm{LEP}_\lambda) \mid \lambda \in \Lambda\}$ with the solution mapping $S : \Lambda \rightrightarrows X$ and assume that at the considered point $\bar{\lambda}$, the solution set $S(\bar{\lambda})$ is nonempty.

Following the lines of investigating $\varepsilon$-solutions to vector optimization problems initiated by Loridan [36], we consider, for each $\varepsilon \in [0;\infty)$, the following approximate problem:

$(\mathrm{LEP}_{\lambda,\varepsilon})$ find $\bar{x} \in K(\lambda)$ such that

$$f(\bar{x},y,\lambda) + \varepsilon e \geq_l 0 \quad \forall y \in K(\lambda),$$

where $e = (0,\ldots,0,1) \in \mathbb{R}^n$. The solution set of $(\mathrm{LEP}_{\lambda,\varepsilon})$ is denoted by $\tilde{S}(\lambda,\varepsilon)$.

We next define the notion of well-posedness for $(\mathbf{LEP})$ and recall continuity-like properties crucial for our analysis in this study.

**Definition 1.** A sequence $\{x_n\}$ with $x_n \in K(\lambda_n)$ is an *approximating sequence* of $(\mathrm{LEP}_{\bar{\lambda}})$ corresponding to a sequence $\{\lambda_n\} \subset \Lambda$ converging to $\bar{\lambda}$ if there is a sequence $\{\varepsilon_n\} \subset (0;\infty)$ converging to 0 such that $x_n \in \tilde{S}(\lambda_n,\varepsilon_n)$ for all $n$.

**Definition 2.** $(\mathbf{LEP})$ is *well posed* at $\bar{\lambda}$ if for any sequence $\{\lambda_n\}$ in $\Lambda$ converging to $\bar{\lambda}$, every corresponding approximating sequence of $(\mathrm{LEP}_{\bar{\lambda}})$ has a subsequence converging to some point of $S(\bar{\lambda})$.

**Definition 3.** $(\mathbf{LEP})$ is *uniquely well posed* at $\bar{\lambda}$ if:

(i) $(\mathrm{LEP}_{\bar{\lambda}})$ has the unique solution $\bar{x}$.
(ii) For any sequence $\{\lambda_n\}$ in $\Lambda$ converging to $\bar{\lambda}$, every corresponding approximating sequence of $(\mathrm{LEP}_{\bar{\lambda}})$ converges to $\bar{x}$.

*Remark 2.* Unfortunately there is no consistency in the literature in the usage of the term "well-posedness." Defining well-posedness here as a kind of "good behavior" of a family of parametric problems, we follow the lines of, e.g., [9, 11, 26]. Other authors, e.g., Bednarczuk [14], use this term as a characterization of a single reference problem. If $f$ in the above setting does not depend on $\lambda$, then the two versions of well-posedness coincide.

**Definition 4 ([13]).** Let $Q : X \rightrightarrows Y$ be a set-valued mapping between metric spaces:

(i) $Q$ is *upper semicontinuous* (usc) at $\bar{x}$ if for any open set $U \supseteq Q(\bar{x})$, there is a neighborhood $N$ of $\bar{x}$ such that $Q(N) \subseteq U$.
(ii) $Q$ is *lower semicontinuous* (lsc) at $\bar{x}$ if for any open subset $U$ of $Y$ with $Q(\bar{x}) \cap U \neq \emptyset$, there is a neighborhood $N$ of $\bar{x}$ such that $Q(x) \cap U \neq \emptyset$ for all $x \in N$.
(iii) $Q$ is *closed* at $\bar{x}$ if for any sequences $\{x_k\} \longrightarrow \bar{x}$ and $\{y_k\} \longrightarrow \bar{y}$ with $y_k \in Q(x_k)$, it holds $\bar{y} \in Q(\bar{x})$.

**Lemma 1 ([13,31]).**

*(i) If $Q$ is usc at $\bar{x}$ and $Q(\bar{x})$ is compact, then for any sequence $\{x_n\} \longrightarrow \bar{x}$, every sequence $\{y_n\}$ with $y_n \in Q(x_n)$ has a subsequence converging to some point in $Q(\bar{x})$. If, in addition, $Q(\bar{x}) = \{\bar{y}\}$ is a singleton, then such a sequence $\{y_n\}$ must converge to $\bar{y}$.*

*(ii) $Q$ is lsc at $\bar{x}$ if and only if for any sequence $\{x_n\} \to \bar{x}$ and any point $y \in Q(\bar{x})$, there is a sequence $\{y_n\}$ with $y_n \in Q(x_n)$ converging to $y$.*

**Definition 5.** Let $g$ be an extended real-valued function on a metric space $X$ and $\varepsilon$ be a real number.

(i) $g$ is *upper $\varepsilon$-level closed* at $\bar{x} \in X$ if for any sequence $\{x_n\} \longrightarrow \bar{x}$,

$$[g(x_n) \geq \varepsilon \quad \forall n] \Rightarrow [g(\bar{x}) \geq \varepsilon].$$

(ii) $g$ is *strongly upper $\varepsilon$-level closed* at $\bar{x} \in X$ if for any sequences $\{x_n\} \longrightarrow \bar{x}$ and $\{v_n\} \subset [0;\infty)$ converging to 0,

$$[g(x_n) + v_n \geq \varepsilon \quad \forall n] \Rightarrow [g(\bar{x}) \geq \varepsilon].$$

*Remark 3.* If $g$ is usc at $\bar{x}$, then it satisfies property (ii) in the last definition, which is obviously stronger than property (i) therein for any real number $\varepsilon$. Property (i) was introduced and investigated in [9, 11]. Property (ii) is a particular case of a more general property also introduced in [9, 11].

We say that a mapping/function satisfies a certain property on a subset of its domain if it is satisfied at every point of this subset.

# 3 Well-Posedness Properties of (LEP)

We are going to establish necessary and/or sufficient conditions for (**LEP**) to be (uniquely) well posed at the reference point $\bar{\lambda} \in \Lambda$. To simplify the presentation, in the sequel, the results will be formulated for the case $n = 2$.

Given $\lambda \in \Lambda$ and $x \in K(\Lambda)$, denote

$$S_1(\lambda) := \{x \in K(\lambda) \mid f_1(x,y,\lambda) \geq 0 \quad \forall y \in K(\lambda)\},$$

$$Z(\lambda,x) := \begin{cases} \{z \in K(\lambda) \mid f_1(x,z,\lambda) = 0\} \text{ if } (\lambda,x) \in \operatorname{gph} S_1, \\ X \text{ otherwise.} \end{cases} \tag{1}$$

$S_1 : \Lambda \rightrightarrows X$ is the solution mapping of the scalar equilibrium problem determined by the real-valued function $f_1$. The set-valued mapping $Z : \Lambda \times K(\Lambda) \rightrightarrows X$ is going to play an important role in our analysis.

Problem $(\text{LEP}_{\lambda,\varepsilon})$ can be equivalently stated as follows:

$(\text{LEP}_{\lambda,\varepsilon})$ find $\bar{x} \in K(\lambda)$ such that

$$\begin{cases} f_1(\bar{x},y,\lambda) \geq 0 & \forall y \in K(\lambda), \\ f_2(\bar{x},z,\lambda) + \varepsilon \geq 0 & \forall z \in Z(\lambda,\bar{x}). \end{cases}$$

This is equivalent to finding $\bar{x} \in S_1(\lambda)$ such that

$$f_2(\bar{x},z,\lambda) + \varepsilon \geq 0 \quad \forall z \in Z(\lambda,\bar{x}).$$

The next lemma is frequently used in the sequel.

**Lemma 2.** *Let $\{x_n\}$ converging to $\bar{x} \in S_1(\bar{\lambda})$ be an approximating sequence of $(\text{LEP}_{\bar{\lambda}})$ corresponding to some sequence $\{\lambda_n\} \longrightarrow \bar{\lambda}$ and assume that $Z$ is lsc at $(\bar{\lambda},\bar{x})$ and $f_2$ is strongly upper 0-level closed on $\{\bar{x}\} \times Z(\bar{\lambda},\bar{x}) \times \{\bar{\lambda}\}$. Then $\bar{x} \in S(\bar{\lambda})$.*

*Proof.* Suppose to the contrary that $\bar{x} \notin S(\bar{\lambda})$. Then, there exists $\bar{z} \in Z(\bar{\lambda},\bar{x})$ such that $f_2(\bar{x},\bar{z},\bar{\lambda}) < 0$. The lower semicontinuity of $Z$ at $(\bar{\lambda},\bar{x})$ ensures the existence, for each $n$, of $z_n \in Z(\lambda_n,x_n)$ such that $\{z_n\} \to \bar{z}$. Due to $x_n \in \tilde{S}(\lambda_n,\varepsilon_n)$, it holds $f_2(x_n,z_n,\lambda_n) + \varepsilon_n \geq 0$ for all $n$. Since $f_2$ is strongly upper 0-level closed at $(\bar{x},\bar{z},\bar{\lambda})$, we get $f_2(\bar{x},\bar{z},\bar{\lambda}) \geq 0$. This yields a contradiction, and, hence, we are done. □

**Theorem 1.** *Suppose that*

*(i) $X$ is compact,*
*(ii) $K$ is lsc and closed at $\bar{\lambda}$,*
*(iii) $Z$ is lsc on $\{\bar{\lambda}\} \times S_1(\bar{\lambda})$,*
*(iv) $f_1$ is upper 0-level closed on $K(\bar{\lambda}) \times K(\bar{\lambda}) \times \{\bar{\lambda}\}$,*
*(v) $f_2$ is strongly upper 0-level closed on $K(\bar{\lambda}) \times K(\bar{\lambda}) \times \{\bar{\lambda}\}$.*

*Then* (**LEP**) *is well posed at $\bar{\lambda}$. Moreover, it is uniquely well posed at this point if $S(\bar{\lambda})$ is a singleton.*

*Proof.* We first prove that $S_1$ is closed at $\bar{\lambda}$. Suppose to the contrary that there are sequences $\{\lambda_n\} \longrightarrow \bar{\lambda}$ and $\{x_n\} \longrightarrow \bar{x}$ with $x_n \in S_1(\lambda_n)$ and $\bar{x} \notin S_1(\bar{\lambda})$. Note that $\bar{x} \in K(\bar{\lambda})$ because $K$ is closed at $\bar{\lambda}$ and $x_n \in K(\lambda_n)$ for all $n$. Then, there exists $\bar{y} \in K(\bar{\lambda})$ satisfying $f_1(\bar{x},\bar{y},\bar{\lambda}) < 0$. The lower semicontinuity of $K$ at $\bar{\lambda}$ ensures that, for each $n$, there is $y_n \in K(\lambda_n)$ such that $\{y_n\} \longrightarrow \bar{y}$. Since $x_n \in S_1(\lambda_n)$, $f_1(x_n,y_n,\lambda_n) \geq 0$. This implies by assumption (iv) that $f_1(\bar{x},\bar{y},\bar{\lambda}) \geq 0$, which yields a contradiction, and hence, $S_1$ is closed at $\bar{\lambda}$.

We next show that $\tilde{S}$ is usc at $(\bar{\lambda},0)$. Indeed, if otherwise, then there is an open set $U \supset \tilde{S}(\bar{\lambda},0)$ along with sequences $\{\lambda_n\} \longrightarrow \bar{\lambda}$, $\{\varepsilon_n\} \downarrow 0$ such that, for each $n$, there is $x_n \in \tilde{S}(\lambda_n,\varepsilon_n) \setminus U$. By the compactness of $X$, we can assume that $(x_n)$ converges to some $\bar{x}$. Since $S_1$ is closed at $\bar{\lambda}$, $\bar{x} \in S_1(\bar{\lambda})$. Thanks to Lemma 2, it holds $\bar{x} \in S(\bar{\lambda}) = \tilde{S}(\bar{\lambda},0)$. This yields a contradiction because $x_n \notin U$ (open) for all $n$. Thus, $\tilde{S}$ is usc at $(\bar{\lambda},0)$.

We finally prove that $S(\bar{\lambda})$ is compact by checking its closedness. Take an arbitrary sequence $\{x_n\}$ in $S(\bar{\lambda})$ converging to $\bar{x}$. It is clear that $\bar{x} \in S_1(\bar{\lambda})$ due to the closedness of $S_1$ at $\bar{\lambda}$. Note that $\{x_n\}$ is, of course, an approximating sequence of (LEP$_{\bar{\lambda}}$). Then, Lemma 2 again implies that $\bar{x} \in S(\bar{\lambda})$ and, hence, $S(\bar{\lambda})$ is compact. Thanks to Lemma 1 (i), we are done. □

*Remark 4.* All assumptions in Theorem 1, except (iii), are formulated in terms of the problem data and normally are not difficult to check. Assumption (iii) involves set-valued mapping $Z$ defined by (1) and can be not so easy to check. Additional research is required to establish verifiable sufficient conditions for lower semicontinuity of $Z$.

The following examples show that none of the assumptions in Theorem 1 can be dropped.

*Example 1 (Compactness of $X$).* Let $X = \Lambda = \mathbb{R}$ (not compact), $K(\lambda) \equiv \mathbb{R}$ (continuous and closed), and $f(x,y,\lambda) = (0,\lambda)$. One can check that $S(\lambda) = S_1(\lambda) = Z(\lambda,x) = \mathbb{R}$ for all $\lambda, x \in \mathbb{R}$. Thus, assumptions (ii)–(v) hold true. However, (**LEP**) is not well posed at $\bar{\lambda} = 0$ because the approximating sequence $\{x_n = n\}$ of (LEP$_{\bar{\lambda}}$) corresponding to $\{\lambda_n = \frac{1}{n}\}$ has no convergent subsequence.

*Example 2 (Lower semicontinuity of $K$).* Let $X = \Lambda = [0;2]$ (compact) and $K$ and $f$ be defined by

$$K(\lambda) := \begin{cases} [0;1] & \text{if } \lambda \neq 0, \\ [0;2] & \text{if } \lambda = 0, \end{cases}$$

$$f(x,y,\lambda) := (x-y,\lambda).$$

One can check that $K$ is closed but not lsc at $\bar{\lambda} = 0$ and

$$S(\lambda) = S_1(\lambda) = \begin{cases} \{1\} & \text{if } \lambda \neq 0, \\ \{2\} & \text{if } \lambda = 0, \end{cases}$$

$$Z(\lambda,x) = \{x\} \quad \forall (\lambda,x) \in \operatorname{gph} S_1.$$

Thus, assumptions (iii)–(v) hold true. However, (**LEP**) is not well posed at $\bar{\lambda}$ because the approximating sequence $\{x_n = 1\}$ of (LEP$_{\bar{\lambda}}$) (corresponding to any sequence $\{\lambda_n\}$) converges to $1 \notin S(\bar{\lambda})$.

*Example 3 (Closedness of $K$).* Let $X = \Lambda = [0;1]$ (compact), $K(\lambda) \equiv (0;1]$ (continuous), and $f(x,y,\lambda) = (0,\lambda)$. It is clear that

$$S(\lambda) = S_1(\lambda) = K(\lambda) \quad \forall \lambda \in \Lambda,$$

$$Z(\lambda,x) = (0;1] \quad \forall (\lambda,x) \in \operatorname{gph} S_1.$$

One can also check that (**LEP**) is not well posed at $\bar{\lambda} = 0$, while all the assumptions of Theorem 1 except the closedness of $K$ at $\bar{\lambda}$ are satisfied.

*Example 4 (Lower semicontinuity of Z).* Let $X = \Lambda = [0;1]$ (compact), $K(\lambda) \equiv [0;1]$ (continuous and closed), $\bar{\lambda} = 0$, and $f(x,y,\lambda) = (\lambda x(x-y), y-x)$. One can check that

$$S_1(\lambda) = \begin{cases} [0;1] & \text{if } \lambda = 0, \\ \{0,1\} & \text{if } \lambda \neq 0, \end{cases}$$

and, for each $(\lambda, x) \in \operatorname{gph} S_1$,

$$Z(\lambda,x) = \begin{cases} [0;1] & \text{if } \lambda = 0 \text{ or } x = 0, \\ \{x\} & \text{if } \lambda \neq 0 \text{ and } x \neq 0. \end{cases}$$

$Z$ is not lsc at $(0,1)$ because by taking $\{(x_n = 1, \lambda_n - \frac{1}{n})\} \longrightarrow (1,0)$, we have $Z(\lambda_n, x_n) = \{1\}$ for all $n$, while $Z(0,1) = [0;1]$. Assumptions (iv) and (v) are obviously satisfied. Finally, we observe that (**LEP**) is not well posed at $\bar{\lambda}$ by calculating the solution mapping $S$ explicitly as follows:

$$S(\lambda) = \begin{cases} \{0\} & \text{if } \lambda = 0, \\ \{0,1\} & \text{if } \lambda \neq 0. \end{cases}$$

*Example 5 (Upper 0-level closedness of $f_1$).* Let $X = \Lambda = [0;1]$ (compact), $K(\lambda) \equiv [0;1]$ (continuous and closed), $\bar{\lambda} = 0$, and

$$f(x,y,\lambda) = \begin{cases} (x-y, \lambda) & \text{if } \lambda = 0, \\ (y-x, \lambda) & \text{if } \lambda \neq 0. \end{cases}$$

One can check that

$$S(\lambda) = S_1(\lambda) = \begin{cases} \{1\} & \text{if } \lambda = 0, \\ \{0\} & \text{if } \lambda \neq 0, \end{cases}$$

$$Z(\lambda,x) = \{x\} \quad \forall (\lambda,x) \in \operatorname{gph} S_1.$$

Hence, all the assumptions except (iv) hold true. However, (**LEP**) is not well posed at $\bar{\lambda}$. Indeed, take sequences $\{\lambda_n = \frac{1}{n}\}$ and $\{x_n = 0\}$ ($x_n \in S(\lambda_n)$). Then, $\{x_n\}$ is an approximating sequence of ($\text{LEP}_{\bar{\lambda}}$) corresponding to $\{\lambda_n\}$, while $\{x_n\} \longrightarrow 0 \notin S(0)$.

Finally, we show that assumption (iv) is not satisfied. Indeed, taking $\{x_n\}$ and $\{\lambda_n\}$ as above and $\{y_n = 1\}$, we have $\{(x_n, y_n, \lambda_n)\} \longrightarrow (0,1,0)$ and $f_1(x_n, y_n, \lambda_n) = 1 > 0$ for all $n$, while $f_1(0,1,0) = -1 < 0$.

*Example 6 (Strong upper 0-level closedness of $f_2$).* Let $X, \Lambda, K, \bar{\lambda}$ be as in Example 5 and

$$f(x,y,\lambda) = \begin{cases} (0, x-y) & \text{if } \lambda = 0, \\ (0, x(x-y)) & \text{if } \lambda \neq 0. \end{cases}$$

One can check that

$$S_1(\lambda) = Z(\lambda, x) = [0;1] \quad \forall x, \lambda \in [0;1],$$

$$S(\lambda) = \begin{cases} \{1\} & \text{if } \lambda = 0, \\ \{0,1\} & \text{if } \lambda \neq 0. \end{cases}$$

Thus, all the assumptions of Theorem 1 except (v) are satisfied. However, it follows from the explicit form of $S$ that (**LEP**) is not well posed at $\bar{\lambda}$. Finally, we show that assumption (v) is not satisfied. Indeed, taking sequences $\{x_n = 0\}$, $\{y_n = 1\}$, $\{\lambda_n = \frac{1}{n}\}$, and $\{\varepsilon_n = \frac{1}{n}\}$, we have $\{(x_n, y_n, \lambda_n, \varepsilon_n)\} \longrightarrow (0,1,0,0)$ and $f_2(x_n, y_n, \lambda_n) + \varepsilon_n > 0$ for all $n$, while $f_2(0,1,0) = -1 < 0$.

In what follows,

$$\mathcal{P}(\bar{\lambda}, \delta, \varepsilon) := \bigcup_{\lambda \in B_\delta(\bar{\lambda})} \tilde{S}(\lambda, \varepsilon),$$

where $B_\delta(\bar{\lambda})$ denotes the closed ball centered at $\bar{\lambda}$ with radius $\delta$. We also use the concept of diameter of a set $A$ in a metric space:

$$\operatorname{diam} A := \sup_{a,b \in A} d(a,b).$$

**Theorem 2.**

(i) *If* (**LEP**) *is uniquely well posed at* $\bar{\lambda}$, *then* $\operatorname{diam} \mathcal{P}(\bar{\lambda}, \delta, \varepsilon) \downarrow 0$ *as* $\delta \downarrow 0$ *and* $\varepsilon \downarrow 0$.

(ii) *Suppose that $X$ is complete and assumptions (ii)–(v) in Theorem 1 hold true. If* $\operatorname{diam} \mathcal{P}(\bar{\lambda}, \delta, \varepsilon) \downarrow 0$ *as* $\delta \downarrow 0$ *and* $\varepsilon \downarrow 0$, *then* (**LEP**) *is uniquely well posed at* $\bar{\lambda}$.

*Proof.*

(i) Let (**LEP**) be uniquely well posed at $\bar{\lambda}$ and $\{\delta_n\} \downarrow 0$, $\{\varepsilon_n\} \downarrow 0$. If $\operatorname{diam} \mathcal{P}(\bar{\lambda}, \delta_n, \varepsilon_n)$ does not converge to 0 as $n \longrightarrow \infty$, then there exists a number $r > 0$ such that for any $n_0 \in \mathbb{N}, \exists n \geq n_0$ with $\operatorname{diam} \mathcal{P}(\bar{\lambda}, \delta_n, \varepsilon_n) > r$. By taking a subsequence if necessary, we can suppose that $\operatorname{diam} \mathcal{P}(\bar{\lambda}, \delta_n, \varepsilon_n) > r$ for all $n$. This implies that, for each $n$, there exist $x_n^1, x_n^2 \in \mathcal{P}(\bar{\lambda}, \delta_n, \varepsilon_n)$ such that

$$d(x_n^1, x_n^2) > \frac{r}{2}. \tag{2}$$

Thus, there are $\lambda_n^1, \lambda_n^2 \in B(\bar{\lambda}, \delta_n)$ such that $x_n^i \in \tilde{S}(\lambda_n^i, \varepsilon_n)$, $i$=1,2. Observe that both $\{\lambda_n^1\}$ and $\{\lambda_n^2\}$ converge to $\bar{\lambda}$ as $n \longrightarrow \infty$, and so $\{x_n^1\}$ and $\{x_n^2\}$ are corresponding approximating sequences of (LEP$_{\bar{\lambda}}$), respectively. Due to the unique well-posedness of (**LEP**) at $\bar{\lambda}$, both $\{x_n^1\}$ and $\{x_n^2\}$ must converge to the only solution $\bar{x}$ to (LEP$_{\bar{\lambda}}$). Hence, $\lim_{n \longrightarrow \infty} d(x_n^1, x_n^2) = 0$. This contradicts (2) and, thus, we are done.

(ii) Suppose that $\{x_n\}$ is an approximating sequence of (LEP$_{\bar{\lambda}}$) corresponding to some sequence $\{\lambda_n\} \longrightarrow \bar{\lambda}$, i.e., there is a sequence $\{\varepsilon_n\} \downarrow 0$ such that $x_n \in \tilde{S}(\lambda_n, \varepsilon_n)$ for all $n$. By setting $\delta_n := d(\lambda_n, \bar{\lambda})$, it holds that $\{\delta_n\} \longrightarrow 0$ as $n \longrightarrow \infty$ and $x_n \in \mathcal{P}(\bar{\lambda}, \delta_n, \varepsilon_n)$ for all $n$. By choosing subsequences if necessary, we can assume that both sequences $\{\delta_n\}$ and $\{\varepsilon_n\}$ are nonincreasing. Thus, $\mathcal{P}(\bar{\lambda}, \delta_n, \varepsilon_n) \supseteq \mathcal{P}(\bar{\lambda}, \delta_m, \varepsilon_m)$ whenever $n \leq m$. From this observation and $\operatorname{diam} \mathcal{P}(\bar{\lambda}, \delta_n, \varepsilon_n) \downarrow 0$ as $n \longrightarrow \infty$, one can directly check that $\{x_n\}$ is a Cauchy sequence and, hence, converges to some point $\bar{x}$ due to the completeness of $X$. Note that assumptions on $K$ and $f_1$ imply the closedness of $S_1$ at $\bar{\lambda}$; see the first reasoning in the proof of Theorem 1. In particular, we have $\bar{x} \in S_1(\bar{\lambda})$, and Lemma 2 then yields $\bar{x} \in S(\bar{\lambda})$.

Finally, we show that $\bar{x}$ is the only solution to (LEP$_{\bar{\lambda}}$). Suppose to the contrary that $S(\bar{\lambda})$ contains also another point $\bar{x}'$ ($\bar{x}' \neq \bar{x}$). It is clear that they both belong to $\mathcal{P}(\bar{\lambda}, \delta, \varepsilon)$ for any $\delta, \varepsilon > 0$. Then, it follows that

$$0 < d(\bar{x}, \bar{x}') \leq \operatorname{diam} \mathcal{P}(\bar{\lambda}, \delta, \varepsilon) \downarrow 0 \text{ as } \delta \downarrow 0 \text{ and } \varepsilon \downarrow 0.$$

This is impossible and, therefore, we are done. □

To weaken the assumption of unique well-posedness in Theorem 2, we are going to use the *Kuratowski measure of noncompactness* of a nonempty set $M$ in a metric space $X$:

$$\mu(M) := \inf \left\{ \varepsilon > 0 \mid M \subseteq \bigcup_{k=1}^{n} M_k,\ M_k \subset X,\ \operatorname{diam} M_k \leq \varepsilon\ \forall k,\ n \in \mathbb{N} \right\}.$$

**Lemma 3 ([38]).** *The following assertions hold true:*

*(i)* $\mu(M) = 0$ *if* $M$ *is compact.*
*(ii)* $\mu(M) \leq \mu(N)$ *whenever* $M \subseteq N$.
*(iii) If* $\mu(M) = 0$, *then* $M$ *is totally bounded, i.e., there are a point* $x_M \in X$ *along with a constant* $\kappa_M > 0$ *such that*

$$d(x, x_M) \leq \kappa_M \quad \forall x \in M.$$

*(iv) If* $\{A_n\}$ *is a sequence of closed subsets in a complete metric space* $X$ *satisfying* $A_{n+1} \subseteq A_n$ *for every* $n \in \mathbb{N}$ *and* $\lim_{n \longrightarrow \infty} \mu(A_n) = 0$, *then* $K := \bigcap_{n \in \mathbb{N}} A_n$ *is a*

*nonempty compact set and* $\lim_{n \longrightarrow \infty} H(A_n, K) = 0$*, where* $H$ *is the Hausdorff distance.*

Recall that the *Hausdorff distance* between two sets $A$ and $B$ in a metric space is defined by

$$H(A,B) := \max\{e(A,B), e(B,A)\},$$

where $e(A,B) := \sup_{a \in A} d(a,B)$ with $d(a,B) := \inf_{b \in B} d(a,b)$.

**Theorem 3.**

*(i) If* (**LEP**) *is well posed at* $\bar{\lambda}$*, then* $\mu(\mathcal{P}(\bar{\lambda}, \delta, \varepsilon)) \downarrow 0$ *as* $\delta \downarrow 0$ *and* $\varepsilon \downarrow 0$*.*

*(ii) Suppose that* $X$ *is complete,* $\Lambda$ *is compact or a finite-dimensional normed space and*

- *(a)* $K$ *is lsc and closed on some neighborhood* $V$ *of* $\bar{\lambda}$*,*
- *(b)* $Z$ *is lsc on* $[V \times X] \cap \operatorname{gph} S_1$*,*
- *(c)* $f_1$ *is upper 0-level closed on* $K(V) \times K(V) \times V$*,*
- *(d)* $f_2$ *is upper a-level closed on* $K(V) \times K(V) \times V$ *for every negative a close to zero.*

*If* $\mu(\mathcal{P}(\bar{\lambda}, \delta, \varepsilon)) \downarrow 0$ *as* $\delta \downarrow 0$ *and* $\varepsilon \downarrow 0$*, then* (**LEP**) *is well posed at* $\bar{\lambda}$*.*

*Proof.*

(i) Suppose that (**LEP**) is well posed at $\bar{\lambda}$. Let $\{x_n\}$ be an arbitrary sequence in $S(\bar{\lambda})$ [and, of course, an approximating sequence of (LEP$_{\bar{\lambda}}$)]. Then, it has a subsequence converging to some point in $S(\bar{\lambda})$. Thus, $S(\bar{\lambda})$ is compact, and so $\mu(S(\bar{\lambda})) = 0$ due to Lemma 3(i). Let any $\varepsilon > 0$ and $S(\bar{\lambda}) \subseteq \bigcup_{k=1}^{n} M_k$ with $\operatorname{diam} M_k \le \varepsilon$ for all $k = \overline{1,n}$. We set

$$N_k = \{y \in X \mid d(y, M_k) \le H(\mathcal{P}(\bar{\lambda}, \delta, \varepsilon), S(\bar{\lambda}))\}$$

and show that $\mathcal{P}(\bar{\lambda}, \delta, \varepsilon) \subseteq \bigcup_{k=1}^{n} N_k$. Pick arbitrary $x \in \mathcal{P}(\bar{\lambda}, \delta, \varepsilon)$. Then $d(x, S(\bar{\lambda})) \le H(\mathcal{P}(\bar{\lambda}, \delta, \varepsilon), S(\bar{\lambda}))$. Due to $S(\bar{\lambda}) \subseteq \bigcup_{k=1}^{n} M_k$, one has

$$d(x, \bigcup_{k=1}^{n} M_k) \le H(\mathcal{P}(\bar{\lambda}, \delta, \varepsilon), S(\bar{\lambda})).$$

Then, there exists $\bar{k} \in \{1, 2, \ldots, n\}$ such that $d(x, M_{\bar{k}}) \le H(\mathcal{P}(\bar{\lambda}, \delta, \varepsilon), S(\bar{\lambda}))$, i.e., $x \in N_{\bar{k}}$. Thus, $\mathcal{P}(\bar{\lambda}, \delta, \varepsilon) \subseteq \bigcup_{k=1}^{n} N_k$.

Because $\mu(S(\bar{\lambda})) = 0$ and

$$\operatorname{diam} N_k = \operatorname{diam} M_k + 2H(\mathcal{P}(\bar{\lambda}, \delta, \varepsilon), S(\bar{\lambda})) \le \varepsilon + 2H(\mathcal{P}(\bar{\lambda}, \delta, \varepsilon), S(\bar{\lambda})),$$

it holds

$$\mu(\mathcal{P}(\bar{\lambda},\delta,\varepsilon)) \leq 2H(\mathcal{P}(\bar{\lambda},\delta,\varepsilon),S(\bar{\lambda})).$$

Note that $H(\mathcal{P}(\bar{\lambda},\delta,\varepsilon),S(\bar{\lambda})) = e(\mathcal{P}(\bar{\lambda},\delta,\varepsilon),S(\bar{\lambda}))$ since $S(\bar{\lambda}) \subseteq \mathcal{P}(\bar{\lambda},\delta,\varepsilon)$ for all $\delta,\varepsilon > 0$.

Now, we claim that $H(\mathcal{P}(\bar{\lambda},\delta,\varepsilon),S(\bar{\lambda})) \downarrow 0$ as $\delta \downarrow 0$ and $\varepsilon \downarrow 0$. Indeed, if otherwise, we can assume that there exist $r > 0$ and sequences $\{\delta_n\} \downarrow 0$, $\{\varepsilon_n\} \downarrow 0$, and $\{x_n\}$ with $x_n \in \mathcal{P}(\bar{\lambda},\delta_n,\varepsilon_n)$ such that

$$d(x_n,S(\bar{\lambda})) \geq r \quad \forall n. \tag{3}$$

Since $\{x_n\}$ is an approximating sequence of (LEP$_{\bar{\lambda}}$) corresponding to some $\{\lambda_n\}$ with $\lambda_n \in B_{\delta_n}(\bar{\lambda})$, it has a subsequence $\{x_{n_k}\}$ converging to some $x \in S(\bar{\lambda})$. Then, $d(x_{n_k},x) < r$ when $n_k$ is sufficiently large. This contradicts (3) and, hence,

$$\mu(\mathcal{P}(\bar{\lambda},\delta,\varepsilon)) \longrightarrow 0 \text{ as } \delta \downarrow 0 \text{ and } \varepsilon \downarrow 0.$$

(ii) Suppose that $\mu(\mathcal{P}(\bar{\lambda},\delta,\varepsilon)) \downarrow 0$ as $\delta \downarrow 0$ and $\varepsilon \downarrow 0$. We firstly show that $\mathcal{P}(\bar{\lambda},\delta,\varepsilon)$ is closed for any $\delta,\varepsilon > 0$. Let $\{x_n\} \in \mathcal{P}(\bar{\lambda},\delta,\varepsilon)$, $\{x_n\} \longrightarrow \bar{x}$. Then, for each $n \in \mathbb{N}$, there exists $\lambda_n \in B_\delta(\bar{\lambda})$ such that $x_n \in \tilde{S}(\lambda_n,\varepsilon)$. Assumption on $\Lambda$ implies that $B_\delta(\bar{\lambda})$ is compact. So, we can assume $\{\lambda_n\}$ converges to some $\lambda \in B_\delta(\bar{\lambda}) \cap V$. Thus, $\bar{x} \in K(\lambda)$ due to the closedness of $K$ at $\lambda$. Assumptions on $K$ and $f_1$ imply that $\bar{x} \in S_1(\lambda)$; see the first reasoning in the proof of Theorem 1. Now, we check that $\bar{x}$ also belongs to $\tilde{S}(\lambda,\varepsilon)$. Indeed, suppose to the contrary that there exists $\bar{z} \in Z(\lambda,\bar{x})$ such that $f_2(\bar{x},\bar{z},\lambda) + \varepsilon < 0$. Then, the lower semicontinuity of $Z$ at $(\lambda,\bar{x})$ ensures that, for each $n$, there is $z_n \in Z(\lambda_n,x_n)$ such that $\{z_n\} \longrightarrow \bar{z}$. Due to the upper $(-\varepsilon)$-level closedness of $f_2$ at $(\bar{x},\bar{z},\lambda)$, $f_2(x_n,z_n,\lambda_n) < -\varepsilon$ when $n$ is sufficiently large. This is a contradiction since $x_n \in \tilde{S}(\lambda_n,\varepsilon)$ for all $n$. Hence, $\bar{x} \in \tilde{S}(\lambda,\varepsilon)$, and so $\bar{x} \in \mathcal{P}(\bar{\lambda},\delta,\varepsilon)$. Therefore, $\mathcal{P}(\bar{\lambda},\delta,\varepsilon)$ is closed for any $\delta,\varepsilon > 0$.

Next, we prove $S(\bar{\lambda}) = \bigcap_{\delta,\varepsilon>0} \mathcal{P}(\bar{\lambda},\delta,\varepsilon)$. We first check that $\bigcap_{\delta>0} \mathcal{P}(\bar{\lambda},\delta,\varepsilon) = \tilde{S}(\bar{\lambda},\varepsilon)$ for any $\varepsilon > 0$. It is clear that $\tilde{S}(\bar{\lambda},\varepsilon) \subseteq \bigcap_{\delta>0} \mathcal{P}(\bar{\lambda},\delta,\varepsilon)$. Now, take any $x \in \bigcap_{\delta>0} \mathcal{P}(\bar{\lambda},\delta,\varepsilon)$. Then, for each sequence $\{\delta_n\} \downarrow 0$, there exists a sequence $\{\lambda_n\}$ with $\lambda_n \in B_{\delta_n}(\bar{\lambda})$ such that $x \in \tilde{S}(\lambda_n,\varepsilon)$ for all $n$. Assumptions on $K$ and $f_1$ again imply $x \in S_1(\bar{\lambda})$. For any $z \in Z(\bar{\lambda},x)$, there exists $z_n \in Z(\lambda_n,x)$, $\{z_n\} \longrightarrow z$, thanks to the lower semicontinuity of $Z$ at $(\bar{\lambda},x)$. As $x \in \tilde{S}(\lambda_n,\varepsilon)$, it holds $f_2(x,z_n,\lambda_n) + \varepsilon \geq 0$ for every $n$. From the upper $(-\varepsilon)$-level closedness of $f_2$ at $(x,z,\bar{\lambda})$, we have $f_2(x,z,\bar{\lambda}) + \varepsilon \geq 0$, i.e., $x \in \tilde{S}(\bar{\lambda},\varepsilon)$. It follows that $\bigcap_{\delta>0} \mathcal{P}(\bar{\lambda},\delta,\varepsilon) \subseteq \tilde{S}(\bar{\lambda},\varepsilon)$ and, thus, $\bigcap_{\delta>0} \mathcal{P}(\bar{\lambda},\delta,\varepsilon) = \tilde{S}(\bar{\lambda},\varepsilon)$. Now, we need to check that $S(\bar{\lambda}) = \bigcap_{\varepsilon>0} \tilde{S}(\bar{\lambda},\varepsilon)$. It is clear that $S(\bar{\lambda}) \subseteq \bigcap_{\varepsilon>0} \tilde{S}(\bar{\lambda},\varepsilon)$. On the other hand, for any $x \in \bigcap_{\varepsilon>0} \tilde{S}(\bar{\lambda},\varepsilon)$, we have $f_2(x,z,\bar{\lambda}) + \varepsilon \geq 0$ for all $z \in Z(\bar{\lambda},x)$ and $\varepsilon > 0$. By letting $\varepsilon$ tend to 0, this implies $f_2(x,z,\bar{\lambda}) \geq 0$ for all $z \in Z(\bar{\lambda},x)$, i.e., $x \in S(\bar{\lambda})$, and, hence, $S(\bar{\lambda}) = \bigcap_{\delta,\varepsilon>0} \mathcal{P}(\bar{\lambda},\delta,\varepsilon)$.

Finally, since $\mu(\mathcal{P}(\bar{\lambda},\delta,\varepsilon)) \downarrow 0$ as $\delta \downarrow 0$ and $\varepsilon \downarrow 0$, Lemma 3(iv) implies the compactness of $S(\bar{\lambda})$ and $H(\mathcal{P}(\bar{\lambda},\delta,\varepsilon),S(\bar{\lambda})) \longrightarrow 0$ as $\delta \downarrow 0$ and $\varepsilon \downarrow 0$. Let $\{x_n\}$ be an approximating sequence of (LEP$_{\bar{\lambda}}$) corresponding to some $\{\lambda_n\} \longrightarrow \bar{\lambda}$. Then, there exists $\{\varepsilon_n\}$ converging to 0 such that $x_n \in \tilde{S}(\lambda_n,\varepsilon_n)$ for all $n$. This means that $x_n \in \mathcal{P}(\bar{\lambda},\delta_n,\varepsilon_n)$, where $\delta_n = d(\bar{\lambda},\lambda_n)$. Note that

$$d(x_n,S(\bar{\lambda})) \leq H(\mathcal{P}(\bar{\lambda},\delta_n,\varepsilon_n),S(\bar{\lambda})) \downarrow 0 \text{ as } n \longrightarrow \infty.$$

Thus, there is $\{\bar{x}_n\} \subset S(\bar{\lambda})$ such that $d(x_n,\bar{x}_n) \downarrow 0$ as $n \longrightarrow \infty$. Since $S(\bar{\lambda})$ is compact, $\{\bar{x}_n\}$ has a subsequence $\{\bar{x}_{n_k}\}$ converging to some $\bar{x} \in S(\bar{\lambda})$ and, hence, $\{x_n\}$ has the corresponding subsequence $\{x_{n_k}\}$ converging to $\bar{x}$. Therefore, (**LEP**) is well posed at $\bar{\lambda}$, and we are done. □

*Remark 5.* Theorem 3 remains valid if the Kuratowski measure is replaced by either Hausdorff or Istrătescu measure. We refer the reader to [23] for further information about these noncompact measures including their equivalence.

Note that when $K(\Lambda)$ is contained in a compact set (in particular, $X$ is compact), the assumption on the measure $\mu$ in Theorem 3 (ii) holds true trivially. Hence, Examples 2–5 again show that assumptions (a)–(c) imposed in Theorem 3(ii) are essential. The following example shows that the upper negative-level closedness of $f_2$ therein is also essential.

*Example 7.* Let $X = \mathbb{R}$ (complete), $\Lambda = [0;1]$ (compact), $K(\lambda) \equiv [-1;1]$ (continuous and closed), $\bar{\lambda} = 0$ and

$$f(x,y,\lambda) := \begin{cases} ((x-y)^2, x-1) & \text{if } \lambda = 0, \\ ((x-y)^2, (x+y)^2) & \text{if } \lambda \neq 0. \end{cases}$$

One can check that

$$S_1(\lambda) = [-1;1] \quad \forall \lambda,$$

$$Z(\lambda,x) = \{x\} \quad \forall (\lambda,x) \in \operatorname{gph} S_1,$$

$$S(\lambda) = \begin{cases} \{1\} & \text{if } \lambda = 0, \\ [-1;1] & \text{if } \lambda \neq 0. \end{cases}$$

We observe that $f_2$ is not 0-level closed at $(-1,1,0)$. Indeed, taking $\{x_n = -1\}$, $\{y_n = 1\}$, and $\{\lambda_n = \frac{1}{n}\}$, we have $\{(x_n,y_n,\lambda_n)\} \longrightarrow (-1,1,0)$ and $f_2(x_n,y_n,\lambda_n) = 0$, while $f_2(-1,1,0) = -2 < 0$. Moreover, all the other assumptions are satisfied, while (**LEP**) is not well posed at $\bar{\lambda}$.

## 4 Applications to Variational Inequalities

In this section, let $\Lambda$ and $K$ be as in the preceding sections, $X$ be a normed space with its dual denoted by $X^*$ and $h_i : X \times \Lambda \longrightarrow X^*$, $i = 1,2$. For each $\lambda \in \Lambda$, we consider the following lexicographic variational inequality:

(LVI$_\lambda$) find $\bar{x} \in K(\lambda)$ such that

$$(\langle h_1(\bar{x},\lambda), y-\bar{x}\rangle, \langle h_2(\bar{x},\lambda), y-\bar{x}\rangle) \geq_l 0 \quad \forall y \in K(\lambda).$$

This is equivalent to finding $\bar{x} \in K(\lambda)$ such that

$$\begin{cases} \langle h_1(\bar{x},\lambda), y-\bar{x}\rangle \geq 0 & \forall y \in K(\lambda), \\ \langle h_2(\bar{x},\lambda), z-\bar{x}\rangle \geq 0 & \forall z \in Z(\lambda,\bar{x}). \end{cases}$$

Here, the set-valued mapping $Z : \Lambda \times K(\Lambda) \rightrightarrows X$ is defined by

$$Z(\lambda,x) := \begin{cases} \{z \in K(\lambda) \mid \langle h_1(x,\lambda), z-x\rangle = 0\} \text{ if } (\lambda,x) \in \operatorname{gph} S_1, \\ X \text{ otherwise,} \end{cases}$$

where $S_1 : \Lambda \rightrightarrows X$ denotes the solution mapping of the scalar variational inequality determined by $h_1$:

$$S_1(\lambda) := \{x \in K(\lambda) \mid \langle h_1(x,\lambda), y-x\rangle \geq 0 \quad \forall y \in K(\lambda)\}.$$

We denote (**LVI**) $:= \{(\text{LVI}_\lambda) \mid \lambda \in \Lambda\}$ with the solution mapping $S : \Lambda \rightrightarrows X$ and assume that at the considered point $\bar{\lambda}$, the solution set $S(\bar{\lambda})$ is nonempty.

Now, for a number $\varepsilon > 0$, we consider the following approximate problem:

(LVI$_{\lambda,\varepsilon}$) find $\bar{x} \in K(\lambda)$ such that

$$\begin{cases} \langle h_1(\bar{x},\lambda), y-\bar{x}\rangle \geq 0 & \forall y \in K(\lambda), \\ \langle h_2(\bar{x},\lambda), z-\bar{x}\rangle + \varepsilon \geq 0 & \forall z \in Z(\lambda,\bar{x}). \end{cases}$$

We also use denotation $\tilde{S}$ for the approximate solution mapping, i.e.,

$$\tilde{S}(\lambda,\varepsilon) := \{x \in S_1(\lambda) \mid \langle h_2(x,\lambda), z-x\rangle + \varepsilon \geq 0 \quad \forall z \in Z(\lambda,x)\}.$$

In the following, we use the concepts defined in Definitions 1–3 with the term "LEP" replaced by "LVI." The next theorems follow from the corresponding results established in Sect. 3 by setting $f_i(x,y,\lambda) := \langle h_i(x,\lambda), y-x\rangle$, $i = 1,2$, therein.

**Theorem 4.** *Suppose that assumptions (i)–(iii) in Theorem 1 are satisfied. Assume, additionally, that*

*(i)* $\{(x,y,\lambda) \in K(\Lambda) \times K(\Lambda) \times \Lambda \mid \langle h_1(x,\lambda), y-x\rangle \geq 0\}$ *is a closed subset of* $K(\Lambda) \times K(\Lambda) \times \Lambda$,

*(ii) the function* $(x,y,\lambda) \mapsto \langle h_2(x,\lambda), y-x\rangle$ *is strongly upper 0-level closed on* $K(\bar{\lambda}) \times K(\bar{\lambda}) \times \{\bar{\lambda}\}$.

*Then* (**LVI**) *is well posed at* $\bar{\lambda}$. *Moreover, it is uniquely well posed at this point if* $S(\bar{\lambda})$ *is a singleton.*

*Remark 6.* Assumptions (i) and (ii) in Theorem 4 are straightforwardly fulfilled when $h_1$ and $h_2$, respectively, are continuous.

**Theorem 5.**

*(i) If* (**LVI**) *is uniquely well posed at* $\bar{\lambda}$, *then* $\operatorname{diam} \mathcal{P}(\bar{\lambda}, \delta, \varepsilon) \downarrow 0$ *as* $\delta \downarrow 0$ *and* $\varepsilon \downarrow 0$.

*(ii) Suppose that X is complete and assumptions (ii)–(iii) in Theorem 1 and (i)–(ii) in Theorem 4 hold true. If* $\operatorname{diam} \mathcal{P}(\bar{\lambda}, \delta, \varepsilon) \downarrow 0$ *as* $\delta \downarrow 0$ *and* $\varepsilon \downarrow 0$, *then* (**LVI**) *is uniquely well posed at* $\bar{\lambda}$.

**Theorem 6.**

*(i) If* (**LVI**) *is well posed at* $\bar{\lambda}$, *then* $\mu(\mathcal{P}(\bar{\lambda}, \delta, \varepsilon)) \downarrow 0$ *as* $\delta \downarrow 0$ *and* $\varepsilon \downarrow 0$.

*(ii) Suppose that X is complete,* $\Lambda$ *is compact or a finite-dimensional normed space, assumptions (a)–(b) in Theorem 3 and assumption (i) in Theorem 4 hold true, and the function* $(x,y,\lambda) \mapsto \langle h_2(x,\lambda), y-x\rangle$ *is upper a-level closed on* $K(V) \times K(V) \times V$ *for every negative a close to zero. If* $\mu(\mathcal{P}(\bar{\lambda}, \delta, \varepsilon)) \downarrow 0$ *as* $\delta \downarrow 0$ *and* $\varepsilon \downarrow 0$, *then* (**LVI**) *is well posed at* $\bar{\lambda}$.

## 5 Acknowledgement

The authors wish to thank the two referees for their helpful remarks and suggestions that helped us significantly improve the presentation of the paper. The research was partially supported by the Australian Research Council, project DP110102011.

## References

[1] Ait Mansour, M., Riahi, H.: Sensitivity analysis for abstract equilibrium problems. J. Math. Anal. Appl. **306**, 684–691 (2005)
[2] Anh, L.Q., Khanh, P.Q.: Semicontinuity of the solution set of parametric multivalued vector quasiequilibrium problems. J. Math. Anal. Appl. **294**, 699–711 (2004)
[3] Anh, L.Q., Khanh, P.Q.: Uniqueness and Hölder continuity of the solution to multivalued equilibrium problems in metric spaces. J. Global Optim. **37**, 449–465 (2007)

[4] Anh, L.Q., Khanh, P.Q.: Semicontinuity of the approximate solution sets of multivalued quasiequilibrium problems. Numer. Funct. Anal. Optim. **29**, 24–42 (2008)
[5] Anh, L.Q., Khanh, P.Q.: Sensitivity analysis for multivalued quasiequilibrium problems in metric spaces: Hölder continuity of solutions. J. Global Optim. **42**, 515–531 (2008)
[6] Anh, L.Q., Khanh, P.Q.: Hölder continuity of the unique solution to quasiequilibrium problems in metric spaces. J. Optim. Theory Appl. **141**, 37–54 (2009)
[7] Anh, L.Q., Khanh, P.Q.: Continuity of solution maps of parametric quasiequilibrium problems. J. Global Optim. **46**, 247–259 (2010)
[8] Anh, L.Q., Khanh, P.Q., Van, D.T.M., Yao, J.C.: Well-posedness for vector quasiequilibria. Taiwan. J. Math. **13**, 713–737 (2009)
[9] Anh, L.Q., Khanh, P.Q., Van, D.T.M.: Well-posedness without semicontinuity for parametric quasiequilibria and quasioptimization. Comput. Math. Appl. **62**, 2045–2057 (2011)
[10] Anh, L.Q., Khanh, P.Q., Tam, T.N.: On Hölder continuity of approximate solutions to parametric equilibrium problems. Nonlinear Anal. **75**, 2293–2303 (2012)
[11] Anh, L.Q., Khanh, P.Q., Van, D.T.M.: Well-posedness under relaxed semicontinuity for bilevel equilibrium and optimization problems with equilibrium constraints. J. Optim. Theory Appl. **153**, 42–59 (2012)
[12] Anh, L.Q., Kruger, A.Y., Thao, N.H.: On Hölder calmness of solution mappings in parametric equilibrium problems. TOP. doi: 10.1007/s11750-012-0259-3
[13] Aubin, J.-P., Frankowska, H.: Set-Valued Analysis. Birkhäuser Boston Inc., Boston (1990)
[14] Bednarczuk, E.: Stability Analysis for Parametric Vector Optimization Problems. Polish Academy of Sciences, Warszawa (2007)
[15] Bianchi, M., Pini, R.: A note on stability for parametric equilibrium problems. Oper. Res. Lett. **31**, 445–450 (2003)
[16] Bianchi, M., Pini, R.: Sensitivity for parametric vector equilibria. Optimization **55**, 221–230 (2006)
[17] Bianchi, M., Kassay, G., Pini, R.: Existence of equilibria via Ekeland's principle. J. Math. Anal. Appl. **305**, 502–512 (2005)
[18] Bianchi, M., Konnov, I.V., Pini, R.: Lexicographic variational inequalities with applications. Optimization **56**, 355–367 (2007)
[19] Bianchi, M., Konnov, I.V., Pini, R.: Lexicographic and sequential equilibrium problems. J. Global Optim. **46**, 551–560 (2010)
[20] Blum, E., Oettli, W.: From optimization and variational inequalities to equilibrium problems. Math. Student **63**, 123–145 (1994)
[21] Burachik, R., Kassay, G.: On a generalized proximal point method for solving equilibrium problems in Banach spaces. Nonlinear Anal. **75**, 6456–6464 (2012)
[22] Carlson, E.: Generalized extensive measurement for lexicographic orders. J. Math. Psych. **54**, 345–351 (2010)
[23] Daneš, J.: On the Istrăţescu's measure of noncompactness. Bull. Math. Soc. Sci. Math. R. S. Roumanie (N.S.) **16**, 403–406 (1974)
[24] Djafari Rouhani, B., Tarafdar, E., Watson, P.J.: Existence of solutions to some equilibrium problems. J. Optim. Theory Appl. **126**, 97–107 (2005)
[25] Emelichev, V.A., Gurevsky, E.E., Kuzmin, K.G.: On stability of some lexicographic integer optimization problem. Control Cybernet. **39**, 811–826 (2010)
[26] Fang, Y.P., Hu, R., Huang, N.J.: Well-posedness for equilibrium problems and for optimization problems with equilibrium constraints. Comput. Math. Appl. **55**, 89–100 (2008)
[27] Flores-Bazán, F.: Existence theorems for generalized noncoercive equilibrium problems: the quasi-convex case. SIAM J. Optim. **11**, 675–690 (2001)
[28] Freuder, E.C., Heffernan, R., Wallace, R.J., Wilson, N.: Lexicographically-ordered constraint satisfaction problems. Constraints **15**, 1–28 (2010)
[29] Hai, N.X., Khanh, P.Q.: Existence of solutions to general quasiequilibrium problems and applications. J. Optim. Theory Appl. **133**, 317–327 (2007)
[30] Iusem, A.N., Sosa, W.: Iterative algorithms for equilibrium problems. Optimization **52**, 301–316 (2003)

[31] Klein, E., Thompson, A.C.: Theory of Correspondences. Wiley Inc., New York (1984)
[32] Konnov, I.V.: On lexicographic vector equilibrium problems. J. Optim. Theory Appl. **118**, 681–688 (2003)
[33] Küçük, M., Soyertem, M., Küçük, Y.: On constructing total orders and solving vector optimization problems with total orders. J. Global Optim. **50**, 235–247 (2011)
[34] Li, X.B., Li, S.J.: Continuity of approximate solution mappings for parametric equilibrium problems. J. Global Optim. **51**, 541–548 (2011)
[35] Li, S.J., Li, X.B., Teo, K.L.: The Hölder continuity of solutions to generalized vector equilibrium problems. Eur. J. Oper. Res. **199**, 334–338 (2009)
[36] Loridan, P.: $\varepsilon$-solutions in vector minimization problems. J. Optim. Theory Appl. **43**, 265–276 (1984)
[37] Mäkelä, M.M., Nikulin, Y., Mezei, J.: A note on extended characterization of generalized trade-off directions in multiobjective optimization. J. Convex Anal. **19**, 91–111 (2012)
[38] Milovanović-Arandjelović, M.M.: Measures of noncompactness on uniform spaces–the axiomatic approach. Filomat 221–225 (2001)
[39] Morgan, J., Scalzo, V.: Discontinuous but well-posed optimization problems. SIAM J. Optim. **17**, 861–870 (2006)
[40] Muu, L.D., Oettli, W.: Convergence of an adaptive penalty scheme for finding constrained equilibria. Nonlinear Anal. **18**, 1159–1166 (1992)
[41] Noor, M.A., Noor, K.I.: Equilibrium problems and variational inequalities. Mathematica **47**, 89–100 (2005)
[42] Sadeqi, I., Alizadeh, C.G.: Existence of solutions of generalized vector equilibrium problems in reflexive Banach spaces. Nonlinear Anal. **74**, 2226–2234 (2011)

# The Best Linear Separation of Two Sets

**V.N. Malozemov and E.K. Cherneutsanu**

**Abstract** Consider the problem of the best approximate separation of two finite sets in the linear case. This problem is reduced to the problem of nonsmooth optimization, analyzing which we use all power of the linear programming theory. Ideologically we follow Bennett and Mangassarian (Optim. Meth. Software 1, 23–34 1992).

**Keywords** The best linear separation • Linear programming

Suppose we have two finite sets in $R^n$

$$A = \{a_i\}_{i=1}^{m} \quad , \quad B = \{b_j\}_{j=1}^{k}.$$

The sets $A$ and $B$ are called *strictly separable*, if there exist a nonzero vector $w \in R^n$ and a real number $\gamma$, such that

$$\langle w, a_i \rangle < \gamma \quad \forall i \in 1:m, \tag{1}$$

$$\langle w, b_j \rangle > \gamma \quad \forall j \in 1:k. \tag{2}$$

If conditions (1) and (2) are satisfied, it is also said that the hyperplane $H$ defined by the equation $\langle w, x \rangle = \gamma$ *strictly separates* the set $A$ from the set $B$.

V.N. Malozemov (✉) • E.K. Cherneutsanu
Saint Petersburg State University, 7–9, Universitetskaya nab., St.Petersburg, 199034, Russia
e-mail: malv@math.spbu.ru; katerinache@yandex.ru

V.F. Demyanov et al. (eds.), *Constructive Nonsmooth Analysis and Related Topics*, Springer Optimization and Its Applications 87, DOI 10.1007/978-1-4614-8615-2_11,

Introduce the function

$$f(g)=\frac{1}{m}\sum_{i=1}^{m}\big[\langle w,a_i\rangle-\gamma+c\big]_+ +\frac{1}{k}\sum_{j=1}^{k}\big[-\langle w,b_j\rangle+\gamma+c\big]_+, \tag{3}$$

where $g=(w,\gamma)$, $c>0$ is a parameter, and $[u]_+=\max\{0,u\}$. In the paper [1], the case $c=1$ was considered. It is clear that $f(g)\geq 0$ for all $g$.

**Theorem 1.** *The sets A and B are strictly separable if and only if there exists a vector $g_*$ such that $f(g_*)=0$.*

*Proof.* Let $f(g_*)=0$ for some vector $g_*=(w_*,\gamma_*)$. First, we show that $w_*\neq 0$. Otherwise

$$f(g_*)=(-\gamma_*+c)_+ +(\gamma_*+c)_+ =\begin{cases}-\gamma_*+c, & \gamma_*\leq -c,\\ 2c, & \gamma_*\in[-c,c],\\ \gamma_*+c, & \gamma_*\geq c.\end{cases}$$

Hence it follows that $f(g_*)\geq 2c$. It contradicts the condition $f(g_*)=0$.

Furthermore, the condition $f(g_*)=0$ guarantees that all the terms

$$\big[\langle w_*,a_i\rangle-\gamma_*+c\big]_+ \quad , \quad \big[-\langle w_*,b_j\rangle+\gamma_*+c\big]_+$$

equal to zero. This is possible only when

$$\langle w_*,a_i\rangle-\gamma_*+c\leq 0 \quad \forall i\in 1:m, \tag{4}$$

$$-\langle w_*,b_j\rangle+\gamma_*+c\leq 0 \quad \forall j\in 1:k. \tag{5}$$

It remains to note that (4) and (5) provide that the conditions of strict separation (1) and (2) are satisfied with $w=w_*$ and $\gamma=\gamma_*$.

Let us prove the converse. Let the conditions (1) and (2) be satisfied. Denote

$$d:=\min_{j\in 1:k}\langle w,b_j\rangle-\max_{i\in 1:m}\langle w,a_i\rangle>0, \tag{6}$$

$$w_*=\Big(\frac{2c}{d}\Big)w\,,\quad \gamma_*=\frac{1}{2}\big[\min_{j\in 1:k}\langle w_*,b_j\rangle+\max_{i\in 1:m}\langle w_*,a_i\rangle\big].$$

According to (6) and the definition of $w_*$

$$\min_{j\in 1:k}\langle w_*,b_j\rangle-\max_{i\in 1:m}\langle w_*,a_i\rangle=2c.$$

We have

$$\max_{i\in 1:m}\langle w_*,a_i\rangle=2\gamma_*-\min_{j\in 1:k}\langle w_*,b_j\rangle=2\gamma_*-2c-\max_{i\in 1:m}\langle w_*,a_i\rangle,$$

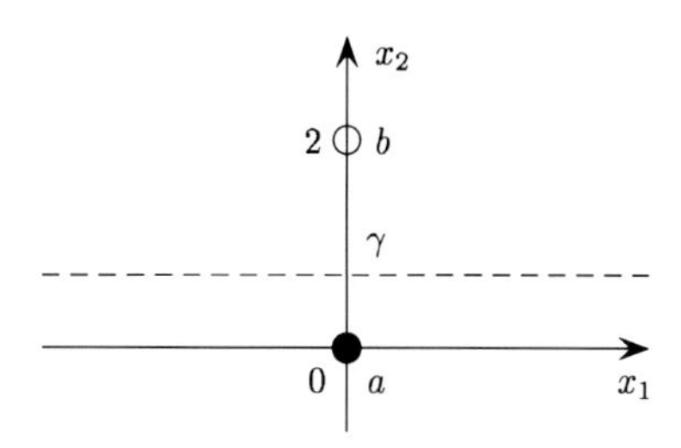


**Fig. 1** A simplest example of strict linear separation

so

$$\max_{i\in 1:m} \langle w_*, a_i \rangle = \gamma_* - c. \tag{7}$$

Similarly

$$\min_{j\in 1:k} \langle w_*, b_j \rangle = 2\gamma_* - \max_{i\in 1:m} \langle w_*, a_i \rangle = 2\gamma_* + 2c - \min_{j\in 1:k} \langle w_*, b_j \rangle,$$

so

$$\min_{j\in 1:k} \langle w_*, b_j \rangle = \gamma_* + c. \tag{8}$$

Let $g_* = (w_*, \gamma_*)$. By (7) and (8) we get $f(g_*) = 0$.

The theorem is proved. □

In the proof of theorem 1 we described a transformation of the vector $g = (w, \gamma)$ that generated a strictly separating hyperplane $H = \{x \mid \langle w, x \rangle = \gamma\}$ into the vector $g_* = (w_*, \gamma_*)$ with the property $f(g_*) = 0$. The point is, given the vector $g$, the value $f(g)$ can be positive (it depends on the parameter $c$).

*Example 1.* Consider two sets $A$ and $B$ on the plane $R^2$, each containing a single point $a = (0,0)$ and $b = (0,2)$, respectively. The vector $g = (w, \gamma)$ with components $w = (0,1)$ and $\gamma \in (0,2)$ generates a line $x_2 = \gamma$ that strictly separates the point $a$ from the point $b$ (see Fig. 1).

At the same time

$$f(g) = [-\gamma + c]_+ + [-2 + \gamma + c]_+.$$

Figure 2 shows a plot of $f(g)$ as a function of $\gamma$ for $c \in (0,1]$.

We see that $f(g) = 0$ for $\gamma \in [c, 2-c]$. At $\gamma \in (0,c) \cup (2-c, 2)$, the line $x_2 = \gamma$ still strictly separates the point $a$ from the point $b$, but $f(g) > 0$.

Consider an extremal problem

$$f(g) \longrightarrow \min, \tag{9}$$

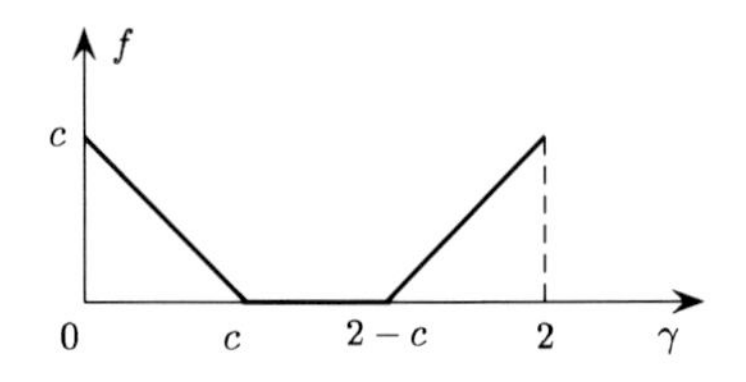


**Fig. 2** A plot of $f(g)$ as a function of $\gamma$

where $f(g)$ is a function of the form (3). This problem is equivalent to a linear programming problem

$$\begin{aligned} \tfrac{1}{m}\textstyle\sum_{i=1}^m y_i + \tfrac{1}{k}\sum_{j=1}^k z_j &\longrightarrow \min, \\ -\langle w, a_i\rangle + \gamma + y_i \geq c, &\quad i \in 1:m; \\ \langle w, b_j\rangle - \gamma + z_j \geq c, &\quad j \in 1:k; \\ y_i \geq 0,\ i \in 1:m; &\quad z_j \geq 0,\ j \in 1:k. \end{aligned} \tag{10}$$

The set of plans of the problem (10) is nonempty (a vector with components $w = 0$, $\gamma = 0$, $y_i \equiv c$, $z_j \equiv c$ is a plan) and the objective function is bounded below by zero. So the problem (10) has a solution. By the equivalence, the problem (9) has a solution too, and the minimum values of the objective functions of these problems are equal. We denote this common value by $\mu$. We also note that if $\big(w_*, \gamma_*, \{u_i^*\}, \{v_j^*\}\big)$ is the solution of (10), then $g_* = \{w_*, \gamma_*\}$ is the solution of (9).

When $\mu = 0$ we get $f(g_*) = 0$. By theorem 1 the vector $g_* = (w_*, \gamma_*)$ generates a hyperplane $H = \big\{x \mid \langle w_*, x\rangle = \gamma_*\big\}$ that strictly separates the set $A$ from the set $B$.

The vector $g_*$ can be reduced to the canonical form. Let

$$\begin{aligned} w_0 &= w_*/\|w_*\|, \\ \gamma_0 &= \frac{1}{2}\Big[\min_{j\in 1:k}\langle w_0, b_j\rangle + \max_{i\in 1:m}\langle w_0, a_i\rangle\Big], \\ c_0 &= \frac{1}{2}\Big[\min_{j\in 1:k}\langle w_0, b_j\rangle - \max_{i\in 1:m}\langle w_0, a_i\rangle\Big], \\ g_0 &= (w_0, \gamma_0). \end{aligned}$$

Then for all $i \in 1:m$

$$\langle w_0, a_i\rangle - \gamma_0 + c_0 = \langle w_0, a_i\rangle - \max_{i\in 1:m}\langle w_0, a_i\rangle \leq 0,$$

and for all $j \in 1:k$

$$-\langle w_0, b_j\rangle + \gamma_0 + c_0 = -\langle w_0, b_j\rangle + \min_{j\in 1:k}\langle w_0, b_j\rangle \leq 0.$$

This means that $f(g_0) = 0$ for $c = c_0$. The hyperplane $H_0 = \big\{x \mid \langle w_0, x\rangle = \gamma_0\big\}$ strictly separates the set $A$ from the set $B$, and the width of the dividing strip is equal to $2c_0$.

As noted earlier, the problem (9) always has a solution. When $\mu > 0$, according to the theorem 1, the sets $A$ and $B$ cannot be strictly separated. In this case we say that the hyperplane $H_* = \{x \mid \langle w_*, x\rangle = \gamma_*\}$ generated by the solution $g_* = (w_*, \gamma_*)$ of the problem (9) is *the best hyperplane approximately separating the set A from the set B* (for a given value of the parameter $c$).

However, there is a catch: there is no guarantee that the component $w_*$ of the vector $g_*$ is nonzero. Let us examine this situation.

**Theorem 2.** *The problem (9) has a solution $g_* = (w_*, \gamma_*)$ with $w_* = 0$ if and only if the following condition holds:*

$$\frac{1}{m}\sum_{i=1}^{m} a_i = \frac{1}{k}\sum_{j=1}^{k} b_j. \tag{11}$$

*Proof.* When $w_* = 0$, it is easy to calculate the extreme value of the objective function of the linear programming problem (10). Indeed,

$$\mu = f(g_*) = \min_{\gamma}\{[-\gamma + c]_+ + [\gamma + c]_+\} = 2c.$$

Of the same extreme value is the linear programming problem that is dual to (10). By the solvability of the dual problem, the following system is consistent:

$$c\left(\textstyle\sum_{i=1}^{m} u_i + \sum_{j=1}^{k} v_j\right) = 2c, \tag{12}$$

$$-\textstyle\sum_{i=1}^{m} u_i a_i + \sum_{j=1}^{k} v_j b_j = 0, \tag{13}$$

$$\textstyle\sum_{i=1}^{m} u_i - \sum_{j=1}^{k} v_j = 0, \tag{14}$$

$$0 \le u_i \le \tfrac{1}{m},\ i \in 1:m; \quad 0 \le v_j \le \tfrac{1}{k},\ j \in 1:k. \tag{15}$$

From (12) and (14) it follows that

$$\sum_{i=1}^{m} u_i = 1, \quad \sum_{j=1}^{k} v_j = 1.$$

Taking into account (15), we conclude that all $u_i$ are equal to $\frac{1}{m}$ and all $v_j$ are equal to $\frac{1}{k}$. Now (13) is equivalent to (11).

Write the problem dual to (10):

$$c\left(\sum_{i=1}^{m} u_i + \sum_{j=1}^{k} v_j\right) \longrightarrow \max$$

subject to constraints (13)–(15). By (11) the set of $u_i \equiv \frac{1}{m}$, $v_j \equiv \frac{1}{k}$ is a plan of this problem. The objective function value is equal to $2c$.

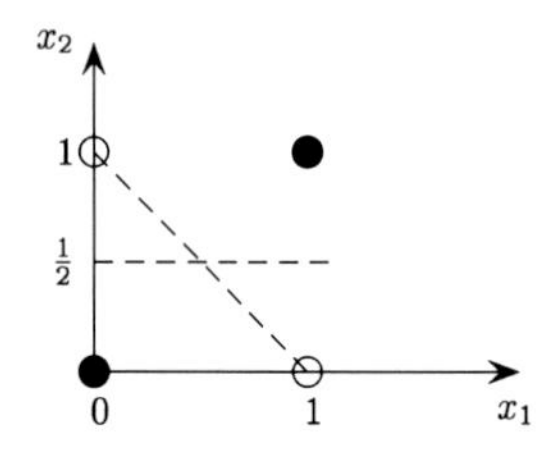


**Fig. 3** An example where condition (11) holds

At the same time, the set

$$w = 0, \quad \gamma = 0, \quad y_i \equiv c, \quad z_j \equiv c \tag{16}$$

is a plan of the problem (10) and the objective function value is also equal to $2c$. Hence it follows that the plan (16) of the problem (10) with $w = 0$ is optimal.

The theorem is proved. □

*Example 2.* Consider two sets on the plane $R^2$:

$$A = \{(0,0)\ ,\ (1,1)\}, \quad B = \{(1,0)\ ,\ (0,1)\}$$

(see Fig. 3). In this case the condition (11) holds. By the theorem 2, the problem (9) has a solution $g_* = (w_*, \gamma_*)$ with $w_* = 0$. In this case $\mu = 2c$.

We will show that the problem (9) has another solution $g_0 = (w_0, \gamma_0)$ with $w_0 \neq 0$.

By (3)

$$f(g) = \frac{1}{2}\{[-\gamma+c]_+ + [w^1+w^2-\gamma+c]_+\} + \frac{1}{2}\{[-w^1+\gamma+c]_+ + [-w^2+\gamma+c]_+\}.$$

Here $w = (w^1, w^2)$. Let

$$w_0 = (c, c), \quad \gamma_0 = c, \quad g_0 = (w_0, \gamma_0).$$

Then $f(g_0) = 2c$. So, a minimum of the function $f(g)$ is attained on the vector $g_0$. The hyperplane $H_0 = \{x \mid x_1 + x_2 = 1\}$ is the best hyperplane approximately separating the set $A$ from the set $B$.

Of the same property are the vector $g_1 = (w_1, \gamma_1)$ with $w_1 = (0, c)$ and $\gamma_1 = \frac{c}{2}$ and the hyperplane $H_1 = \{x \mid x_2 = \frac{1}{2}\}$ (see Fig. 3).

The peculiarity noted in example 2 is of general nature.

**Theorem 3.** *When $\mu > 0$, the problem (9) has a solution $g_0 = (w_0, \gamma_0)$ with $w_0 \neq 0$.*

*Proof.* Assume that the solution $g_* = (w_*, \gamma_*)$ of the problem (9) has zero component $w_*$. We will construct another solution $g_0 = (w_0, \gamma_0)$ with $w_0 \neq 0$.

By theorem 2 the relation (11) holds and $\mu = 2c$. Take any nonzero vector $h \in R^n$ and consider a linear programming problem

$$\langle h, w \rangle \longrightarrow \min, \tag{17}$$

$$-\tfrac{1}{m}\sum_{i=1}^{m} y_i - \tfrac{1}{k}\sum_{j=1}^{k} z_j = -2c;$$

$$-\langle w, a_i \rangle + \gamma + y_i \geq c, \quad i \in 1:m;$$

$$\langle w, b_j \rangle - \gamma + z_j \geq c, \quad j \in 1:k;$$

$$y_i \geq 0,\ i \in 1:m; \quad z_j \geq 0,\ j \in 1:k.$$

The vector (16) satisfies the constraints of the problem (17), so it is its plan. We will show that this plan cannot be optimal.

Indeed, if the plan (16) is optimal, then the problem dual to (17) must have a plan with the same (i.e., zero) value of the objective function. Thus, the following system must be consistent:

$$c\left(\sum_{i=1}^{m} u_i + \sum_{j=1}^{k} v_j - 2\zeta\right) = 0, \tag{18}$$

$$-\sum_{i=1}^{m} u_i a_i + \sum_{j=1}^{k} v_j b_j = h, \tag{19}$$

$$\sum_{i=1}^{m} u_i - \sum_{j=1}^{k} v_j = 0, \tag{20}$$

$$0 \leq u_i \leq \tfrac{1}{m}\zeta,\ i \in 1:m; \quad 0 \leq v_j \leq \tfrac{1}{k}\zeta,\ j \in 1:k. \tag{21}$$

However, it can be shown that this system is inconsistent.

From (18) and (20) it follows that

$$\sum_{i=1}^{m} u_i = \zeta, \quad \sum_{j=1}^{k} v_j = \zeta.$$

By (21) we obtain $u_i \equiv \frac{1}{m}\zeta$, $v_j \equiv \frac{1}{k}\zeta$. Equality (19) takes the form

$$\zeta\left(-\frac{1}{m}\sum_{i=1}^{m} a_i + \frac{1}{k}\sum_{j=1}^{k} b_j\right) = h.$$

But this contradicts to (11) (recall that $h \neq 0$).

It is ascertained that the plan (16) of the problem (17) with a zero value of the objective function is not optimal. Hence, there exist a plan

$$\left(w_0,\ \gamma_0,\ \{u_i^0\},\ \{v_j^0\}\right) \tag{22}$$

with a negative value of the objective function. Such a plan must be with $w_0 \neq 0$.

Now we note that the plan (22) of the problem (17) satisfies the constraints of (10) and on it the objective function of the problem (10) takes the smallest possible value equal to $2c$ (recall that $\mu = 2c$). By the equivalence of the problems (9) and (10) the vector $g_0 = (w_0, \gamma_0)$ with $w_0 \neq 0$ is a solution of the problem (9).

The theorem is proved. □

*Remark 1.* As a nonzero vector $h$ we can take, for example, any nonzero difference $b_{j_0} - a_{i_0}$. In this case, a set of plans of the problem dual to the problem (17), which is defined by (19)– (21), will not be empty. Together with the nonempty set of plans of the problem (17) this guarantees the existence of the optimal plan of the problem (17).

When $\mu > 0$ the solution $g_0 = (w_0, \gamma_0)$ of the problem (9) with $w_0 \neq 0$ can be reduced to the canonical form. Set

$$w_1 = w_0/\|w_0\|,$$

$$\gamma_1 = \frac{1}{2}\big[\min_{j\in 1:k}\langle w_1, b_j\rangle + \max_{i\in 1:m}\langle w_1, a_i\rangle\big],$$

$$c_1 = \frac{1}{2}\big[\min_{j\in 1:k}\langle w_1, b_j\rangle - \max_{i\in 1:m}\langle w_1, a_i\rangle\big],$$

$$g_1 = (w_1, \gamma_1).$$

In this case $c_1 \leq 0$. When $c_1 = 0$ the hyperplane $H_1 = \{x \mid \langle w_1, x\rangle = \gamma_1\}$ nonstrictly separates the set $A$ from the set $B$. When $c_1 < 0$ the same hyperplane $H_1$ is the best approximately separating the set $A$ from the set $B$.

By definition of $w_1$, $\gamma_1$, $c_1$ we have

$$\langle w_1, a_i\rangle - \gamma_1 + c_1 \leq 0, \quad i \in 1:m$$

$$-\langle w_1, b_j\rangle + \gamma_1 + c_1 \leq 0, \quad j \in 1:k.$$

When $c_1 < 0$ these inequalities define a "mixed strip"

$$c_1 \leq \langle w_1, x\rangle - \gamma_1 \leq -c_1,$$

which contains both the points of the set $A$ and the points of the set $B$. The width of the mixed strip is equal to $2|c_1|$.

The example of the best approximate separation of two sets is illustrated in Fig. 4.

To emphasize the dependence on the parameter $c$, we will write $f(g,\, c)$ and $\mu(c)$ instead of $f(g)$ and $\mu$. It is obvious that for all $c > 0$ the following formula holds:

$$f(cg,\, c) = cf(g,\, 1).$$

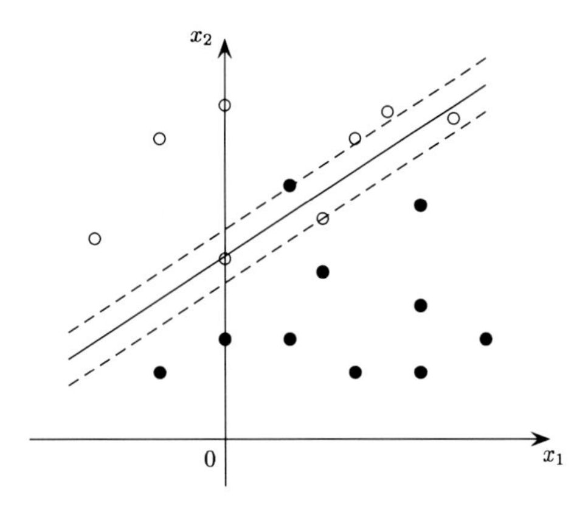


**Fig. 4** An example of the best approximate linear separation

So

$$\mu(c) = \min_{g} f(g, c) = \min_{g} f(cg, c) = c \min_{g} f(g, 1) = c\mu(1).$$

Moreover, if $g_1$ is a solution of the problem (9) with $c = 1$, then the vector $g_c = cg_1$ will be a solution of the problem (9) for arbitrary $c > 0$. Thus, the additive parameter $c > 0$ plays the role of a normalizing factor.

## Reference

[1] Bennett, K.P., Mangassarian, O.L.: Robust linear programming discrimination of two linearly inseparable sets. Optim. Meth. Software **1**, 23–34 (1992)

# Alternance Form of Optimality Conditions in the Finite-Dimensional Space

**V.F. Demyanov and V.N. Malozemov**

**Abstract** In solving optimization problems, necessary and sufficient optimality conditions play an outstanding role. They allow, first of all, to check whether a point under study satisfies the conditions, and, secondly, if it does not, to find a "better" point. This is why such conditions should be "constructive" letting to solve the above-mentioned problems. For the class of directionally differentiable functions in $\mathbb{R}^n$, a necessary condition for an unconstrained minimum requires for the directional derivative to be non-negative in all directions. This condition becomes efficient for special classes of directionally differentiable functions. For example, in the case of convex and max-type functions, the necessary condition for a minimum takes the form $0_n \in C$ where $C \subset \mathbb{R}^n$ is a convex compact set. The problem of verifying this condition is reduced to that of finding the point of $C$ which is the nearest to the origin. If the origin does not belong to $C$, we easily find the steepest descent direction and are able to construct a numerical method. For the classical Chebyshev approximation problem (the problem of approximating a function $f(t) : G \longrightarrow \mathbb{R}$ by a polynomial $P(t)$), the condition for a minimum takes the so-called alternance form: for a polynomial $P^*(t)$ to be a solution to the Chebyshev approximation problem, a collection of points $\{t_i \mid t_i \in G\}$ should exist at which the difference $P^*(t) - f(t)$ attains its maximal absolute value with alternating signs. This condition can easily be verified, and if it does not hold, one can find a "better" polynomial. In the present paper, it will be demonstrated that the alternance form of the necessary conditions is valid not only for Chebyshev approximation problems but also in the general

V.F. Demyanov
Applied Mathematics Department, Saint Petersburg State University, Peterhof, Saint-Petersburg, 198504, Russia
e-mail: vfd@ad9503.spb.edu

V.N. Malozemov
Mathematics and Mechanics Department, Saint Petersburg State University, Universitetskii prospekt 28, Peterhof, Saint-Petersburg, 198504, Russia
e-mail: malv@math.spbu.ru

V.F. Demyanov et al. (eds.), *Constructive Nonsmooth Analysis and Related Topics*, Springer Optimization and Its Applications 87, DOI 10.1007/978-1-4614-8615-2_12,

case of directionally differentiable functions. Here only unconstrained optimization problems are discussed. In many cases a constrained optimization problem can be reduced (via Exact Penalization Techniques) to an unconstrained one. In the paper, optimality conditions are first formulated in terms of directional derivatives. Next, the notions of upper and lower exhausters are introduced, and optimality conditions are stated by means of upper and lower exhausters. In all these cases the optimality conditions are presented in the form $0_n \in C$ where $C \subset \mathbb{R}^n$ is a convex closed bounded set (or a family of such sets). It is proved that the condition $0_n \in C$ can be formulated in the alternance form. The result obtained is applied to deduce the well-known Chebyshev alternation rule in the problem of Chebyshev approximation of a function by a polynomial. The problem of Chebyshev approximation of several functions by a polynomial is also discussed, and optimality conditions are stated in the alternance form.

**Keywords** Necessary optimality conditions • Alternance form • Directionally differentiable functions

## 1 Introduction and Statement of the Problem

In solving optimization problems, necessary and sufficient optimality conditions play an outstanding role [11–14, 16]. They allow, first of all, to check whether a point under study satisfies the optimality conditions, and, secondly, if it does not, to find a "better" point.

For the class of directionally differentiable functions in the $n$-dimensional space, a necessary condition for an unconstrained minimum is

$$f'(x^*,g) \geq 0 \quad \forall g \in \mathbb{R}^n \tag{1}$$

while a necessary condition for an unconstrained maximum is

$$f'(x^*,g) \leq 0 \quad \forall g \in \mathbb{R}^n$$

where $f'(x,g)$ is the directional derivative of the function $f$ at a point $x \in \mathbb{R}^n$. The above stated conditions become efficient for special classes of directionally differentiable functions. For example, in the case of convex and max-type functions, $f'(x,g)$ takes the form

$$f'(x,g) = \max_{v \in \partial f(x)} (v,g),$$

where $\partial f(x)$ is the subdifferential of the function $f$ at a point $x \in \mathbb{R}^n$ (see [15]). Then the condition (1) is equivalent to

$$0_n \in \partial f(x^*) \tag{2}$$

If a point $x_0 \in \mathbb{R}^n$ is given, to check condition (2), one can find

$$\min_{z\in\partial f(x_0)} ||z|| = ||z(x_0)||.$$

If $||z(x_0)|| = 0$, it means that $x_0$ satisfies the necessary condition (2). Otherwise, the direction $g_0 = -\frac{z(x_0)}{||z(x_0)||}$ is the steepest descent direction, while the value $||z(x_0)||$ is just the rate of steepest descent. Different types of numerical methods can be constructed making use of steepest descent directions (see, e.g., [7, 8, 10]).

Consider now the classical Chebyshev approximation problem. Let

$$F(A) = \max_{t\in G} |P(t,A) - f(t)|$$

where

$$P(t,A) = a_0 + a_1t + \cdots + a_nt^n, \quad A = (a_0,\ldots,a_n) \in \mathbb{R}^{n+1},$$

$G \in \mathbb{R}$ is a given closed bounded set, the function $f$ is continuous on $G$. It is required to find

$$\min_{A\in\mathbb{R}^{n+1}} F(A).$$

The condition for a minimum (1) takes the so-called alternance form: for a point $A^* \in \mathbb{R}^{n+1}$ to be an unconstrained minimizer of the function $F(A)$, it is necessary and sufficient that there exist $n+2$ points $\{t_i \mid t_i \in G\}$ such that

$$t_1 < t_2 < \cdots < t_{n+2}, \quad |P(t_i,A^*) - f(t_i)| = F(A^*)$$

and the signs of the function $P(t,A^*) - f(t)$ are alternating at these points:

$$P(t_i,A^*) - f(t_i) = f(t_{i-1}) - P(t_{i-1},A^*) \quad \forall i \in 2 : (n+2).$$

This condition allows one not only to check optimality conditions but also to construct numerical methods based on the alternance approach (see, e.g., [5]). It turned out that the alternance form of necessary conditions is valid not only for Chebyshev approximation problems but also for a wide class of mathematical programming problems (see a survey of these results in [3]). In the present paper, an alternance form of necessary conditions different from (but, of course, equivalent to) the one in [3] is described for the general class of directionally differentiable functions. In some cases it may be more convenient to use the conditions in form (2), in other ones – the condition in the alternance form.

In the paper we discuss only unconstrained optimization problems in the finite-dimensional space. In many cases a constrained optimization problem can be reduced (via Exact Penalization Techniques) to an unconstrained one. The paper is organized as follows. In Sect. 2 optimality conditions are formulated in terms of directional derivatives. In Sect. 3 the notions of upper and lower exhausters are introduced, and optimality conditions are stated by means of upper and lower exhausters. In Subsection 3.2 it is shown how to find the upper and lower exhausters in the case of quasidifferentiable functions. In all these cases the optimality conditions are presented in the form $0_n \in C$ where $C \subset \mathbb{R}^n$ is a convex closed bounded set (or a family of such sets). In Sect. 4 it is proved that the condition $0_n \in C$ can be formulated in the alternance form. The result obtained is applied to deduce the well-known Chebyshev alternation rule in the problem of Chebyshev approximation of a function by a polynomial (Sect. 5). The problem of Chebyshev approximation of several functions by a polynomial is considered in Sect. 6, where optimality conditions are stated in the alternance form.

## 2 Directionally Differentiable Functions

Let us consider the class of directionally differentiable (d.d.) functions. Recall that a function $f : \mathbb{R}^n \to \mathbb{R}$ is called *Dini directionally differentiable* $(D.-d.d.)$ at $x \in \mathbb{R}^n$ if the limit

$$f'_D(x;g) := \lim_{\alpha \downarrow 0} \frac{f(x+\alpha g) - f(x)}{\alpha} \tag{3}$$

exists and is finite for every $g \in \mathbb{R}^n$. The quantity $f'_D(x,g)$ is called the *Dini derivative* of $f$ at $x$ in the direction $g$.

A function $f : \mathbb{R}^n \to \mathbb{R}$ is called *Hadamard directionally differentiable* $(H.-d.d.)$ at $x \in \mathbb{R}^n$ if the limit

$$f'_H(x;g) := \lim_{[\alpha,g'] \to [+0,g]} \frac{f(x+\alpha g') - f(x)}{\alpha} \tag{4}$$

exists and is finite for every $g \in \mathbb{R}^n$. The quantity $f'_H(x;g)$ is called the *Hadamard derivative* of $f$ at $x$ in the direction $g$.

The Hadamard directional differentiability implies the Dini directional differentiability, while the opposite is not true. Since the functions $f'_D(x;g)$ and $f'_H(x;g)$ are positively homogeneous (p.h.) (as functions of $g$) of degree one, it is sufficient to consider only $g \in S_1 := \{g \in \mathbb{R}^n \mid ||g|| = 1\}$. (Recall that a function $h(g)$ is p.h. if $h(\lambda g) = \lambda h(g) \quad \forall\, \lambda > 0$.) If $f$ is $H.-d.d.$, then the function $h(g) = f'_H(x;g)$ is continuous.

The directional derivatives allow us to formulate first-order necessary conditions for an extremum:

**Theorem 1.** *For a point $x^* \in \mathbb{R}^n$ to be a (local or global) minimizer of $f$ on $\mathbb{R}^n$ it is necessary that*

$$f'_D(x^*;g) \geq 0 \quad \forall g \in \mathbb{R}^n, \tag{5}$$

$$f'_H(x^*;g) \geq 0 \quad \forall g \in \mathbb{R}^n. \tag{6}$$

*If*

$$f'_H(x^*;g) > 0 \quad \forall g \in \mathbb{R}^n \setminus \{0_n\}, \tag{7}$$

*then $x^*$ is a strict local minimizer of $f$ on $\mathbb{R}^n$.*

**Theorem 2.** *For a point $x^{**} \in \mathbb{R}^n$ to be a (local or global) maximizer of $f$ on $\mathbb{R}^n$ it is necessary that*

$$f'_D(x^{**};g) \leq 0 \quad \forall g \in \mathbb{R}^n, \tag{8}$$

$$f'_H(x^{**};g) \leq 0 \quad \forall g \in \mathbb{R}^n. \tag{9}$$

*If*

$$f'_H(x^{**};g) < 0 \quad \forall g \in \mathbb{R}^n \setminus \{0_n\}, \tag{10}$$

*then $x^{**}$ is a strict local maximizer of $f$ on $\mathbb{R}^n$.*

Note that the necessary conditions for a minimum (5) and (6) and for a maximum (8) and (9) *do not coincide* any more, and the sufficient conditions (7) and (10) have no equivalence in the smooth case, *they are just impossible.*

A point $x^* \in \mathbb{R}^n$, satisfying (5) and ((6)), is called a *$D(H)$-inf-stationary point* of $f$ on $\mathbb{R}^n$. A point $x^{**} \in \mathbb{R}^n$, satisfying (8) and ((9)), is called a *$D(H)$-sup-stationary point* of $f$ on $\mathbb{R}^n$.

# 3 Exhausters

## *3.1 Exhausters of a Positively Homogeneous Function*

Let a function $f : \mathbb{R}^n \longrightarrow \mathbb{R}$ be directionally differentiable (in the sense of Dini or Hadamard) and let $h(g) = f'(x,g)$ be the corresponding ($D$ or $H$)-derivative of the function $f$ at the point $x$ in a direction $g$. Fix $x \in \mathbb{R}$. If the function $h$ is upper semicontinuous as a function of $g$, then (see [8]) $h(g)$ can be expressed in the form

$$h(g) = \inf_{C \in E^*} \max_{v \in C}(v,g), \tag{11}$$

where $E^* = E^*(x)$ is a family of convex closed bounded sets of the space $\mathbb{R}^n$, and if $h(g) = f'(x,g)$ is lower semicontinuous as a function of $g$, then $h(g)$ can be written as

$$h(g) = \sup_{C \in E_*} \min_{w \in C} (w, g), \tag{12}$$

where $E_* = E_*(x)$ is also a family of convex closed bounded sets of the space $\mathbb{R}^n$.

The family of sets $E^*$ is called an *upper exhauster* of the function $f$ at the point $x$ (respectively, in the sense of Dini or Hadamard), and the family $E_*$ is called a *lower exhauster* of the function $f$ at the point $x$ (in the sense of Dini or Hadamard).

For an arbitrary p.h. function $h$ for which the relation (11) holds, the family $E^*$ is called an *upper exhauster* (of the function $h$). If the relation (12) is valid, then the family $E_*$ is called a *lower exhauster* of $h$.

If $h$ is continuous in $g$, then both representations (11) and (12) are true. If the function $h$ is Lipschitz, then (see [2]) (11) can be written in the form

$$h(g) = h_1(g) = \min_{C \in E^*} \max_{v \in C} (v, g) \quad \forall g \in \mathbb{R}^n, \tag{13}$$

and (12) – in the form

$$h(g) = h_2(g) = \max_{C \in E_*} \min_{w \in C} (w, g) \quad \forall g \in \mathbb{R}^n, \tag{14}$$

where the families $E^*$ and $E_*$ are totally bounded. The functions $h_1$ and $h_2$ are positively homogeneous (p.h.) as functions of direction $g$ and therefore represent positively homogeneous approximations of the increment of the function $f$ in a vicinity of the point $x$.

Exhausters were introduced in [4, 9]. Some properties and applications of exhausters are discussed in [1, 6, 17]. Theorems 1 and 2 imply the following optimality conditions.

**Theorem 3.** *Let $E^*$ be an upper exhauster of the function $f$ at a point $x^* \in \mathbb{R}^n$. Then for the point $x^*$ to be an unconstrained (global or local) minimizer of $f$, it is necessary that*

$$0_n \in C \quad \forall\, C \in E^*. \tag{15}$$

*Let the function $h$ be continuous. If there exists a $\delta > 0$ such that*

$$B_\delta \subset C \quad \forall\, C \in E^*, \tag{16}$$

*where $B_\delta = \{x \in \mathbb{R}^n \mid ||x|| \le \delta\}$, then $x^*$ is a strict local minimizer of the function $f$.*

**Theorem 4.** *Let $E_*$ be a lower exhauster of the function $f$ at a point $x^{**}$. Then for the point $x^{**}$ to be an unconstrained (global or local) maximizer of $f$, it is necessary that*

$$0_n \in C \quad \forall\, C \in E_*. \tag{17}$$

*Let the function h be continuous. If there exists a $\delta > 0$ such that*

$$B_\delta \subset C \quad \forall\, C \in E_*, \tag{18}$$

*then $x^{**}$ is a strict local maximizer of the function $f$.*

*Remark 1.* In the case of Lipschitz directionally differentiable functions (and if the representations (13) and (14) are used) the conditions (16) and (18) can be replaced, respectively, by the conditions

$$0_n \in \operatorname{int} C \quad \forall C \in E^* \tag{19}$$

and

$$0_n \in \operatorname{int} C \quad \forall C \in E_*. \tag{20}$$

To illustrate this Remark, consider the function $f(x) = -||x||^2$ where $x \in \mathbb{R}^n$. This function is Lipschitz around the point $x_0 = 0_n$. At $x_0$ it is directionally differentiable and

$$h(g) = f'(x_0, g) = 0\; \forall g \in \mathbb{R}^n.$$

Clearly,

$$h(g) = \inf_{C \in E^*} \max_{v \in C} (v, g) \quad \forall g \in \mathbb{R}^n, \tag{21}$$

where $E^* = \{B_\delta \mid \delta > 0\}$ is an upper exhauster of $h$ at $x_0$. But it is possible to get the following presentation

$$h(g) = \min_{C \in \hat{E}^*} \max_{v \in C} (v, g) \quad \forall g \in \mathbb{R}^n, \tag{22}$$

where $\hat{E}^* = \{B_\delta \mid \delta \geq 0\}$ is also an upper exhauster of $h$ at $x_0$. For the exhauster $E^*$ the condition (19) holds, however, the point $x_0$ is not a local minimizer.

Thus, conditions for a minimum are expressed by an upper exhauster, and conditions for a maximum are formulated via a lower one. If $h$ is a continuous function, it has the both exhausters: an upper exhauster and a lower one. An upper exhauster is called a proper one for the minimization problem (and adjoint for the maximization problem) while a lower exhauster is called a proper one for the maximization problem (and adjoint for the minimization problem). Upper and lower exhausters are not uniquely defined.

### 3.2 Quasidifferentiable Functions

A function $f : \mathbb{R}^n \longrightarrow \mathbb{R}$ is called (see [7, 8, 10]) quasidifferentiable in the sense of Dini (Hadamard) at a point $x \in \mathbb{R}^n$, if it is directionally differentiable in the sense of Dini (Hadamard) at the point $x$ and if its directional derivative (respectively in the sense of Dini or Hadamard) $f'(x,g)$ can be represented in the form

$$f'(x;g) = \max_{v \in \underline{\partial} f(x)} (v,g) + \min_{w \in \overline{\partial} f(x)} (w,g) \qquad \forall g \in \mathbb{R}^n, \tag{23}$$

where $\underline{\partial} f(x) \subset \mathbb{R}^n$, $\overline{\partial} f(x) \subset \mathbb{R}^n$ are convex compact sets. The pair of sets $\mathcal{D}f(x) = [\underline{\partial} f(x), \overline{\partial} f(x)]$ is called a quasidifferential of the function $f$ at the point $x$. Quasidifferentials, like exhausters, are not uniquely defined.

From (23), it is not difficult to get an upper exhauster and a lower one. Indeed,

$$h(g) = f'(x;g) = \min_{w \in \overline{\partial} f(x)} \max_{v \in \underline{\partial} f(x)} (v+w,g) =$$

$$= \min_{w \in \overline{\partial} f(x)} \max_{v \in [w+\underline{\partial} f(x)]} (v,g) = \min_{C \in E^*} \max_{v \in C} (v,g),$$

where

$$E^* = \{C = w + \underline{\partial} f(x) \mid \ w \in \overline{\partial} f(x)\}. \tag{24}$$

Hence, the family of sets (24) is an upper exhauster of the function $f$ at the point $x$.

Analogously,

$$h(g) = f'(x;g) = \max_{v \in \underline{\partial} f(x)} \min_{w \in \overline{\partial} f(x)} (v+w,g) =$$

$$= \max_{v \in \underline{\partial} f(x)} \min_{w \in [v+\overline{\partial} f(x)]} (w,g) = \max_{C \in E_*} \min_{w \in C} (w,g)$$

where

$$E_*(h) = \{C = v + \overline{\partial} f(x) \mid v \in \underline{\partial} f(x)\}. \tag{25}$$

It means that the family of sets (25) represents a lower exhauster of $f$ at $x$.

Therefore, for quasidifferentiable functions upper and lower exhausters are easily constructed by means of a quasidifferential.

## 4 The Alternance Form of the Condition $0_n \in C$

Optimality conditions in Theorems 3 and 4 are expressed in the form $0_n \in C$ which should be checked for one or several (may be, infinitely many) sets $C$. Thus, let us discuss the condition

$$0_n \in C, \tag{26}$$

where $C$ is the convex closed bounded set. In some important practical cases, the set $C$ is a convex hull of a finite or infinite number of points. For such a case the following equivalent formulation of the condition (26) can be useful. Let $\hat{C}$ be an arbitrary set of $n$ linearly independent vectors of $\mathbb{R}^n$.

Let the condition (26) hold where $C = \text{co}\{C'\}$, and $C'$ is a closed bounded set (not necessarily convex) of points in $\mathbb{R}^n$.

**Theorem 5.** *The condition (26) is equivalent to the following one:*
*There exist $n+1$ points $V_1, V_2, \ldots, V_{n+1}$ such that*

$$V_i \in C' \quad \forall i \in 1:r,$$

$$V_i \in \hat{C} \quad \forall i \in (r+1):(n+1),$$

*where $0 < r \leq n+1$ and the determinants $\Delta_i$ of the n-th order, constructed from the vectors $V_1, V_2, \ldots, V_{n+1}$ by consecutive deleting one of these vectors have the following property:*

$$\Delta_i \neq 0, \quad \forall i \in 1:r; \qquad sign\ \Delta_{i+1} = -sign\ \Delta_i \quad \forall i \in 1:(r-1), \tag{27}$$

$$\Delta_i = 0 \quad \forall i \in (r+1):(n+1). \tag{28}$$

*Proof.* Let the condition (26) hold. Then there exist points $V_1, V_2, \ldots, V_m$ and coefficients $\alpha_1, \ldots, \alpha_m$, where $0 < m \leq n+1$, such that

$$V_i \in C', \ \alpha_i > 0, \ \sum_{i \in 1:m} \alpha_i = 1, \quad \sum_{i \in 1:m} \alpha_i V_i = 0_n. \tag{29}$$

If $m = 1$, then, clearly, $V_1 = 0_n$, and the statement of the theorem holds if one takes the entire set of vectors $\hat{C}$ (containing $n$ points) as points $V_2, \ldots, V_{n+1}$.

If $m > 1$, then

$$\sum_{i \in 2:m} \beta_i V_i = -V_1$$

where $\beta_i = \frac{\alpha_i}{\alpha_1} > 0$. Thus, the point $V_1$ is a point of the conic hull of the points $V_2, \ldots, V_m$. Every point of a conic hull can be represented as a conical combination of not more than $n$ linearly independent points. Let it be points $V_2, \ldots, V_r$ (with $r \leq m$). We have the relation

$$\sum_{i \in 2:r} \gamma_i V_i = -V_1$$

where $\gamma_i > 0$. Now, take vectors $V_{r+1}, \ldots, V_{n+1}$ from the set $\hat{C}$ such that the vectors $V_2, \ldots, V_{n+1}$ are linearly independent. Then the following expansion holds

$$\sum_{i\in 2:(n+1)} \gamma_i V_i = -V_1 \tag{30}$$

where $\gamma_i > 0 \quad \forall i \in 2:r, \qquad \gamma_i = 0 \quad \forall i \in (r+1):(n+1)$. The relation (30) is a system (of $n$ linear equations with $n$ unknowns) with the nonzero determinant $\Delta_1$ (since the vectors $V_2,\ldots,V_{n+1}$ are linearly independent). By the Cramer formula,

$$\gamma_i = \frac{\Delta_i'}{\Delta_1} \qquad \forall i \in 2:(n+1), \tag{31}$$

where $\Delta_i'(i \in 2:(n+1))$ is the determinant $\Delta_i'$, constructed from the vectors $V_2,\ldots,V_{n+1}$ by replacing the vector $V_i$ with the vector $-V_1$. Note that $\operatorname{sign} \Delta_i' = \operatorname{sign} \Delta_1 \quad \forall i \in 2:r$ (since $\gamma_i > 0$). Clearly,

$$\Delta_i' = (-1)^{i-1}\Delta_i \qquad \forall i \in 2:(n+1). \tag{32}$$

Hence,

$$\operatorname{sign} \Delta_i = (-1)^{i-1}\operatorname{sign} \Delta_1 \qquad \forall i \in 2:r. \tag{33}$$

This implies the relation (27).

The relation (28) is based on the formulas (31) and (32) and the fact that $\gamma_i = 0$ for $i \in (r+1):(n+1)$.

Let us prove the inverse statement. Consider the system (30) where the relations (27) and (28) are valid. Due to (31) and (32),

$$\gamma_i = (-1)^{i-1}\frac{\Delta_i}{\Delta_1} \qquad \forall i \in 2:(n+1). \tag{34}$$

This relation and the relations (27) and (28) yield

$$\gamma_i = 0 \text{ if } i \in r+1:(n+1); \qquad \gamma_i > 0 \text{ if } i \in 2:r.$$

Thus,

$$\sum_{i=2}^{r} \gamma_i V_i + V_1 = 0_n. \tag{35}$$

The relation (35) implies

$$\sum_{i\in 1:r} \alpha_i V_i = 0_n, \tag{36}$$

where

$$\alpha_1 = \frac{1}{\gamma_0}, \quad \alpha_i = \frac{\gamma_i}{\gamma_0} \quad \forall i \in 2:r, \quad \gamma_0 = 1 + \sum_{i\in 2:r} \gamma_i.$$

Clearly,

$$\alpha_i > 0, \sum_{i \in 1:r} \alpha_i = 1. \tag{37}$$

It follows from (36) and (37) that $0_n \in C$. This completes the proof.

**Corollary 1.** *If $r = 1$, then $V_1 = 0_n$.*
*If $r = n+1$, then $0_n \in int\, C$.*

**Corollary 2.** *It follows from (34) and (35) that*

$$\sum_{i \in 1:r} (-1)^{i-1} \Delta_i V_i = 0_n.$$

*This is equivalent to the condition*

$$\sum_{i \in 1:r} |\Delta_i| V_i = 0_n. \tag{38}$$

## 5 Chebyshev Approximation of a Function by a Polynomial

The problem of approximation of a given function $f : G \longrightarrow \mathbb{R}$ on a set $G \subset \mathbb{R}$ by a polynomial of a fixed degree is the well-known *Chebyshev polynomial approximation problem* ([5]).

Let a closed bounded set of points $G \subset \mathbb{R}$ and a continuous function $f : G \longrightarrow \mathbb{R}$ be given. On $\mathbb{R}$ let us consider polynomials $P(t,A)$ of degree $n$

$$P(t,A) = a_0 + a_1 t + \cdots + a_n t^n = (A, B(t)), \tag{39}$$

where

$$A, B(t) \in \mathbb{R}^{n+1}, \ A = (a_0, \ldots, a_n), \ B(t) = (1, t, \ldots, t^n), \tag{40}$$

and $(A,B)$ denotes the scalar product of vectors $A$ and $B$.

Put

$$\sigma(t,A) = P(t,A) - f(t), \quad \eta(t,A) = \frac{1}{2}(\sigma(t,A))^2, \tag{41}$$

$$F(A) = \max_{t \in G} \eta(t,A). \tag{42}$$

It is required to find a point $A^* \in \mathbb{R}^{n+1}$ such that

$$F(A^*) = \min_{A \in \mathbb{R}^{n+1}} F(A) = F^*. \tag{43}$$

Since

$$F(A) \geq 0 \quad \forall A \in \mathbb{R}^{n+1},$$

then, clearly, $F^* \geq 0$.

It is assumed in the sequel that $F^* > 0$ (if it is not the case, the approximated function $f$ coincides with some polynomial on the set $G$). Then, in particular, $|G| \geq n+2$ (it follows from the interpolation arguments).

The problem (43) is equivalent to the problem of minimizing the functional

$$\overline{F}(A) = \max_{t \in G} |P(t,A) - f(t)| = \max_{t \in G} |\sigma(t,A)|.$$

The function $F : \mathbb{R}^{n+1} \longrightarrow \mathbb{R}$ is a max-type function, and its directional derivative can be expressed in the form

$$F'(A,g) = \max_{t \in R(A)} (\eta'_A(t,A), g) = \max_{V \in \underline{\partial} F(A)} (V,g), \tag{44}$$

where

$$\underline{\partial} F(A) = \text{co}\{\sigma(t,A)B(t) \mid t \in R(A)\}, \tag{45}$$

$$R(A) = \{t \in G \mid \eta(t,A) = F(A)\}, \quad g \in \mathbb{R}^{n+1}.$$

The following optimality condition for an unconstrained minimum of the function $F$ holds.

**Theorem 6.** *For a point $A^* \in \mathbb{R}^{n+1}$ to be a global minimizer of $F$ on $\mathbb{R}^{n+1}$ it is necessary and sufficient that*

$$0_{n+1} \in \underline{\partial} F(A^*). \tag{46}$$

*Remark 2.* The condition (46) is sufficient due to the convexity of $F$.

Now, let us employ Theorem 5. Note that $A^* \in \mathbb{R}^{n+1}$. In our case (see (45)) $C' = \{\sigma(t,A^*)B(t) \mid t \in R(A^*)\}$. By Theorem 5, there exist $n+2$ vectors $\sigma(t_1,A^*)B(t_1),\ldots,\sigma(t_r,A^*)B(t_r),V_{r+1},\ldots,V_{n+2}$, satisfying (27) and (28). By Corollary 2,

$$\sum_{i \in 1:r} (-1)^{i-1} \Delta_i \sigma(t_i,A^*)B(t_i) = 0_{n+1}. \tag{47}$$

If $r < n+1$ and $t_i \neq t_j \; \forall i \neq j$, the vectors $B(t_1),\ldots,B(t_r)$ are linearly independent; therefore the condition (47) can be satisfied only if $r = n+2$. If, in addition, the points $t_1,\ldots,t_{r+2}$ are ordered: $t_1 < t_2 < \cdots < t_{k+2}$, Theorem 6 can be reformulated as follows:

**Theorem 7.** *For a point $A^* \in \mathbb{R}^{n+1}$ to be a global minimizer of $F$ on $\mathbb{R}^{n+1}$ it is necessary and sufficient that there exist $n+2$ points $\{t_i \in G \mid i \in 1:(n+2)\}$ such that*

$$t_1 < t_2 < \cdots < t_{n+2},$$

*at the points $t_i$ the function $|P(t,A^*) - f(t)| = |\sigma(t,A^*)|$ attains its maximal value on $G$ and the signs of $\sigma(t,A^*)$ are alternating:*

$$\sigma(t_{i+1},A^*) = -\sigma(t_i,A^*) \quad \forall i \in 1:(n+1).$$

*Proof.* Indeed,

$$\Delta_i = \Delta_i' \prod_{j \neq i} \sigma(t_j,A^*) = \frac{\Delta_i'}{\sigma(t_i,A^*)} \prod_{j=1}^{n+2} \sigma(t_j,A^*),$$

where $\Delta_i'$ is the Vandermonde determinant of the matrix composed by the columns $B(t_1),\ldots,B(t_{i-1}),B(t_{i+1}),\ldots,B(t_{n+2})$. Since the points of the alternance $\{t_i\}$ are arranged in the increasing order, then all $\Delta_i'$ are positive. Hence, the alternation of the signs of the determinants $\Delta_i$ is equivalent to the alternation of the signs of the values $\sigma(t_i,A^*)$.

Theorem 7 is the famous Chebyshev theorem. The set of points $\{t_i\}$ satisfying the conditions of Theorem 7 is called the *Chebyshev alternance*.

# 6 Chebyshev Approximation of Several Functions by a Polynomial

## 6.1 *Introduction and Statement of the Problem*

Let closed bounded sets of points $G_k \subset \mathbb{R}$, $k \in 1:p$, and continuous functions $f_k : G_k \longrightarrow \mathbb{R}$, $k \in 1:p$, be given. The sets $G_k$ may intersect. On $\mathbb{R}$ let us consider polynomials $P(t,A)$ of degree $n$

$$P(t,A) = a_0 + a_1 t + \cdots + a_n t^n = (A, B(t)), \tag{48}$$

where, as above,

$$A, B(t) \in \mathbb{R}^{n+1}, \ A = (a_0,\ldots,a_n), \ B(t) = (1,t,\ldots,t^n), \tag{49}$$

and $(A,B)$ denotes the scalar product of vectors $A$ and $B$.

Put

$$\sigma_k(t,A) = P(t,A) - f_k(t), \quad \eta_k(t,A) = \frac{1}{2}\sigma_k^2(t,A), \tag{50}$$

$$F_k(A) = \max_{t \in G_k} \eta_k(t,A). \tag{51}$$

Now, let us construct the functionals

$$F(A) = \max_{k \in 1:p} F_k(A), \tag{52}$$

$$\Phi(A) = \sum_{k \in 1:p} F_k(A) \tag{53}$$

and consider the following problems:

**Problem 1.** Find a point $A^* \in \mathbb{R}^{n+1}$ such that

$$F(A^*) = \min_{A \in \mathbf{R}^{n+1}} F(A). \tag{54}$$

**Problem 2.** Find a point $A^* \in \mathbb{R}^{n+1}$ such that

$$\Phi(A^*) = \min_{A \in \mathbf{R}^{n+1}} \Phi(A). \tag{55}$$

The problem (54) is equivalent to the problem of minimizing the functional

$$\overline{F}(A) = \max_{k \in 1:p} \max_{t \in G_k} |P(t,A) - f_k(t)|.$$

In the following two subsections, we discuss Problems 1 and 2 in the case $p = 2$.

## *6.2 Optimality Conditions for Problem 1*

Put

$$\sigma_k(t,A) = P(t,A) - f_k(t), \quad \eta_k(t,A) = \frac{1}{2}\sigma_k^2(t,A) \quad \forall k \in 1:2, \tag{56}$$

$$F_1(A) = \max_{t \in G_1} \eta_1(t,A), \quad F_2(A) = \max_{t \in G_2} \eta_2(t,A), \tag{57}$$

$$F(A) = \max\{F_1(A),\ F_2(A)\}. \tag{58}$$

It is required to find a point $A^* \in \mathbb{R}^{n+1}$ such that

$$F(A^*) = \min_{A \in \mathbf{R}^{n+1}} F(A). \tag{59}$$

If the sets $G_1$ and $G_2$ do not intersect, Problem 1 becomes the one discussed in Sect. 5.

The functions $F_1,\ F_2 : \mathbb{R}^{n+1} \longrightarrow \mathbb{R}$ are max-type functions, and their directional derivatives can be expressed in the form

$$F_1'(A,g) = \max_{t\in R_1(A)} (\eta_{1A}'(t,A),g) = \max_{V\in\underline{\partial}F_1(A)} (V,g), \tag{60}$$

$$F_2'(A,g) = \max_{t\in R_2(A)} (\eta_{2A}'(t,A),g) = \max_{V\in\underline{\partial}F_2(A)} (V,g), \tag{61}$$

where

$$\underline{\partial}F_1(A) = \mathrm{co}\{\sigma_1(t,A)B(t) \mid t\in R_1(A)\}, \tag{62}$$

$$\underline{\partial}F_2(A) = \mathrm{co}\{\sigma_2(t,A)B(t) \mid t\in R_2(A)\}, \tag{63}$$

$$R_1(A) = \{t\in G_1 \mid \eta_1(t,A) = F_1(A)\},$$

$$R_2(A) = \{t\in G_2 \mid \eta_2(t,A) = F_2 A)\}, \quad g\in\mathbb{R}^{n+1}.$$

The function $F : \mathbb{R}^{n+1} \longrightarrow \mathbb{R}$ is also a max-type function, and therefore its directional derivative takes the form

$$F'(A,g) = \max_{V\in\underline{\partial}F(A)} (V,g), \tag{64}$$

$$\underline{\partial}F(A) = \begin{cases} \underline{\partial}F_1(A), & \text{if } F_1(A) > F_2(A), \\ \underline{\partial}F_2(A), & \text{if } F_1(A) < F_2(A), \\ \mathrm{co}\left\{\underline{\partial}F_1(A),\ \underline{\partial}F_2(A)\right\}, & \text{if } F_1(A) = F_2(A). \end{cases}$$

The following optimality condition for a minimum of the function $F$ holds.

**Theorem 8.** *For a point $A^*\in\mathbb{R}^{n+1}$ to be a global minimizer of $F$ on $\mathbb{R}^{n+1}$ it is necessary and sufficient that*

$$0_{n+1}\in\underline{\partial}F(A^*). \tag{65}$$

In the case $F_1(A^*)\neq F_2(A^*)$, this is just Theorem 6 where $F(A^*)$ is replaced by $F_1(A^*)$ or $F_2(A^*)$, respectively. In the case $F_1(A^*) = F_2(A^*)$, Theorem 8 implies the following result.

**Theorem 9.** *Let $F_1(A^*) = F_2(A^*)$. Then for a point $A^*\in\mathbb{R}^{n+1}$ to be a global minimizer of $F$ on $\mathbb{R}^{n+1}$ it is necessary and sufficient that*

*(1) Either there exist a point $t^*$ such that $t^*\in R_1(A^*)\bigcap R_2(A^*)$ and*

$$\sigma_1(t^*,A^*) = -\sigma_2(t^*,A^*) \tag{66}$$

*(2) Or there exist* $n+2$ *points* $\{t_i \in R_1(A^*) \bigcup R_2(A^*) \mid i \in 1:(n+2)\}$ *such that*

$$t_1 < t_2 < \cdots < t_{n+2},$$

*and at the points* $t_i \in R_1(A^*)$ *the function* $|\sigma_1(t,A^*)|$ *attains its maximal in* $t$ *value (on the set* $G_1$*), while at the points* $t_i \in R_2(A^*)$ *the function* $|\sigma_2(t,A^*)|$ *attains its maximal value (on the set* $G_2$*), and the signs of values*

$$\xi(t_i,A^*) = \begin{cases} \sigma_1(t_i,A^*), \ \text{if } t_i \in R_1(A^*), \\ \sigma_2(t_i,A^*), \ \text{if } t_i \in R_2(A^*) \end{cases}$$

*are alternating.*

Note that the maximal values of the functions $|\sigma_1(t,A^*)|$ and $|\sigma_2(t,A^*)|$ on the corresponding sets $G_1$ and $G_2$ coincide.

*Proof.* N e c e s s i t y. Denote $R(A^*) = R_1(A^*) \bigcap R_2(A^*)$. The statement is trivial if there exists a point $t^* \in R(A^*)$ at which the condition (66) holds. Therefore one can assume that either $R(A^*) = \emptyset$ or at all points $t^*$ from $R(A^*)$ the equality $\sigma_1(t^*,A^*) = \sigma_2(t^*,A^*)$ is valid. In particular, the points $\sigma_1(t^*,A^*)B(t^*)$ and $\sigma_2(t^*,A^*)B(t^*)$ coincide (they are common points of $\underline{\partial} F_1(A^*)$ and $\underline{\partial} F_2(A^*)$). In such a case, as well as in the case $R(A^*) = \emptyset$, the existence of $(n+2)$-point alternance is proved as in Sect. 5.

In the proof of s u f f i c i e n c y one should note that in the case 1) the vectors $\sigma_1(t^*,A^*)B(t^*)$ and $\sigma_2(t^*,A^*)B(t^*)$ belong to $\underline{\partial} F(A^*)$. Their half-sum, which is equal to $0_{n+1}$, also belongs to $\underline{\partial} F(A^*)$, i.e., $0_{n+1} \in \underline{\partial} F(A^*)$.

*Remark 3.* The condition (65) is sufficient due to the convexity of $F$.

### *6.3 Optimality Conditions for Problem 2*

Put

$$\sigma_k(t,A) = P(t,A) - f_k(t), \quad \eta_k(t,A) = \frac{1}{2}\sigma_k^2(t,A) \quad \forall k \in 1:2, \tag{67}$$

$$F_1(A) = \max_{t \in G_1} \eta_1(t,A), \quad F_2(A) = \max_{t \in G_2} \eta_2(t,A), \tag{68}$$

$$\Phi(A) = F_1(A) + F_2(A)\}. \tag{69}$$

It is required to find a point $A^* \in \mathbb{R}^{n+1}$ such that

$$\Phi(A^*) = \min_{A \in \mathbb{R}^{n+1}} \Phi(A). \tag{70}$$

The functions $F_1,\ F_2 : \mathbb{R}^{n+1} \longrightarrow \mathbb{R}$ are max-type functions, and their directional derivatives were expressed in the form (60) and (61). The function $\Phi : \mathbb{R}^{n+1} \longrightarrow \mathbb{R}$

is the sum of two max-type functions, and therefore its directional derivative takes the form

$$\Phi'(A,g) = \max_{V \in \underline{\partial}\Phi(A)} (V,g), \tag{71}$$

where

$$\underline{\partial}\Phi(A) = \underline{\partial}F_1(A) + \underline{\partial}F_2(A).$$

The following necessary condition for a minimum of the function $F$ holds.

**Theorem 10.** *For a point $A^* \in \mathbb{R}^{n+1}$ to be a global minimizer of $F$ on $\mathbb{R}^{n+1}$ it is necessary and sufficient that*

$$0_{n+1} \in \underline{\partial}\Phi(A^*). \tag{72}$$

The condition (72) means that there exist points $\{t_{1i} \in R_1(A^*) \mid i \in 1:r_1\}$ and coefficients

$$\{\alpha_i \mid \alpha_i > 0, \sum_{1:r_1} \alpha_i = 1\},\ r_1 \le n+2,$$

and also points $\{t_{2j} \in R_2(A^*) \mid j \in 1:r_2\}$ and coefficients

$$\{\beta_j \mid \beta_j > 0, \sum_{1:r_2} \beta_j = 1,\ r_2 \le n+2\}$$

such that

$$\sum_{i \in 1:r_1} \alpha_i \sigma_1(t_{1i},A^*)B(t_{1i}) + \sum_{j \in 1:r_2} \beta_j \sigma_2(t_{2j},A^*)B(t_{2j}) = 0_{n+1}. \tag{73}$$

Assume that $R_1(A^*) \bigcap R_2(A^*) = \emptyset$ (this is the case, e.g., if $G_1 \bigcap G_2 = \emptyset$). Then all vectors $B(t_{1i})$ and $B(t_{2j})$ in (73) are different, and any collection of them consisting of less than $n+2$ points represents a set of linearly independent vectors. Therefore the relation (73) is possible if and only if $r_1 + r_2 = n+2$.

Let

$$V_i = \begin{cases} \sigma_1(t_{1i},A^*)B(t_{1i}),\ t_{1i} \in R_1(A^*), \\ \sigma_2(t_{2j},A^*)B(t_{2j}),\ t_{2j} \in R_2(A^*) \end{cases}$$

be the vectors taking part in the representation (73). By $\Delta_i$ denote the determinants of the $(n+1)$-th order, constructed from the vectors $V_1, V_2, \ldots, V_{n+2}$ by consecutive deleting one of these vectors. It follows from (38) that

$$\sum_{i \in 1:n+2} |\Delta_i| V_i = 0_{n+1}. \tag{74}$$

The condition (72) is equivalent to the following one.

**Theorem 11.** *For a point $A^* \in \mathbb{R}^{n+1}$ to be a global minimizer of $\Phi$ on $\mathbb{R}^{n+1}$ it is necessary and sufficient that*

*(1) There exist $n+2$ points $\{t_i \in R_1(A^*) \bigcup R_2(A^*) \mid i \in 1:(n+2)\}$ such that*

$$t_1 < t_2 < \cdots < t_{n+2},$$

*at the points $t_i \in R_1(A^*)$ the function $|\sigma_1(t,A^*)|$ attains its maximal value (on the set $G_1$), while at the points $t_i \in R_2(A^*)$ the function $|\sigma_2(t,A^*)|$ attains its maximal value (on the set $G_2$), and the signs of values*

$$\xi(t_i,A^*) = \begin{cases} \sigma_1(t_i,A^*), \text{ if } t_i \in R_1(A^*), \\ \sigma_2(t_i,A^*), \text{ if } t_i \in R_2(A^*) \end{cases}$$

*are alternating.*

*(2) The equality*

$$\sum_{t_i \in R_1(A^*)} |\Delta_i| = \sum_{t_i \in R_2(A^*)} |\Delta_i| \tag{75}$$

*holds.*

*Proof.* The proof of property (1) is similar to that of Theorem 7. Property (2) follows from the relations (73) and (74).

*Remark 4.* The condition (72) is sufficient due to the convexity of the function $\Phi$.

*Remark 5.* Observe that to check whether a point $A^*$ is a minimizer in Problem 1, it is sufficient to find $n+2$ alternating points. For the Problem 2, one has to find an $(n+2)$-alternance and, in addition, to check condition (75).

**Acknowledgements** The authors are thankful to two unknown Referees whose remarks and suggestions were very fruitful and useful. The work was supported by the Russian Foundation for Basic Research (RFFI) under Grant No 12-01-00752.

## References

[1] Abbasov, M.E., Demyanov, V.F.: Extremum Conditions for a Nonsmooth Function in Terms of Exhausters and Coexhausters, (In Russian) Trudy Instituta Matematiki i Mekhaniki UrO RAN, 2009, Vol. 15, No. 4. Institute of Mathematics and Mechanics Press, Ekaterinburg. English translation: Proceedings of the Steklov Institute of Mathematics, 2010, Suppl. 2, pp. S1–S10. Pleiades Publishing, Ltd. (2010)

[2] Castellani, M.: A dual representation for proper positively homogeneous functions, J. Global Optim. **16** (4), 393–400 (2000)

[3] Daugavet, V.A., Malozemov, V.N.: Nonlinear approximation problems. In: Moiseev, N.N. (ed.) The State-of-the-Art of Operations Research Theory. (in Russian) pp. 336–363, Moscow, Nauka Publishers (1979)

[4] Demyanov, V.F.: Exhausters and convexificators—new tools in nonsmooth analysis. In: Demyanov, V.F., Rubinov, A.M. (eds.) Quasidifferentiability and Related Topics, pp. 85–137. Kluwer, Dordrecht (2000)
[5] Demyanov, V.F., Malozemov, V.N.: Introduction to Minimax, p. 368. Nauka, Moscow (1972) [English translation: John Wiley and Sons, 1974. vii+307 p. Second Edition by Dover, New York (1990)]
[6] Demyanov, V.F., Roshchina, V.A.: Exhausters and subdifferentials in non-smooth analysis. Optimization **57**(1), 41–56 (2008)
[7] Demyanov, V.F., Rubinov, A.M.: Quasidifferential Calculus. Springer—Optimization Software, New York (1986)
[8] Demyanov, V.F., Rubinov, A.M.: Constructive Nonsmooth Analysis. Peter Lang Verlag, Frankfurt a/M (1995)
[9] Demyanov, V.F., Rubinov, A.M.: Exhaustive families of approximations revisited. In: Gilbert, R.P., Panagiotopoulos, P.D., Pardalos, P.M. (eds.) From Convexity to Nonconvexity, pp. 43–50. Kluwer, Dordrecht (2001)
[10] Demyanov, V. F., Vasilev, L. V.: Nondifferentiable Optimization, Translated from the Russian by Tetsushi Sasagawa. Translation Series in Mathematics and Engineering, pp. xvii+452. Optimization Software, Inc., Publications Division, New York (1985)
[11] Hiriart-Urruty, J.-B., Lemarechal, C.: Convex Analysis and Minimization Algorithms, Parts I and II. Springer, Berlin (1993)
[12] Ioffe, A.D., Tikhomirov, V.M.: Theory of Extremal Problems, p. 480.Nauka Publishers, Moscow (1974). [English transl. by North-Holland, 1979]
[13] Polyak, B.T.: Introduction to Optimization, (In Russian) p. 384. Nauka Publishers, Moscow (1983)
[14] Pschenichnyi, B.N.: Convex Analysis and Extremal Problems. Nauka, Moscow (1980)
[15] Rockafellar, R.T.: Convex Analysis. p. xviii+451. Princeton University Press, Princeton (1970)
[16] Shor N.Z.: Methods for Minimizing Nondifferentiable Functions and Their Applications. p. 200. Naukova dumka, Kiev (1979)
[17] Uderzo, A.: Convex approximators, convexificators and exhausters: applications to constrained extremum problems. In: Demyanov, V.F., Rubinov, A.M. (eds.) Quasidifferentiability and Related Topics, pp. 297–327. Kluwer, Dordrecht (2000)

# Optimal Multiple Decision Statistical Procedure for Inverse Covariance Matrix

**Alexander P. Koldanov and Petr A. Koldanov**

**Abstract** A multiple decision statistical problem for the elements of inverse covariance matrix is considered. Associated optimal unbiased multiple decision statistical procedure is given. This procedure is constructed using the Lehmann theory of multiple decision statistical procedures and the conditional tests of the Neyman structure. The equations for thresholds calculation for the tests of the Neyman structure are analyzed.

**Keywords** Inverse covariance matrix • Tests of the Neyman structure • Multiple decision statistical procedure • Generating hypothesis

## 1 Introduction

A market network is constructed by means of some similarity measure between every pairs of stocks . The most popular measure of similarity between stocks of a market is the correlation between them [1–4]. The analysis of methods of market graph construction [2] from the statistical point of view was started in [5]. In [5] multiple decision statistical procedure for market graph construction based on the Pearson test is suggested. The authors of [5] note that a procedure of this type can be made optimal in the class of unbiased multiple decision statistical procedures if one uses the tests of the Neyman structure for generating hypothesis. In the present paper we use the partial correlations as a measure of similarity between stocks. In this case the elements of inverse covariance matrix are the weights of links between stocks in the market network [6].

A.P. Koldanov • P.A. Koldanov (✉)
National Research University, Higher School of Economics,
Bolshaya Pecherskaya 25/12, Nizhny Novgorod, 603155, Russia
e-mail: akoldanov@hse.ru; pkoldanov@hse.ru

V.F. Demyanov et al. (eds.), *Constructive Nonsmooth Analysis and Related Topics*,
Springer Optimization and Its Applications 87, DOI 10.1007/978-1-4614-8615-2_13,

The main goal of the paper is investigation of the problem of identification of inverse covariance matrix as a multiple decision statistical problem. As a result an optimal unbiased multiple decision statistical procedure for identification of inverse covariance matrix is given. This procedure is constructed using the Lehmann theory of multiple decision statistical procedures and the conditional tests of the Neyman structure. In addition the equations for thresholds calculation for the tests of the Neyman structure are analyzed.

The paper is organized as follows. In Sect. 2 we briefly recall the Lehmann theory of multiple decision statistical procedures. In Sect. 3 we describe the tests of the Neyman structure. In Sect. 4 we formulate the multiple decision problem for identification of inverse covariance matrix. In Sect. 5 we construct and study the optimal tests for testing of generating hypothesis for the elements of an inverse covariance matrix. In Sect. 6 we construct the multiple statistical procedure and consider some particular cases. In Sect. 7 we summarize the main results of the paper.

## 2 Lehmann Multiple Decision Theory

In this section we recall for the sake of completeness the basic idea of the Lehmann theory following the paper [5].

Suppose that the distribution of a random vector $R$ is taken from the family $f(r,\theta):\theta\in\Omega$, where $\theta$ is a parameter, $\Omega$ is the parametric space, $r$ is an observation of $R$. We need to construct a statistical procedure for the selection of one from the set of $L$ hypotheses, which in the general case can be stated as:

$$\begin{aligned} &H_i:\theta\in\Omega_i, i=1,\dots,L,\\ &\Omega_i\cap\Omega_j=\emptyset,\ i\neq j,\ \textstyle\bigcup_{i=1}^{L}\Omega_i=\Omega. \end{aligned} \tag{1}$$

The most general theory of a multiple decision procedure is the Lehmann theory [7]. The Lehmann theory is based on three concepts: generating hypothesis, generating hypothesis testing and compatibility conditions, and additivity condition for the loss function.

The multiple decision problem (1) of selecting one from the set of $L$ hypotheses$H_i:\theta\in\Omega_i; i=1,\dots,L$ is equivalent to a family of M two decision problems:

$$H'_j:\theta\in\omega_j \,\text{vs}\, K'_j:\theta\in\omega_j^{-1}, j=1,\dots,M, \tag{2}$$

with

$$\bigcup_{i=1}^{L}\Omega_i=\omega_j\cup\omega_j^{-1}=\Omega.$$

This equivalence is given by the relations:

$$\Omega_i = \bigcap_{j=1}^{M} \omega_j^{\chi_{i,j}}, \quad \omega_j = \bigcup_{\{i:\chi_{i,j}=1\}} \Omega_i,$$

where

$$\chi_{i,j} = \begin{cases} 1, & \Omega_i \cap \omega_j \neq \emptyset, \\ -1, & \Omega_i \cap \omega_j = \emptyset. \end{cases} \tag{3}$$

Hypotheses $H'_j : (j = 1,\ldots,M)$ are called generating hypotheses for the problem (1).

The equivalence between problem (1) and the family of problems (2) reduces the multiple decision problem to the testing of generating hypothesis. Any statistical procedure $\delta_j$ for hypothesis testing of $H'_j$ can be written in the following form

$$\delta_j(r) = \begin{cases} \partial_j, & r \in X_j, \\ \partial_j^{-1}, & r \in X_j^{-1}, \end{cases} \tag{4}$$

where $\partial_j$ is the decision of acceptance of $H'_j$ and $\partial_j^{-1}$ is the decision of acceptance of $K'_j$, $X_j$ is the acceptance region of $H'_j$ and $X_j^{-1}$ is the acceptance region of $K'_j$ (rejection region of $H'_j$) in the sample space. One has $X_j \cap X_j^{-1} = \emptyset$, $X_j \cup X_j^{-1} = X$, $X$ being the sample space.

Define the acceptance region for $H_i$ by

$$D_i = \bigcap_{j=1}^{M} X_j^{\chi_{i,j}}, \tag{5}$$

where $\chi_{i,j}$ are defined by (3), and put $X_j^1 = X_j$. Note that $\bigcup_{i=1}^{L} D_i \subset X$, but it is possible that $\bigcup_{i=1}^{L} D_i \neq X$.

Therefore, if $D_1, D_2, \ldots, D_L$ is a partition of the sample space $X$, then one can define the statistical procedure $\delta(r)$ by

$$\delta(r) = \begin{cases} d_1, & r \in D_1, \\ d_2, & r \in D_2, \\ \ldots & \ldots \ldots \\ d_L, & r \in D_L. \end{cases} \tag{6}$$

According to [8,9] we define the conditional risk for multiple decision statistical procedure by

$$risk(\theta, \delta) = E_\theta w(\theta, \delta(R)) = \sum_{k=1}^{L} w(\theta, d_k) P_\theta(\delta(R) = d_k), \tag{7}$$

where $E_\theta$ is the expectation according to the density function $f(x,\theta)$ and $w(\theta,d_k)$ is the loss from decision $d_k$ under the condition that $\theta$ is true, $\theta \in \Omega$. Under the additivity condition of the loss function (see [5] for more details) the conditional risk can be written as

$$risk(\theta,\delta) = \sum_{j=1}^{M} risk(\theta,\delta_j), \quad \theta \in \Omega. \tag{8}$$

We call statistical procedure optimal in a class of statistical procedures if it has minimal conditional risk for all $\theta \in \Omega$ in this class. The main result of the Lehmann theory (see [7]) states that if Eq. (8) is satisfied and statistical procedures (4) are optimal in the class of unbiased statistical tests, then the associated multiple decision statistical procedure (6) is optimal in the class of unbiased multiple decision procedures.

## 3 Unbiasedness and Tests of the Neyman Structure

The class of unbiased multiple decision statistical procedures according to Lehmann [9, 10] is defined by:

$$E_\theta w(\theta,\delta(R)) \leq E_\theta w(\theta',\delta(R)) \, for\, any\, \theta,\theta' \in \Omega.$$

Let $f(r;\theta)$ be the density of the exponential family:

$$f(r;\theta) = c(\theta)exp(\sum_{j=1}^{M} \theta_j T_j(r))h(r), \tag{9}$$

where $c(\theta)$ is a function defined in the parameters space, $h(r)$, $T_j(r)$ are functions defined in the sample space, and $T_j(R)$ are the sufficient statistics for $\theta_j$, $j = 1,\ldots,M$.

Suppose that generating hypotheses (2) has the form:

$$H'_j : \theta_j = \theta_j^0 \text{ vs } K'_j : \theta_j \neq \theta_j^0, \ \ j = 1,2,\ldots,M, \tag{10}$$

where $\theta_j^0$ are fixed. For a fixed $j$ the parameter $\theta_j$ is called information or structural parameter and $\theta_k, k \neq j$ are called nuisance parameters. According to [9] the optimal unbiased tests for generating hypotheses (10) are:

$$\delta_j = \begin{cases} \partial_j, & c_j^1(t_1,\ldots,t_{j-1},t_{j+1},\ldots,t_M) < t_j < c_j^2(t_1,\ldots,t_{j-1},t_{j+1},\ldots,t_M), \\ \partial_j^{-1}, & \text{otherwise}, \end{cases} \tag{11}$$

where $t_i = T_i(r), i = 1, \ldots, M$ and constants $c_j^1(t_1, \ldots, t_{j-1}, t_{j+1}, \ldots, t_M)$, $c_j^2(t_1, \ldots, t_{j-1}, t_{j+1}, \ldots, t_M)$ are defined from the equations

$$\int_{c_j^1}^{c_j^2} f(t_j, \theta_j^0 | T_i = t_i, i = 1, \ldots, M; i \neq j) dt_j = 1 - \alpha_j, \tag{12}$$

and

$$\begin{aligned} &\int_{-\infty}^{c_j^1} t_j f(t_j, \theta_j^0 | T_i = t_i, i = 1, \ldots, M; i \neq j) dt_j \\ &+ \int_{c_j^2}^{+\infty} t_j f(t_j, \theta_j^0 | T_i = t_i, i = 1, \ldots, M; i \neq j) dt_j \\ &= \alpha_j \int_{-\infty}^{+\infty} t_j f(t_j, \theta_j^0 | T_i = t_i, i = 1, \ldots, M; i \neq j) dt_j, \end{aligned} \tag{13}$$

where $f(t_j, \theta_j^0 | T_i = t_i, i = 1, \ldots, M; i \neq j)$ is the density of conditional distribution of statistics $T_j$ and $\alpha_j$ is the level of significance of the test.

A test satisfying (12) is said to have Neyman structure. This test is characterized by the fact that the conditional probability of rejection of $H_j'$ (under the assumption that $H_j'$ is true) is equal to $\alpha_j$ on each of the surfaces $\bigcap_{k \neq j} (T_k(x) = t_k)$. Therefore the multiple decision statistical procedure associated with the tests of the Neyman structure (11), (12), and (13) is optimal in the class of unbiased multiple decision procedures.

## 4 Problem of Identification of Inverse Covariance Matrix

In this section we formulate the multiple decision problem for elements of inverse covariance matrix.

Let N be the number of stocks on a financial market, and let n be the number of observations. Denote by $r_i(t)$ the daily return of the stock $i$ for the day $t$ ($i = 1, \ldots, N; t = 1, \ldots, n$). We suppose $r_i(t)$ to be an observation of the random variable $R_i(t)$. We use the standard assumptions: the random variables $R_i(t)$, $t = 1, \ldots, n$ are independent and have all the same distribution as a random variable $R_i (i = 1, \ldots, N)$. The random vector $(R_1, R_2, \ldots, R_N)$ describes the joint behavior of the stocks. We assume that the vector $(R_1, R_2, \ldots, R_N)$ has a multivariate normal distribution with covariance matrix $\|\sigma_{ij}\|$ where $\sigma_{ij} = cov(R_i, R_j) = E(R_i - E(R_i))(R_j - E(R_j)), \rho_{ij} = (\sigma_{ij})/(\sqrt{\sigma_{ii}\sigma_{jj}}), i, j = 1, \ldots, N$, $E(R_i)$ is the expectation of the random variable $R_i$. We define a sample space as $R^{N \times n}$ with the elements $(r_i(t))$. Statistical estimation of $\sigma_{ij}$ is $s_{ij} = \Sigma_{t=1}^n (r_i(t) - \overline{r_i})(r_j(t) - \overline{r_j})$ where $\overline{r_i} = (1/n) \sum_{t=1}^n r_i(t)$. The sample correlation between the stocks $i$ and $j$ is defined by $r_{ij} = (s_{ij})/(\sqrt{s_{ii}s_{jj}})$.

It is known [11] that for a multivariate normal vector the statistics $(\overline{r_1}, \overline{r_2}, \ldots, \overline{r_N})$ and $\|s_{ij}\|$ (matrix of sample covariances) are sufficient.

Let $\sigma^{ij}$ be the elements of inverse covariance matrix $\|\sigma^{ij}\|$. Then the problem of identification of inverse covariance matrix can be formulated as a multiple decision problem of the selection of one from the set of hypotheses:

$$
\begin{aligned}
&H_1 : \sigma^{ij} = \sigma_0^{ij},\ i,j = 1,\ldots,N,\ i<j,\\
&H_2 : \sigma^{12} \neq \sigma_0^{12}, \sigma^{ij} = \sigma_0^{ij}, i,j = 1,\ldots,N, (i,j) \neq (1,2),\ i<j,\\
&H_3 : \sigma^{13} \neq \sigma_0^{13}, \sigma^{ij} = \sigma_0^{ij}, i,j = 1,\ldots,N, (i,j) \neq (1,3),\ i<j,\\
&H_4 : \sigma^{12} \neq \sigma_0^{12}, \sigma^{13} \neq \sigma_0^{13}, \sigma^{ij} = \sigma_0^{ij}, i,j = 1,\ldots,N, (i,j) \neq (1,2),\ (i,j) \neq (1,3),\\
&\ldots\\
&H_L : \sigma^{ij} \neq \sigma_0^{ij}, i,j = 1,\ldots,N,\ i<j,
\end{aligned}
\tag{14}
$$

where $L = 2^M$ with $M = N(N-1)/2$.

Multiple decision problem (14) is a particular case of the problem (1). The parameter space $\Omega$ is the space of positive semi-definite matrices $||\sigma^{ij}||$, $\Omega_k$ is a domain in the parameters space associated with the hypothesis $H_k$ from the set (14) $k = 1,\ldots,L$. For the multiple decision problem (14) we introduce the following set of generating hypotheses:

$$h_{i,j} : \sigma^{ij} = \sigma_0^{ij} \text{ vs } k_{i,j} : \sigma^{ij} \neq \sigma_0^{ij},\ i,\ j = 1,2,\ldots,N,\ i<j, \tag{15}$$

We use the following notations: $\partial_{i,j}$ is the decision of acceptance of the hypothesis $h_{i,j}$ and $\partial_{i,j}^{-1}$ is the decision of rejection of $h_{i,j}$.

## 5 Tests of the Neyman Structure for Testing of Generating Hypothesis

Now we construct the optimal test in the class of unbiased tests for generating hypothesis (15). To construct these tests we use the sufficient statistics $s_{ij}$ with the following Wishart density function [11]:

$$f(\{s_{k,l}\}) = \frac{[\det(\sigma^{kl})]^{n/2} \times [\det(s_{kl})]^{(n-N-2)/2} \times \exp[-(1/2)\sum_k \sum_l s_{k,l}\sigma^{kl}]}{2^{(Nn/2)} \times \pi^{N(N-1)/4} \times \Gamma(n/2)\Gamma((n-1)/2)\cdots\Gamma((n-N+1)/2)}$$

if the matrix $(s_{kl})$ is positive definite, and $f(\{s_{kl}\}) = 0$ otherwise. One has for a fixed $i<j$:

$$f(\{s_{kl}\}) = C(\{\sigma^{kl}\}) \times \exp[-\sigma^{ij}s_{ij} - \frac{1}{2} \sum_{(k,l)\neq(i,j);(k,l)\neq(j,i)} s_{kl}\sigma^{kl}] \times h(\{s_{kl}\}),$$

where

$$C(\{\sigma^{kl}\}) = \frac{1}{q}\det(\sigma^{kl})]^{n/2},$$

$$q = 2^{(Nn/2)} \times \pi^{N(N-1)/4} \times \Gamma(n/2)\Gamma((n-1)/2)\cdots\Gamma((n-N+1)/2),$$

$$h(\{s_{kl}\}) = [\det(s_{kl})]^{(n-N-2)/2}.$$

Therefore, this distribution belongs to the class of exponential distributions with parameters $\sigma^{kl}$.

The optimal tests of the Neyman structure (11), (12), and (13) for generating hypothesis (15) take the form:

$$\delta_{i,j}(\{s_{kl}\}) = \begin{cases} \partial_{i,j}, & c^1_{i,j}(\{s_{kl}\}) < s_{ij} < c^2_{i,j}(\{s_{kl}\}), \ (k,l) \neq (i,j), \\ \partial^{-1}_{i,j}, & s_{ij} \leq c^1_{i,j}(\{s_{kl}\}) \text{ or } s_{ij} \geq c^2_{i,j}(\{s_{kl}\}), \ (k,l) \neq (i,j), \end{cases} \tag{16}$$

where the critical values are defined from the equations

$$\frac{\int_{I\cap[c^1_{i,j};c^2_{i,j}]} \exp[-\sigma_0^{ij} s_{ij}][\det(s_{kl})]^{(n-N-2)/2} ds_{ij}}{\int_I \exp[-\sigma_0^{ij} s_{ij}][\det(s_{kl})]^{(n-N-2)/2} ds_{ij}} = 1 - \alpha_{i,j}, \tag{17}$$

$$\begin{aligned} &\int_{I\cap[-\infty;c^1_{i,j}]} s_{ij}\exp[-\sigma_0^{ij} s_{ij}][\det(s_{kl})]^{(n-N-2)/2} ds_{ij} \\ &+ \int_{I\cap[c^2_{i,j};+\infty]} s_{ij}\exp[-\sigma_0^{ij} s_{ij}][\det(s_{kl})]^{(n-N-2)/2} ds_{ij} \\ &= \alpha_{i,j}\int_I s_{ij}\exp[-\sigma_0^{ij} s_{ij}][\det(s_{kl})]^{(n-N-2)/2} ds_{ij}, \end{aligned} \tag{18}$$

where $I$ is the interval of values of $s_{ij}$ such that the matrix $(s_{kl})$ is positive definite and $\alpha_{ij}$ is the level of significance of the tests.

Consider Eqs. (17) and (18). Note that $\det(s_{kl})$ is a quadratic polynomial of $s_{ij}$. Let $\det(s_{k,l}) = -C_1 s^2_{i,j} + C_2 s_{i,j} + C_3 = C_1(-s^2_{i,j} + A s_{i,j} + B)$, $C_1 > 0$ then the positive definiteness of the matrix $(s_{k,l})$ for a fixed $s_{k,l}$, $(k,l) \neq (i,j)$ gives the following interval for the value of $s_{i,j}$:

$$I = \{x : \frac{A}{2} - \sqrt{\frac{A^2}{4} + B} < x < \frac{A}{2} + \sqrt{\frac{A^2}{4} + B}\}. \tag{19}$$

Now we define the functions

$$\Psi_1(x) = \int_0^x (-t^2 + At + B)^K \exp(-\sigma_0^{ij} t) dt, \tag{20}$$

$$\Psi_2(x) = \int_0^x t(-t^2 + At + B)^K \exp(-\sigma_0^{ij} t) dt, \tag{21}$$

where $K = (n-N-2)/2$. One can calculate the critical values of $c^1_{i,j}, c^2_{i,j}$ from the equations:

$$\Psi_1(c^2_{ij}) - \Psi_1(c^1_{ij}) = (1-\alpha_{ij})(\Psi_1(\frac{A}{2}+\sqrt{\frac{A^2}{4}+B}) - \Psi_1(\frac{A}{2}-\sqrt{\frac{A^2}{4}+B})), \quad (22)$$

$$\Psi_2(c^2_{ij}) - \Psi_2(c^1_{ij}) = (1-\alpha_{ij})(\Psi_2(\frac{A}{2}+\sqrt{\frac{A^2}{4}+B}) - \Psi_2(\frac{A}{2}-\sqrt{\frac{A^2}{4}+B})). \quad (23)$$

The test (16) can be written in terms of sample correlations. First note that $\det(s_{kl}) = \det(r_{kl})s_{11}s_{22}\dots s_{NN}$ where $r_{kl}$ are the sample correlations. One has

$$\frac{\int_{J\cap[e^1_{i,j};e^2_{i,j}]}\exp[-\sigma_0^{ij}r_{ij}\sqrt{s_{ii}s_{jj}}][\det(r_{kl})]^{(n-N-2)/2}dr_{ij}}{\int_J \exp[-\sigma_0^{ij}r_{ij}\sqrt{s_{ii}s_{jj}}][\det(r_{kl})]^{(n-N-2)/2}dr_{ij}} = 1-\alpha_{i,j}, \quad (24)$$

$$\begin{aligned}&\int_{J\cap[-\infty;e^1_{i,j}]} r_{ij}\exp[-\sigma_0^{ij}r_{ij}\sqrt{s_{ii}s_{jj}}][\det(r_{kl})]^{(n-N-2)/2}dr_{ij}\\ &+\int_{J\cap[e^2_{i,j};+\infty]} r_{ij}\exp[-\sigma_0^{ij}r_{ij}\sqrt{s_{ii}s_{jj}}][\det(r_{kl})]^{(n-N-2)/2}dr_{ij}\\ &=\alpha_{i,j}\int_J r_{ij}\exp[-\sigma_0^{ij}r_{ij}\sqrt{s_{ii}s_{jj}}][\det(r_{kl})]^{(n-N-2)/2}dr_{ij},\end{aligned} \quad (25)$$

where $I = J\sqrt{s_{ii}s_{jj}}$, $c^k_{i,j} = e^k_{i,j}\sqrt{s_{ii}s_{jj}}; k=1,2$. Therefore, the tests (16) take the form

$$\delta_{i,j}(\{r_{kl}\}) = \begin{cases} \partial_{i,j}, & e^1_{i,j}(s_{ii},s_{jj},\{r_{kl}\}) < r_{ij} < e^2_{i,j}(s_{ii},s_{jj},\{r_{kl}\}),\\ & (k,l)\neq(i,j),\\ \partial^{-1}_{i,j}, & r_{ij}\le e^1_{i,j}(s_{ii},s_{jj}\{r_{kl}\}) \text{ or } r_{ij}\ge e^2_{i,j}(s_{ii},s_{jj}\{r_{kl}\}),\\ & (k,l)\neq(i,j).\end{cases} \quad (26)$$

It means that the tests of the Neyman structure for generating hypothesis (15) do not depend on $s_{k,k}$, $k\neq i$, $k\neq j$. In particular for $N=3$, $(i,j)=(1,2)$ one has

$$\delta_{1,2}(\{r_{k,l}\}) = \begin{cases} \partial_{1,2}, & e^1_{12}(s_{11},s_{22},r_{13},r_{23}) < r_{12} < e^2_{12}(s_{11},s_{22},r_{13},r_{23}),\\ \partial^{-1}_{1,2}, & r_{12}\le e^1_{12}(s_{11},s_{22},r_{13},r_{23}) \text{ or } r_{12}\ge e^2_{12}(s_{11},s_{22},r_{13},r_{23}).\end{cases} \quad (27)$$

To emphasize the peculiarity of the constructed test we consider some interesting particular cases.

$n-N-2=0$, $\sigma_0^{ij}\neq 0$. In this case expressions (20) and (21) can be simplified. Indeed one has in this case

$$\Psi_1(x) = \int_0^x \exp(-\sigma_0^{ij}t)dt = \frac{1}{\sigma_0^{ij}}(1-\exp(-\sigma_0^{ij}x)),$$

$$\Psi_2(x) = \int_0^x t\exp(-\sigma_0^{ij}t)dt = -\frac{x}{\sigma_0^{ij}}\exp(-\sigma_0^{ij}x) - \frac{1}{(\sigma_0^{ij})^2}\exp(-\sigma_0^{ij}x) + \frac{1}{(\sigma_0^{ij})^2}.$$

Finally one has the system of two equations for defining constants $c_{ij}^1, c_{ij}^2$:

$$\begin{aligned}&\exp(-\sigma_0^{ij}c_{ij}^1) - \exp(-\sigma_0^{ij}c_{ij}^2)\\ &= (1-\alpha_{ij})\{\exp(-\sigma_0^{ij}(\frac{A}{2} - \sqrt{\frac{A^2}{4}+B})) - \exp(-\sigma_0^{ij}(\frac{A}{2} + \sqrt{\frac{A^2}{4}+B}))\},\end{aligned} \quad (28)$$

$$\begin{aligned}&c_{ij}^1\exp(-\sigma_0^{ij}c_{ij}^1) - c_{ij}^2\exp(-\sigma_0^{ij}c_{ij}^2)\\ &= (1-\alpha_{ij})\{(\frac{A}{2} - \sqrt{\frac{A^2}{4}+B})\exp(-\sigma_0^{ij}(\frac{A}{2} - \sqrt{\frac{A^2}{4}+B}))\\ &-(\frac{A}{2} + \sqrt{\frac{A^2}{4}+B})\exp(-\sigma_0^{ij}(\frac{A}{2} + \sqrt{\frac{A^2}{4}+B}))\}.\end{aligned} \quad (29)$$

$\sigma_0 = 0$. In this case the critical values are defined by the system of algebraic equations (22) and (23) where the functions $\Psi_1$, $\Psi_2$ are defined by

$$\Psi_1(x) = \int_0^x (-t^2 + At + B)^K dt,$$

$$\Psi_2(x) = \int_0^x t(-t^2 + At + B)^K dt,$$

In this case the tests of the Neyman structure have the form

$$\delta_{i,j}(\{r_{k,l}\}) = \begin{cases} \partial_{i,j}, & e_{i,j}^1(\{r_{kl}\} < r_{ij} < e_{i,j}^2(\{r_{kl}\}, \ (k,l) \neq (i,j), \\ \partial_{i,j}^{-1}, & r_{ij} \leq e_{i,j}^1(\{r_{kl}\} \text{ or } r_{ij} \geq e_{i,j}^1(\{r_{kl}\}, \ (k,l) \neq (i,j). \end{cases} \quad (30)$$

$n - N - 2 = 0,\ \sigma_0 = 0$. In this case one has

$$\begin{aligned} c_{i,j}^1 &= \tfrac{A}{2} - (1-\alpha_{ij})\sqrt{\tfrac{A^2}{4}+B},\\ c_{i,j}^2 &= \tfrac{A}{2} + (1-\alpha_{ij})\sqrt{\tfrac{A^2}{4}+B}.\end{aligned} \quad (31)$$

## 6 Multiple Statistical Procedure Based on the Tests of the Neyman Structure

Now it is possible to construct the multiple decision statistical procedure for problem (14) based on the tests of Neyman structure. Then the multiple decision statistical procedure (6) takes the form:

$$\delta = \begin{cases} d_1, \; c^1_{ij}(\{s_{kl}\}) < r_{ij} < c^2_{ij}(\{s_{kl}\}), \\ \qquad i < j, i, j, k, l = 1, \dots, N, (k,l) \neq (i,j), \\ d_2, \; r_{12} \leq c^1_{12}(\{s_{kl}\}) \text{ or } r_{12} \geq c^2_{12}(\{s_{kl}\}), c^1_{ij}(\{s_{kl}\}) < r_{ij} < c^2_{ij}(\{s_{kl}\}), \\ \qquad i < j, i, j, k, l = 1, \dots, N, (k,l) \neq (i,j), \\ \dots \; \dots\dots\dots \\ d_L, \; r_{ij} \leq c^1_{ij}(\{s_{kl}\}) \text{ or } r_{ij} \geq c^2_{ij}(\{s_{kl}\}), \\ \qquad i < j, i, j, k, l = 1, \dots, N, (k,l) \neq (i,j), \end{cases} \tag{32}$$

where $c_{ij}(\{s_{kl}\})$ are defined from the Eqs. (17) and (18).

One has $D_k = \{r \in R^{N \times n} : \; \delta(r) = d_k\}$, $k = 1, 2, \dots, L$. It is clear that

$$\bigcup_{k=1}^{L} D_k = R^{N \times n}.$$

Then $D_1, D_2, \dots, D_L$ is a partition of the sample space $R^{N \times n}$. The tests of the Neyman structure for generating hypothesis (15) are optimal in the class of unbiased tests. Therefore if the condition of the additivity (8) of the loss function is satisfied, then the associated multiple decision statistical procedure is optimal. For discussion of additivity of the loss function see [5].

We illustrate statistical procedure (32) with an example.

Let $N = 3$. In this case problem (14) is the problem of the selection of one from eight hypotheses:

$$\begin{aligned} H_1 &: \sigma^{12} = \sigma_0^{12}, \; \sigma^{13} = \sigma_0^{13}, \; \sigma^{23} = \sigma_0^{23}, \\ H_2 &: \sigma^{12} \neq \sigma_0^{12}, \; \sigma^{13} = \sigma_0^{13}, \; \sigma^{23} = \sigma_0^{23}, \\ H_3 &: \sigma^{12} = \sigma_0^{12}, \; \sigma^{13} \neq \sigma_0^{13}, \; \sigma^{23} = \sigma_0^{23}, \\ H_4 &: \sigma^{12} = \sigma_0^{12}, \; \sigma^{13} = \sigma_0^{13}, \; \sigma^{23} \neq \sigma_0^{23}, \\ H_5 &: \sigma^{12} \neq \sigma_0^{12}, \; \sigma^{13} \neq \sigma_0^{13}, \; \sigma^{23} = \sigma_0^{23}, \\ H_6 &: \sigma^{12} = \sigma_0^{12}, \; \sigma^{13} \neq \sigma_0^{13}, \; \sigma^{23} \neq \sigma_0^{23}, \\ H_7 &: \sigma^{12} \neq \sigma_0^{12}, \; \sigma^{13} = \sigma_0^{13}, \; \sigma^{23} \neq \sigma_0^{23}, \\ H_8 &: \sigma^{12} \neq \sigma_0^{12}, \; \sigma^{13} \neq \sigma_0^{13}, \; \sigma^{23} \neq \sigma_0^{23}. \end{aligned} \tag{33}$$

Generating hypotheses are:

$h_{1,2} : \sigma^{12} = \sigma_0^{12}$ vs $k_{1,2} : \sigma^{12} \neq \sigma_0^{12}$, $\sigma^{13}, \sigma^{23}$ are the nuisance parameters.

$h_{1,3} : \sigma^{13} = \sigma_0^{13}$ vs $k_{1,3} : \sigma^{13} \neq \sigma_0^{13}$, $\sigma^{12}, \sigma^{23}$ are the nuisance parameters.

$h_{2,3} : \sigma^{23} = \sigma_0^{23}$ vs $k_{2,3} : \sigma^{23} \neq \sigma_0^{23}$, $\sigma^{12}, \sigma^{13}$ are the nuisance parameters.

In this case multiple statistical procedure for problem (33) (if $\sigma_0 \neq 0$) is:

$$\delta = \begin{cases} d_1, & c_{12}^1 < r_{12} < c_{12}^2, & c_{13}^1 < r_{13} < c_{13}^2, & c_{23}^1 < r_{23} < c_{23}^2, \\ d_2, & r_{12} \leq c_{12}^1 \text{ or } r_{12} \geq c_{12}^2, & c_{13}^1 < r_{13} < c_{13}^2, & c_{23}^1 < r_{23} < c_{23}^2, \\ d_3, & c_{12}^1 < r_{12} < c_{12}^2, & r_{13} \leq c_{13}^1 \text{ or } r_{13} \geq c_{13}^2, & c_{23}^1 < r_{23} < c_{23}^2, \\ d_4, & c_{12}^1 < r_{12} < c_{12}^2, & c_{13}^1 < r_{13} < c_{13}^2, & r_{23} \leq c_{23}^1 \text{ or } r_{23} \geq c_{23}^2, \\ d_5, & r_{12} \leq c_{12}^1 \text{ or } r_{12} \geq c_{12}^2, & r_{13} \leq c_{13}^1 \text{ or } r_{13} \geq c_{13}^2, & c_{23}^1 < r_{23} < c_{23}^2, \\ d_6, & c_{12}^1 < r_{12} < c_{12}^2, & r_{13} \leq c_{13}^1 \text{ or } r_{13} \geq c_{13}^2, & r_{23} \leq c_{23}^1 \text{ or } r_{23} \geq c_{23}^2, \\ d_7, & r_{12} \leq c_{12}^1 \text{ or } r_{12} \geq c_{12}^2, & c_{13}^1 < r_{13} < c_{13}^2, & r_{23} \leq c_{23}^1 \text{ or } r_{23} \geq c_{23}^2, \\ d_8, & r_{12} \leq c_{12}^1 \text{ or } r_{12} \geq c_{12}^2, & r_{13} \leq c_{13}^1 \text{ or } r_{13} \geq c_{13}^2, & r_{23} \leq c_{23}^1 \text{ or } r_{23} \geq c_{23}^2. \end{cases} \tag{34}$$

The critical values $c_{12}^k = c_{12}^k(r_{13}, r_{23}, s_{11}s_{22})$, $c_{13}^k = c_{13}^k(r_{12}, r_{23}, s_{11}s_{33})$, $c_{23}^k = c_{23}^k(r_{12}, r_{13}, s_{22}s_{33})$; $k - 1,2$ are defined from Eqs. (24) and (25). If $n = 5$; $\sigma_0^{ij} \neq 0$; $i, j = 1,2,3$, then the critical values $c_{ij}^k$; $k = 1,2$ are defined from (28) and (29).

If $\sigma_0^{ij} = 0, \forall i, j$ and $n = 5$, then tests (30) for generating hypothesis depend on the sample correlation only. Therefore the corresponding multiple statistical procedure with $L$ decisions depends only on the sample correlation too. This procedure is (34) where constants $c_{12}^k = c_{12}^k(r_{13}, r_{23}), c_{13}^k = c_{13}^k(r_{12}, r_{23}), c_{23}^k = c_{23}^k(r_{12}, r_{13}); k = 1,2$.

In this case

$$I_{1,2} = (r_{13}r_{23} - G_{1,2}; r_{13}r_{23} + G_{1,2}),$$
$$I_{1,3} = (r_{12}r_{23} - G_{1,3}; r_{12}r_{23} + G_{1,3}),$$
$$I_{2,3} = (r_{12}r_{13} - G_{2,3}; r_{12}r_{13} + G_{2,3}),$$

where

$$G_{1,2} = \sqrt{(1 - r_{13}^2)(1 - r_{23}^2)},$$
$$G_{1,3} = \sqrt{(1 - r_{12}^2)(1 - r_{23}^2)},$$
$$G_{2,3} = \sqrt{(1 - r_{12}^2)(1 - r_{13}^2)},$$

and the critical values are

$$c_{12}^1 = r_{13}r_{23} - (1 - \alpha_{12})G_{1,2},$$
$$c_{12}^2 = r_{13}r_{23} + (1 - \alpha_{12})G_{1,2},$$
$$c_{13}^1 = r_{12}r_{23} - (1 - \alpha_{13})G_{1,3},$$
$$c_{13}^2 = r_{12}r_{23} + (1 - \alpha_{13})G_{1,3},$$

$$c_{23}^1 = r_{12}r_{13} - (1 - \alpha_{23})G_{2,3},$$

$$c_{23}^2 = r_{12}r_{13} + (1 - \alpha_{23})G_{2,3}.$$

Note that in this case test (34) has a very simple form.

## 7 Concluding Remarks

Statistical problem of identification of elements of inverse covariance matrix is investigated as multiple decision problem. Solution of this problem is developed on the base of the Lehmann theory of multiple decision procedures and theory of tests of the Neyman structure. It is shown that this solution is optimal in the class of unbiased multiple decision statistical procedures. Obtained results can be applied to market network analysis with partial correlations as a measure of similarity between stocks returns.

**Acknowledgements** The authors are partly supported by National Research University, Higher School of Economics, Russian Federation Government grant, N. 11.G34.31.0057

## References

[1] Mantegna, R.N.: Hierarchical structure in financial market. Eur. Phys. J. series B **11**, 193–197 (1999)
[2] Boginsky V., Butenko S., Pardalos P.M.: On structural properties of the market graph. In: Nagurney, A. (ed.) Innovations in Financial and Economic Networks. pp. 29–45. Edward Elgar Publishing Inc., Northampton (2003)
[3] Boginsky V., Butenko S., Pardalos P.M.: Statistical analysis of financial networks. Comput. Stat. Data. Anal. **48**, 431–443 (2005)
[4] M. Tumminello, T. Aste, T. Di Matteo, R.N. Mantegna, H.A.: Tool for Filtering Information in Complex Systems. Proc. Natl. Acad. Sci. Uni. States Am. **102** 30, 10421–10426 (2005)
[5] Koldanov, A.P., Koldanov, P.A., Kalyagin, V.A., Pardalos, P.M.: Statistical Procedures for the Market Graph Construction. Comput. Stat. Data Anal. **68**, 17–29, DOI: 10.1016/j.csda.2013.06.005. (2013)
[6] Hero, A., Rajaratnam, B.: Hub discovery in partial correlation graphical models. IEEE Trans. Inform. Theor. **58**–I(9), 6064–6078 (2012)
[7] Lehmann E.L.: A theory of some multiple decision procedures 1. Ann. Math. Stat. **28**, 1–25 (1957)
[8] Wald, A.: Statistical Decision Function. Wiley, New York (1950)
[9] Lehmann E.L., Romano J.P.: Testing Statistical Hypothesis. Springer, New York (2005)
[10] Lehmann E.L.: A general concept of unbiasedness. Ann. Math. Stat. **22**, 587–597 (1951)
[11] Anderson T.W.: An Introduction to Multivariate Statistical Analysis. 3 edn. Wiley-Interscience, New York (2003)

# Conciliating Generalized Derivatives

**Jean-Paul Penot**

**Abstract** As a new illustration of the versatility of abstract subdifferentials we examine their introduction in the field of analysis of second-order generalized derivatives. We also consider some calculus rules for some of the various notions of such derivatives and we give an account of the effect of the Moreau regularization process.

**Keywords** Calculus rules • Coderivatives • Epi-derivatives • Generalized derivatives • Subdifferentials • Variational analysis

## 1 Introduction

It is likely that many mathematicians and users have been disoriented by the abundance of concepts in nonsmooth analysis. The following quotation of F.H. Clarke's review of the book [45] attests that: "In recent years, the subject has known a period of intense abstract development which has led to a rather bewildering array of competing and unclearly related theories." Even in leaving apart some important approaches such as Demyanov's quasidifferentials [7, 8], Jeyakumar and Luc's generalized Jacobians [20], Warga's derivate containers [46] and in focussing on subdifferentials, one is faced with very different constructs. It may be useful to be aware of their specific features in order to choose the concept adapted to the problem one has to solve. That is the point of view adopted in the author's book [38]. It appears there that even if the approaches are different, a number of important properties are shared by all usual subdifferentials, at least on suitable spaces.

Thus, several researchers endeavored to give a synthetic approach to generalized derivatives in proposing a list of properties that can be taken as axioms allowing

J.-P. Penot (✉)
Laboratoire Jacques-Louis Lions, Université Pierre et Marie Curie (Paris 6), Paris, France
e-mail: penot@ann.jussieu.fr

V.F. Demyanov et al. (eds.), *Constructive Nonsmooth Analysis and Related Topics*, Springer Optimization and Its Applications 87, DOI 10.1007/978-1-4614-8615-2_14,

to ignore the various constructions. Such a list may be more or less complete depending on the needs or the aims. Among the many attempts in this direction, we quote the recent papers [18, 41] as they gather the most complete lists to date; see also their references.

Among the results using a general subdifferential $\partial$, we note mean value theorems [36], characterizations of convexity [5, 6], or generalized convexity [33, 35, 37, 42], optimality conditions using Chaney type second derivatives [40]. It is the purpose of the present paper to show that a general approach using an arbitrary subdifferential can also be workable for general second-order derivatives.

It is amazing that second-order derivatives can be considered for functions that are not even differentiable. In fact, many approaches can be adopted and it is a second aim of the present paper to relate such derivatives. Since the methods are so diverse (primal, dual, using second-order expansions or not), such an aim is not obvious. In fact, several results are already known; we leave them apart. For instance, for the links between parabolic second derivatives with epi-derivatives, we refer to [13–15, 30, 39] and [45, Sect. 13.J]. We do not consider generalized hessians and generalized Jacobians of the derivative that have been the object of a strong attention during the last decades (see [29] and its references). We also leave apart second-order derivatives involving perturbations of the nominal point as in [4, 25], [45, Proposition 13.56], [47], and second-order derivatives aimed at convex functions [10–12].

## 2 Calculus of Second-Order Epi-derivatives

The (lower) *second-order epi-derivative* of $f : X \to \overline{\mathbb{R}}$ at $x \in f^{-1}(\mathbb{R})$ relative to $x^* \in X^*$ has been introduced by Rockafellar in [43]; it is given by

$$\forall u \in X \qquad f''(x,x^*,u) := \liminf_{(t,u')\to(0_+,u)} \frac{2}{t^2}(f(x+tu') - f(x) - \langle x^*, tu'\rangle).$$

We also write $f''_{x,x^*}(u) = f''(x,x^*,u)$. When the directional derivative $f'(x)$ of $f$ at $x$ exists, we write $f''_x$ instead of $f''_{x,x^*}$ with $x^* := f'(x)$. Obviously, if for all $u \in X$ one has $f''(x,x^*,u) > -\infty$, one has $x^* \in \partial_D f(x)$, the directional subdifferential of $f$ at $x$. Moreover, if $f$ is twice directionally differentiable at $x$ in the sense that there exist some $x^* := f'(x) \in X^*$ and some $D^2 f(x) \in L^2(X,X;\mathbb{R})$ such that for all $u \in X$ one has

$$f'(x)(u) = \lim_{(t,u')\to(0_+,u)} \frac{f(x+tu') - f(x)}{t},$$

$$D^2 f(x)(u,u) = \lim_{(t,u')\to(0_+,u)} \frac{f(x+tu') - f(x) - \langle x^*, tu'\rangle}{t^2/2},$$

then $f''(x,x^*,u) = D^2 f(x)(u,u)$. More generally, let us say that a continuous linear map $H : X \to X^*$ is a *directional subhessian* of $f$ at $(x,x^*) \in X \times X^*$ if there exists a function $r : \mathbb{R} \times X \to \mathbb{R}$ such that $\lim_{(t,u') \to (0_+,u)} t^{-2} r(t,u') = 0$ for all $u \in X$ and

$$\forall u' \in X,\ \forall t \in \mathbb{P} \qquad f(x+tu') \geq f(x) + \langle x^*, tu' \rangle + t^2 \langle Hu', u' \rangle + r(t,u').$$

If there exists a function $s : X \to \mathbb{R}$ such that $r(t,u) = s(tu)$ for all $(t,u) \in \mathbb{R} \times X$ and $\|x\|^{-2} s(x) \to 0$ as $x \to 0$ in $X \backslash \{0\}$ we say that $H$ is a *firm subhessian* or a Fréchet subhessian. Then one sees that $H$ is a directional subhessian of $f$ at $(x,x^*)$ if, and only if, for all $u \in X$ one has

$$f''(x,x^*,u) \geq \langle Hu, u \rangle.$$

We refer to [32] for the links between subhessians and conjugacy and to [19] for calculus rules for (limiting) subhessians.

Second-order epi-derivatives are useful to get second-order optimality conditions (see [43–45]).

The calculus rules for such derivatives are not simple (see [16, 17, 27, 31, 34, 45]). However, some estimates can be easily obtained and have some usefulness. Clearly, if $f = g + h$ and if $x^* = y^* + z^*$, one has

$$f''(x,x^*,u) \geq g''(x,y^*,u) + h''(x,z^*,u)$$

for all $u \in \mathrm{dom} g''_{x,y^*} \cap \mathrm{dom} h''_{x,z^*}$. Suppose $f := h \circ g$, where $g : X \to Y$ is twice directionally differentiable at $x$ and $h : Y \to \overline{\mathbb{R}}$ is finite at $y := g(x)$, given $y^* \in Y^*$ and $x^* := y^* \circ g'(x)$, one has

$$f''(x,x^*,u) \geq h''(y,y^*,g'(x)u) + \langle y^*, D^2 g(x)(u,u) \rangle,$$

as shows the fact that

$$w(t,u') := \frac{2}{t^2}\left(g(x+tu') - g(x) - tg'(x)u' - \frac{t^2}{2} D^2 g(x)(u,u)\right) \underset{(t,u') \to (0_+,u)}{\to} 0,$$

so that $g'(x)u' + (1/2)t(D^2 g(x)(u,u) + w(t,u')) \to g'(x)u$ as $(t,u') \to (0_+,u)$.

Let us consider the relationships of such derivatives with graphical derivatives of the subdifferential. Recall that the (outer) *graphical derivative* at $(x,y) \in \mathrm{gph}(F)$ of a multimap $F : X \rightrightarrows Y$ between two normed spaces is the multimap $DF(x,y)$ whose graph is the (weak) tangent cone at $(x,y)$ to the graph of $F$. In other terms,

$$\forall u \in X \qquad DF(x,y)(u) := \text{seq}-\text{weak}-\limsup_{(t,u') \to (0_+,u)} \Delta_t F(x,y)(u'),$$

where

$$\Delta_t F(x,y)(u') := \frac{1}{t}(F(x+tu') - y).$$

We say that $F$ has a *graphical derivative* at $(x,y)$ or that $F$ is (weakly) proto-differentiable at $(x,y)$ if the sequential weak limsup of $\Delta_t F(x,y)$ as $t \to 0_+$ coincides with $\liminf_{t\to 0_+} \Delta_t F(x,y)$.

For $f : X \to \overline{\mathbb{R}}$ finite at $x$ and $x^* \in \partial f(x)$, let us introduce the differential quotient

$$\forall u \in X \qquad \Delta_t^2 f_{x,x^*}(u) := \frac{2}{t^2}\left[f(x+tu) - f(x) - \langle x^*, tu\rangle\right].$$

We say that $f : X \to \overline{\mathbb{R}}$ finite at $x$ has a *second-order epi-derivative* at $(x,x^*)$ with $x^* \in \partial f(x)$ (in the sense of Mosco) if the sequential weak limsup as $t \to 0_+$ of the epigraph of $\Delta_t^2 f_{x,x^*}$ coincides with the liminf as $t \to 0_+$ of the epigraph of $\Delta_t^2 f_{x,x^*}$.

The function $f : X \to \overline{\mathbb{R}}$ is said to be *paraconvex* around $x$ if it is lower semicontinuous, finite at $x$ and if there exist $c > 0$, $r > 0$ such that the function $f + (c/2)\left\|\cdot\right\|^2$ is convex on the ball $B(x,r)$ with center $x$ and radius $r$.

Assertion (a) of the next proposition completes [45, Lemma 13.39]. Assertion (b) extends [45, Theorem 13.40] to the infinite dimensional case and [9] to a nonconvex case. Our assumptions are slightly different from the assumptions of [23] dealing with primal lower-nice functions, a class that is larger than the class of paraconvex functions but requires a more complex analysis. See also [3, 21–23, 28].

**Proposition 1.**

(a) *For any function $f : X \to \overline{\mathbb{R}}$ on a normed space $X$ that is finite at $x \in X$ and for all $x^* \in \partial f(x)$ one has*

$$D(\partial f)(x,x^*)(u) = \text{seq} - \text{weak} - \limsup_{(t,u')\to(0_+,u)} \partial\left(\frac{1}{2}\Delta_t^2 f_{x,x^*}\right)(u'). \tag{1}$$

(b) *If $X$ is a Hilbert space and if $f$ is paraconvex around $x$, then $\partial f$ has a graphical derivative $D(\partial f)(x,x^*)$ at $(x,x^*)$ if, and only if, $f$ has a second-order epi-derivative $f''_{x,x^*}$ at $(x,x^*)$. Then*

$$D(\partial f)(x,x^*)(u) = \frac{1}{2}\partial f''_{x,x^*}(u).$$

Let us observe that when $f$ has a second-order epi-derivative at $(x,x^*)$ that is quadratic and continuous this relation ensures that $D(\partial f)(x,x^*)$ is the linear and continuous map from $X$ into $X^*$ corresponding to this quadratic form.

*Proof.*

(a) We first observe that the calculus rules of subdifferentials ensure that

$$\partial\left(\frac{1}{2}\Delta_t^2 f_{x,x^*}\right)(\cdot) = \Delta_t(\partial f)(x,x^*)(\cdot).$$

Passing to the (sequential weak) limsup as $t \to 0_+$ yields equality on the graphs and relation (1).

(b) Changing $f$ into $g := f + (c/2)\|\cdot\|^2$ for some appropriate $c > 0$ and observing that $(c/2)\|\cdot\|^2$ is twice continuously differentiable, we reduce the proof to the case $f$ is convex. Now we note that for all $t > 0$ we have $(0,0) \in \partial(\frac{1}{2}\Delta_t^2 f_{x,x^*})$ and that $(\Delta_t^2 f_{x,x^*})(0) = 0$. Thus we can apply the equivalence of Attouch's theorem ([1, Theorem 3.66]) and get the conclusion. □

## 3 Epi-derivatives and Regularization

Let us consider the interplay between second-order epi-differentiation and the Moreau regularization (or Moreau envelope). The *Moreau envelope* ${}^r f := e_r f$ of a function $f : X \to \mathbb{R}_\infty := \mathbb{R} \cup \{\infty\}$ on a Hilbert space $X$ given by

$$(e_r f)(x) = \inf_{w \in X} \left( f(w) + \frac{1}{2r} \|x - w\|^2 \right).$$

It is well known (see [2, Proprosition 12.29], [38, Theorem 4.124] for instance) that if $f$ is paraconvex around $x \in X$ and *quadratically minorized* in the sense that for some $a \in \mathbb{R}$, $b \in \mathbb{R}$ one has $f \geq b - a\|\cdot\|^2$, then there exists some $\bar{r} > 0$ such that for all $r \in ]0, \bar{r}[$ the Moreau envelope $e_r f$ is differentiable at $x$ with gradient $\nabla(e_r f)(x) = x_r := (1/r)(x - p_r)$ where $p_r := p_r(x) \in X$ is such that $(e_r f)(x) = f(p_r) + (1/2r)\|x - p_r\|^2$. The following result has been obtained in [9, Theorem 4.3] for a lower semicontinuous, proper, convex function $f$ under the additional assumption that $f$ is twice epi-differentiable at $p_r$. The method of [9, Theorem 4.3] uses duality and in particular Attouch's theorem, whereas we take a direct approach. We denote by $J : X \to X^*$ the Riesz isomorphism characterized by $\langle J(x), w \rangle = \langle x \mid w \rangle$, where $\langle \cdot \mid \cdot \rangle$ is the scalar product of $X$.

**Proposition 2.** *Let $X$ be a Hilbert space and let $f : X \to \overline{\mathbb{R}}$ be paraconvex around $x \in X$ and quadratically minorized. Then, there exists some $\bar{r} > 0$ such that for all $r \in ]0, \bar{r}[$ the Moreau envelope $e_r f$ is differentiable at $x$ with gradient $\nabla(e_r f)(x) = x_r := (1/r)(x - p_r)$ where $p_r := p_r(x) \in X$ is such that $(e_r f)(x) = f(p_r) + (1/2r)\|x - p_r\|^2$. Moreover, the sequential weak (lower) second derivative of the Moreau envelope of $f$ is related to the Moreau envelope of the (lower) sequential weak second derivative of $f$ by the following relation in which $x_r^* := J(x_r) \in \partial f(p_r)$ is the derivative of $e_r f$ at $x$:*

$$\forall u \in X \qquad \frac{1}{2}(e_r f)''(x, x_r^*, u) = e_r \left( \frac{1}{2} f''(p_r, x_r^*, \cdot) \right)(u).$$

In particular, when $f$ is lower semicontinuous, proper, convex, and twice epi-differentiable at $p_r$ for $x_r^*$, the function $(e_r f)''(x, x_r^*, \cdot)$ is continuous (and even of class $C^1$): in such a case $(1/2)f''(p_r, x_r^*, \cdot)$ is lower semicontinuous, proper, convex, so that $e_r((1/2)f''(p_r, x_r^*, \cdot))$ is of class $C^1$. For the proof we need a result of independent interest about the interchange of minimization with a sequential weak lower epi-limit.

**Lemma 1.** *Let $M$ be a metric space, let $S \subset M$, $\overline{s} \in \mathrm{cl}(S)$ and let $Z$ be a reflexive Banach space endowed with its weak topology. Let $g : S \times Z \to \overline{\mathbb{R}}$ be coercive in its second variable, uniformly for $s \in S \cap V$, where $V$ is a neighborhood of $t$ in $M$. Then*

$$\liminf_{s(\in S)\to\overline{s}} (\inf_{z\in Z} g(s,z)) = \inf_{z\in Z} (\mathrm{seq} - \liminf_{(s,z')(\in S\times Z)\to(\overline{s},z)} g(s,z'))$$

*Proof.* The inequality

$$\liminf_{s(\in S)\to\overline{s}} (\inf_{z\in Z} g(s,z)) \leq \inf_{z\in Z} (\mathrm{seq} - \liminf_{(s,z')(\in S\times Z)\to(\overline{s},z)} g(s,z'))$$

stems from the fact that for all $\overline{z} \in Z$, $(z_n) \to \overline{z}$ for the weak topology and $(s_n) \to \overline{s}$ in $S$ one has $\inf_{z\in Z} g(s_n,z) \leq g(s_n,z_n)$ for all $n \in \mathbb{N}$, hence $\liminf_n (\inf_{z\in Z} g(s,z)) \leq \liminf_n g(s_n,z_n)$.

Let $r > \liminf_{s(\in S)\to\overline{s}} (\inf_{z\in Z} g(s,z))$. Given a sequence $(\varepsilon_n) \to 0_+$ one can find some $s_n \in B(\overline{s},\varepsilon_n) \cap S$, $z_n \in Z$ such that $r > g(s_n,z_n)$. We may suppose $B(\overline{s},\varepsilon_n) \subset V$ for all $n$. The coercivity assumption ensures that $(z_n)$ is bounded. Taking a subsequence if necessary, we may suppose $(z_n)$ has a weak limit $z$. Then we get

$$r \geq \liminf_n g(s_n,z_n) \geq \mathrm{seq} - \liminf_{(s,z')(\in S\times Z)\to(\overline{s},z)} g(s,z')$$

and $r \geq \inf_{z\in Z} (\mathrm{seq} - \liminf_{(s,z')(\in S\times Z)\to(\overline{s},z)} g(s,z'))$. □

For $(t,u) \in \mathbb{P} \times X$, assuming the classical fact that ${}^r f = e_r f$ is differentiable at $x$ with derivative $x_r^*$, we adopt the simplified notation

$$\Delta_t^2 (e_r f)_x(u) := \frac{2}{t^2}((e_r f)(x+tu) - (e_r f)(x) - \langle x_r^*, tu\rangle).$$

Then we have the following exchange property generalizing [45, Lemma 13.39].

**Lemma 2.** *With the preceding notation, one has*

$$\frac{1}{2}\Delta_t^2 (e_r f)_x = e_r \left( \frac{1}{2}\Delta_t^2 f_{p_r,x_r^*} \right).$$

*Proof.* Setting $w = p_r + tz$ with $t \in \mathbb{P}$, $z \in X$, for all $u \in X$ we have $\langle x_r^*, tu\rangle = \frac{t}{r}\langle x - p_r \mid u\rangle$, $(e_r f)(x) = f(p_r) + \frac{1}{2r}\|x - p_r\|^2$,

$$\begin{aligned}(e_r f)(x+tu) &= \inf_{w\in X}\left( f(w) + \frac{1}{2r}\|x+tu-w\|^2 \right) \\ &= \inf_{z\in X}\left( f(p_r+tz) + \frac{1}{2r}\|x - p_r + t(u-z)\|^2 \right),\end{aligned}$$

$$\frac{1}{2r}\|x-p_r+t(u-z)\|^2-\frac{1}{2r}\|x-p_r\|^2-\frac{t}{r}\langle x-p_r \mid u\rangle$$
$$=-\frac{t}{r}\langle x-p_r \mid z\rangle+\frac{t^2}{2r}\|u-z\|^2$$

hence

$$\begin{aligned}
&\frac{1}{2}\Delta_t^2(e_rf)_x(u)\\
&=\frac{1}{t^2}\inf_{w\in X}\left[f(w)-f(p_r)+\frac{1}{2r}\|x+tu-w\|^2-\frac{1}{2r}\|x-p_r\|^2-\langle x_r^*,tu\rangle\right]\\
&=\frac{1}{t^2}\inf_{z\in X}\left[f(p_r+tz)-f(p_r)-\frac{t}{r}\langle x-p_r \mid z\rangle+\frac{t^2}{2r}\|u-z\|^2\right]\\
&=\inf_{z\in X}\left[\frac{1}{t^2}(f(p_r+tz)-f(p_r)-\langle x_r^*,tz\rangle)+\frac{1}{2r}\|u-z\|^2\right]\\
&=e_r\left(\frac{1}{2}\Delta_t^2 f_{p_r,x_r^*}\right)(u).
\end{aligned}$$

*Proof of Proposition 2.* The first assertion is deduced from the convex case by considering $f+c\|\cdot\|^2$ for some appropriate $c>0$ (see [2], [45, Theorem 2.26]). For the second one, in Lemma 1 let us set $Z:=X$, $S:=\mathbb{R}\times X$, $M:=\mathbb{P}\times X$, $s:=(t,u')$, $\overline{s}:=(0,u)$, $g:=h+k$ with

$$h(s,z):=\Delta_t^2 f(p_r,x_r^*,z):=(1/t^2)(f(p_r+tz)-f(p_r)-\langle x_r^*,tz\rangle),$$
$$k(s,z):=(1/2r)\left\|u'-z\right\|^2.$$

It remains to show that for all $z\in Z$ one has

$$\text{seq}-\liminf_{(s,z')(\in S\times Z)\to(\overline{s},z)} g(s,z')=\frac{1}{2}f''(p_r,x_r^*,z)+\frac{1}{2r}\|u-z\|^2. \tag{2}$$

For every sequences $(t_n)\to 0_+$, $(u_n)\to u$, $(z_n)\to z$ in the weak topology of $Z=X$, one has

$$\begin{aligned}
\liminf_n g((t_n,u_n),z_n)&\geq \liminf_n h((t_n,u_n),z_n)+\liminf_n k((t_n,u_n),z_n)\\
&\geq \frac{1}{2}f''(p_r,x_r^*,z)+\frac{1}{2r}\|u-z\|^2.
\end{aligned}$$

Let $(t_n)\to 0_+$, $(z_n)\to z$ (in the weak topology) be such that $(\Delta_{t_n}^2 f(p_r,x_r^*,z_n))\to \frac{1}{2}f''(p_r,x_r^*,z)$. Setting $u_n:=u+z_n-z$, we note that $(u_n)\to u$ in the weak topology and $\|u_n-z_n\|^2=\|u-z\|^2$ for all $n$, so that

$$\frac{1}{2}f''(p_r,x_r^*,z)+\frac{1}{2r}\|u-z\|^2 = \lim_n h((t_n,u_n),z_n)+\lim_n k((t_n,u_n),z_n)$$
$$= \lim_n g((t_n,u_n),z_n) \geq \text{seq}-\liminf_{(s,z')(\in S\times Z)\to(\overline{s},z)} g(s,z').$$

Since the reverse inequality stems from the preceding estimate, we get relation (2). □

## 4 Second-Order Derivatives via Coderivatives

Let us recall that the *coderivative* of a multimap (or set-valued map) $F : U \rightrightarrows V$ between two normed spaces at $(\overline{u},\overline{v}) \in F$ (identified with its graph $\mathrm{gph}(F)$) is the multimap $D^*F(\overline{u},\overline{v}) : V^* \rightrightarrows U^*$ defined by

$$D^*F(\overline{u},\overline{v})(v^*) := \{u^* \in U^* : (u^*,-v^*) \in N(F,(\overline{u},\overline{v}))\}.$$

If $\varphi : U \to V$ is a map of class $C^1$ between two normed spaces, the coderivative of $\varphi$ at $(\overline{u},\varphi(\overline{u}))$, denoted by $D^*\varphi(\overline{u})$ rather than $D^*\varphi(\overline{u},\varphi(\overline{u}))$, is

$$D^*\varphi(\overline{u}) = \varphi'(\overline{u})^T : V^* \to U^*,$$

a single-valued (linear) map rather than a multimap. When $\varphi = f'$ is the derivative of a function $f : U \to \mathbb{R}$ (so that $V = U^*$), denoting by $D^2f : U \to L(U,U^*)$ the derivative of $\varphi := f'$, we obtain that $D^*f'(\overline{u}) = D^2f(\overline{u})^T$ maps $U^{**}$ into $U^*$ and

$$D^*f'(\overline{u})(u^{**}) = u^{**} \circ \varphi'(\overline{u}) = u^{**} \circ D^2f(\overline{u}) \in U^*.$$

Denoting by $A \mapsto A^b$ the isomorphism from $L(U,U^*)$ onto the space $L^2(U,U;\mathbb{R})$ of continuous bilinear forms on $U$, we see that the restriction of $\partial^2 f(\overline{u}) := D^*f'(\overline{u})$ to $U$ considered as a subspace of $U^{**}$ is $D^2f(\overline{u})^b$ :

$$\forall u \in U \qquad \partial^2 f(\overline{u})(u) := D^*f'(\overline{u})(u) = D^2f(\overline{u})^b(u,\cdot) \in U^*.$$

Calculus rules for coderivatives are presented in [24, 26]. In [27] calculus rules for coderivatives of subdifferentials are given in the finite dimensional case. Here we look for extensions to the infinite dimensional case.

Since the coderivative of a multimap is defined through the normal cone to its graph, it is natural to expect that calculus rules for second-order derivatives of functions depend on calculus rules for normal cones under images and inverse images. Let us recall such rules (see [38, Proposition 2.108, Theorem 2.111]).

**Proposition 3.** *Let $U$, $V$, $W$ be Banach spaces, and let $j : U \to V$, $k : U \to W$ be maps of class $C^1$, $E \subset U$, $H \subset W$, $\overline{u} \in E$, $\overline{v} := j(\overline{u}) \in V$, $\overline{w} := k(\overline{u}) \in W$.*

*If* $E := k^{-1}(H)$, *and if* $k'(\overline{u})(U) = W$, *then* $N(E,\overline{u}) = k'(\overline{u})^T(N(H,\overline{w}))$.
*If* $F := j(E)$, *then* $N(F,\overline{v}) \subset (j'(\overline{u})^T)^{-1}(N(E,\overline{u}))$.
*If* $F := j(k^{-1}(H))$, *and if* $k'(\overline{u})(U) = W$, *then, for all* $v^* \in N(F,\overline{v})$, *there exists a unique* $w^* \in N(H,\overline{w})$ *such that* $j'(\overline{u})^T(v^*) = k'(\overline{u})^T(w^*)$.

For the second assertion it is not necessary to suppose $U$ and $V$ are complete. Moreover, the differentiability assumptions on $j$ and $k$ can be adapted to the specific subdifferentials that are used. But since we wish to adopt a general approach, we ignore such refinements.

Let us consider a composite function $f := h \circ g$, where $X$, $Y$ are Banach spaces, $g : X_0 \to Y$ is of class $C^2$ on some open subset $X_0$ of $X$ whose derivative at $x$ around $\overline{x} \in X_0$ is surjective and $h : Y \to \overline{\mathbb{R}}$ is lower semicontinuous and finite around $g(\overline{x})$. For all usual subdifferentials one has the formula

$$\partial f(x) = g'(x)^T(\partial h(g(x))),$$

where $g'(x)^T$ is the transpose of the derivative $g'(x)$ of $g$ at $x$.

In order to compute $\partial^2 f(\overline{x},\overline{x}^*) := D^*\partial f(\overline{x},\overline{x}^*)$, where $\overline{x}^* \in \partial f(\overline{x})$, $\overline{x}^* = g'(\overline{x})^T(\overline{y}^*)$ with $\overline{y}^* \in \partial h(g(\overline{x}))$, let us denote by $F$ (resp. $H$) the graph of $\partial f$ (resp. $\partial h$) and set

$$E := \{(x,y^*) \in X \times Y^* : \ (g(x),y^*) \in H\},$$
$$j := I_X \times g'(x)^T : (x,y^*) \mapsto (x, y^* \circ g'(x)),$$
$$k := g \times I_{Y^*} : (x,y^*) \mapsto (g(x),y^*).$$

Then, for $U := X \times Y^*$, $V := X \times X^*$, $W := Y \times Y^*$, one has

$$E = k^{-1}(H), \qquad F = j(E).$$

**Proposition 4.** *With the preceding assumptions, one has* $x^* \in \partial^2 f(\overline{x},\overline{x}^*)(x^{**})$ *if, and only if, for* $y^{**} = g'(\overline{x})^{TT}(x^{**})$ *and some* $y^* \in \partial^2 h(g(\overline{x}),\overline{y}^*)(y^{**})$, *one has*

$$x^* = g'(\overline{x})^T(y^*) + \langle x^{**}, \overline{y}^* \circ (D^2 g(\overline{x})(\cdot))\rangle. \tag{3}$$

Note that for all $u \in X$ one has $D^2 g(\overline{x})(u) \in L(X,Y)$ and $\overline{y}^* \circ (D^2 g(\overline{x})(u)) \in X^*$. If $x^{**}$ is the image of some $x \in X$ through the canonical injection of $X$ into $X^{**}$, then for $y := g'(\overline{x})(x)$, $y^* \in \partial^2 h(g(\overline{x}),\overline{y}^*)(y)$, one gets

$$x^* = y^* \circ g'(\overline{x}) + \overline{y}^* \circ (D^2 g(\overline{x})(x)),$$

a formula akin the classical formula of the twice differentiable case.

*Proof.* The derivatives of $j$ and $k$ at $\overline{u} := (\overline{x},\overline{y}^*)$ are given by

$$j'(\overline{x},\overline{y}^*)(x,y^*) = (x, \overline{y}^* \circ (D^2 g(\overline{x})(x)) + y^* \circ g'(\overline{x})),$$
$$k'(\overline{x},\overline{y}^*)(x,y^*) = (g'(\overline{x})x, y^*),$$

so that $k'(\overline{x},\overline{y}^*)(X \times Y^*) = Y \times Y^*$ and we can apply the last assertion of Proposition 3. Thus, the relation $j'(\overline{u})^T(v^*) = k'(\overline{u})^T(w^*)$ with $v^* := (x^*, -x^{**})$, $w^* := (y^*, -y^{**})$ can be transcribed as

$$\langle (x^*, -x^{**}), (x, \overline{y}^* \circ (D^2 g(\overline{x})(x)) + y^* \circ g'(\overline{x})) \rangle = \langle (y^*, -y^{**}), (g'(\overline{x})x, y^*) \rangle.$$

for all $(x,y^*) \in X \times Y^*$.Taking successively $x = 0$ and then $y^* = 0$, we get

$$\forall y^* \in Y^* \qquad \langle x^{**}, g'(\overline{x})^T y^* \rangle = \langle y^{**}, y^* \rangle,$$
$$\forall x \in X \qquad \langle x^*, x \rangle - \langle x^{**}, \overline{y}^* \circ (D^2 g(\overline{x})(x)) \rangle = \langle y^*, g'(\overline{x})x \rangle$$

or $g'(\overline{x})^{TT}(x^{**}) = y^{**}$, $x^* = g'(\overline{x})^T(y^*) + \langle x^{**}, \overline{y}^* \circ (D^2 g(\overline{x})(\cdot)) \rangle$. □

## 5 Coderivative of the Gradient Map of a Moreau Envelope

Let us turn to the coderivative of the gradient map of the Moreau envelope of a paraconvex function $f$ on a Hilbert space $X$ such that for some $a \in \mathbb{R}$, $b \in \mathbb{R}$ one has $f \geq b - a\|\cdot\|^2$. We recalled that there exists some $\overline{r} > 0$ such that for all $r \in ]0, \overline{r}[$ the Moreau envelope $e_r f = {}^r f$ is differentiable with gradient $\nabla(e_r f)(x) = (1/r)(x - p_r(x))$ where $p_r(x) \in X$ is such that $e_r(x) = f(p_r(x)) + (1/2r)\|x - p_r(x)\|^2$ and $x - p_r(x) \in r\partial^\nabla f(p_r(x))$, where the *subgradient* $\partial^\nabla g(x)$ of $g$ at $x$ is the inverse image of $\partial g(x)$ under the Riesz isomorphism $J : X \to X^*$. Here we abridge $(e_r f)(x)$ into $e_r(x)$, whereas we write $p_r(x)$ instead of $p_r$ in order to take into account the dependence on $x$. We have to apply calculus rules for coderivatives. We do not need the full pictures of [24, 26]. A simple lemma is enough. Here we take the coderivatives associated with the directional, the firm (Fréchet), the limiting, and the Clarke subdifferentials, and we simply write $\partial$ instead of $\partial_D$, $\partial_F$, $\partial_L$, $\partial_C$, respectively.

**Lemma 3.**

(a) *If* $F^{-1} : Y \rightrightarrows X$ *is the inverse multimap of* $F : X \rightrightarrows Y$, *then for all* $(y,x) \in F^{-1}$ *and all* $x^* \in X^*$ *one has* $y^* \in D^*F^{-1}(y,x)(x^*)$ *if and only if* $-x^* \in D^*F(x,y)(-y^*)$.
(b) $u^* \in D^*(-F)(x,-y)(v^*)$ *if, and only if,* $u^* \in D^*F(x,y)(-v^*)$.
(c) *If* $F := I + G$, *where* $G : X \rightrightarrows X$ *and* $I$ *is the identity map, then for all* $v^* \in X^*$ *one has* $D^*F(x,y)(v^*) = v^* + D^*G(x, y-x)(v^*)$.

*Proof.* We give the proof in the case the subdifferential is the directional subdifferential $\partial_D$. The cases of the firm, the limiting, and the Clarke subdifferentials are also simple applications of the definitions.

(a) We observe that $N(F^{-1}, (y,x)) = (N(F,(x,y))^{-1}$. Thus, for $x^* \in X^*$ one has $y^* \in D^*F^{-1}(y,x)(x^*)$ if, and only if, $(-x^*, y^*) \in N(F,(x,y))$, if, and only if, $-x^* \in D^*F(x,y)(-y^*)$.
(b) The equivalence is a consequence of the fact that

$$(u^*, -v^*) \in N(\text{gph}(-F), (x, -y)) \Longleftrightarrow (u^*, v^*) \in N(\text{gph}(F), (x, y)).$$

(c) It is easy to see that $v \in DF(x,y)(u)$ for $(u,v) \in X \times X$ if, and only if, there exists $w \in DG(x, y-x)(u)$ such that $v = u + w$, if, and only if, $v - u \in DG(x, y-x)(u)$. Then $(u^*, -v^*) \in N(F, (x,y))$ if, and only if, for all $u \in X$, $w \in DG(x, y-x)(u)$, one has

$$\langle u^*, u \rangle + \langle -v^*, u + w \rangle \leq 0,$$

if, and only if, $(u^* - v^*, -v^*) \in N(G, (x, y-x))$ or $u^* = v^* + (u^* - v^*) \in v^* + D^*G(x, y-x)(v^*)$. □

Applying the preceding lemma to the map $p_r = (I + r\partial^\nabla f)^{-1}$, we get, since $\nabla e_r(x) = r^{-1}(x - p_r(x))$

$$u^* \in D^*\nabla e_r(x)(v^*) \Leftrightarrow u^* \in r^{-1}(v^* + D^*p_r(x)(-v^*)),$$

$$w^* \in D^*p_r(x)(-v^*) \Leftrightarrow v^* \in D^*(I + r\partial^\nabla f)(p_r(x), x)(-w^*)$$

$$\Leftrightarrow v^* \in -w^* + rD^*\partial^\nabla f(p_r(x), x)(-w^*).$$

Therefore, $u^* \in D^*\nabla e_r(x)(v^*)$ if, and only if, for $w^* := ru^* - v^* \in D^*p_r(x)(-v^*)$ one has $v^* + w^* \in rD^*\partial^\nabla f(p_r(x), x)(-w^*)$ or $ru^* \in rD^*\partial^\nabla f(p_r(x), x)(-w^*)$ or $u^* \in D^*\partial^\nabla f(p_r(x), x)(v^* - ru^*)$. We have obtained an expression of $D^*\nabla e_r f$ in terms of $D^*\partial^\nabla f$. We state it as follows.

**Proposition 5.** *If $f$ is paraconvex around $x$ and such that for some $a \in \mathbb{R}$, $b \in \mathbb{R}$ one has $f \geq b - a\|\cdot\|^2$, then for $r > 0$ small enough, one has $u^* \in D^*\nabla(e_r f)(x)(v^*)$ if, and only if, $u^* \in D^*\partial^\nabla f(p_r(x), x)(v^* - ru^*)$*

$$D^*\nabla(e_r f)(x) = (rI + (D^*\partial^\nabla f(p_r(x), x))^{-1})^{-1}.$$

## 6 Conclusion

The choice of notation and terminology is not a simple matter. It has some importance in terms of use. The famous example of the quarrel between Leibniz and Newton shows that the choice of the mathematician that is often considered as the first discoverer of a concept is not always the choice that remains in use. For what concerns terminology, in mathematics as in other sciences, both names of scientists and descriptive names are in use. The difference is that mathematicians prefer ordinary names such as "field,""group,"and "ring" to sophisticated names issued from Greek or Latin. For what concerns notation, there is a trade-off between simplicity and clarity. Both are desirable, but avoiding ambiguity is crucial. Nonetheless, local abuses of notation are often tolerated. Here, because our aim was limited to a few glances at second-order generalized derivatives, we have avoided a heavy notation involving upper epi-limits and we have often omitted the mention that weak convergence is involved. That is not always suitable.

Let us observe that the use of a generic subdifferential avoids delicate choices of notation for subdifferentials. In his first contributions the author endeavored to use symbols like $f^0(x,\cdot)$, $f^{\uparrow}(x,\cdot)$ rather than letters like $f^C(x,\cdot)$ to denote generalized derivatives. But the disorder among such symbols led him to adopt a more transparent choice in [38] and elsewhere. Also, he tried to combine the advantages of descriptive terms and authors' names. More importantly, he chose to adopt the simplest notation for simple, fundamental objects such as Fréchet normal cones $N_F$ and Fréchet subdifferentials $\partial_F$ that are used in most proofs rather than affect them with decorations such as $\hat{N}$ or $\overline{N}$ usually kept for completions or limiting constructions. The fact that limiting constructions can be performed by using different convergences comforts that choice.

It would be interesting to decide whether the influences of dominant philosophies (behaviorism in America, Cartesianism, Kantianism, positivism, existentialism, structuralism in Europe) play some role in the different choices present in the mathematical literature. But that question is outside the scope of the present contribution.

## References


[1] Attouch, H.: Variational Convergence for Functions and Operators. Pitman, Boston (1984)

[2] Bauschke, H.H., Combettes, P.-L.: Convex Analysis and Monotone Operator Theory in Hilbert Spaces. CMS Books in Mathematics. Springer, New York (2011)

[3] Bernard, F., Thibault, L., Zlateva, N.: Prox-regular sets and epigraphs in uniformly convex Banach spaces: various regularities and other properties. Trans. Amer. Math. Soc. **363**(4), 2211–2247 (2011)

[4] Cominetti, R., Correa, R.: A generalized second-order derivative in nonsmooth optimization. SIAM J. Control Optim. **28**(4), 789–809 (1990)

[5] Correa, R., Jofré, A., Thibault, L., Characterization of lower semicontinuous convex functions. Proc. Amer. Math. Soc. **116**(1), 67–72 (1992)

[6] Correa, R., Jofre, A., Thibault, L.: Subdifferential characterization of convexity. In: Du, D.Z., Qi, L., Womersley, R.S. (eds.) Recent Advances in Nonsmooth Optimization, pp. 18–23. World Scientific, Singapore (1995)

[7] Demyanov, V.F., Rubinov, A.M.: Constructive Nonsmooth Analysis. Approximation and Optimization, vol. 7. Peter Lang, Frankfurt am Main, Berlin (1995)

[8] Demyanov, V.F., Stavroulakis, G.E., Polyakova, L.N., Panagiotopoulos, P.D.: Quasidifferentiability and Nonsmooth Modelling in Mechanics, Engineering and Economics. Nonconvex Optimization and Its Applications, vol. 10. Kluwer Academic Publishers, Dordrecht (1996)

[9] Do, C.N.: Generalized second-order derivatives of convex functions in reflexive Banach spaces. Trans. Amer. Math. Soc. **334**(1), 281–301 (1992)

[10] Hiriart-Urruty, J.-B.: Limiting behaviour of the approximate first order and second order directional derivatives for a convex function. Nonlinear Anal. **6**(12), 1309–1326 (1982)

[11] Hiriart-Urruty, J.-B., Seeger, A.: Calculus rules on a new set-valued second order derivative for convex functions. Nonlinear Anal. **13**(6), 721–738 (1989)

[12] Hiriart-Urruty, J.-B., Seeger, A.: The second-order subdifferential and the Dupin indicatrices of a nondifferentiable convex function. Proc. London Math. Soc. (3) **58**(2), 351–365 (1989)
[13] Huang, L.R., Ng, K.F.: Second-order necessary and sufficient conditions in nonsmooth optimization. Math. Program. **66**, 379–402 (1994)
[14] Huang, L.R., Ng, K.F.: On second-order directional derivatives in nonsmooth optimization. In: Du, D.Z., Qi, L., Womersley, R.S. (eds.) Recent Advances in Nonsmooth Optimization, pp. 159–171. World Scientific, Singapore (1995)
[15] Huang, L.R., Ng, K.F.: On some relations between Chaney's generalized second order directional derivative and that of Ben-Tal and Zowe. SIAM J. Control Optim. **34**(4), 1220–1235 (1996)
[16] Ioffe, A.: On some recent developments in the theory of second order optimality conditions. In: Optimization (Varetz, 1988). Lecture Notes in Mathematics, vol. 1405, pp. 55–68. Springer, Berlin (1989)
[17] Ioffe, A.D.: Variational analysis of a composite function: a formula for the lower second-order epi-derivative. J. Math. Anal. Appl. **160**(2), 379–405 (1991)
[18] Ioffe, A.D.: Theory of subdifferentials. Adv. Nonlinear Anal. **1**, 47–120 (2012)
[19] Ioffe, A.D., Penot, J.-P.: Limiting subhessians, limiting subjets and their calculus. Trans. Amer. Math. Soc. **349**, 789–807 (1997)
[20] Jeyakumar, V., Luc, D.T.: Nonsmooth Vector Functions and Continuous Optimization. Springer Optimization and Its Applications, vol. 10. Springer, New York (2008)
[21] Jourani, A., Thibault, L., Zagrodny, D.: $C^{1,\omega(\cdot)}$ -regularity and Lipschitz-like properties of subdifferential. Proc. Lond. Math. Soc. (3) **105**(1), 189–223 (2012)
[22] Levy, A.B.: Second-order epi-derivatives of composite functionals, in optimization with data perturbations II. Ann. Oper. Res. **101**, 267–281 (2001)
[23] Levy, A.B., Poliquin, R., Thibault, L.: Partial extension of Attouch's theorem with applications to proto-derivatives of subgradient mappings. Trans. Amer. Math. Soc. **347**, 1269–1294 (1995)
[24] Li, S.J., Penot, J.-P., Xue, X.: Codifferential calculus. Set-Valued Var. Anal. **19**(4), 505–536 (2011)
[25] Michel, Ph., Penot, J.-P.: Second-order moderate derivatives. Nonlinear Anal. Theor. Methods Appl. **22**(7), 809–821 (1994)
[26] Mordukhovich, B.S.: Variational Analysis and Generalized Differentiation I. Grundlehren der Mathematischen Wissenchaften, vol. 330. Springer, Berlin (2006)
[27] Mordukhovich, B., Rockafellar, R.T.: Second-order subdifferential calculus with applications to tilt stability in optimization. SIAM J. Optim. **22**(3), 953–986 (2012)
[28] Ndoutoumé, J., Théra, M.: Generalized second-order derivatives of convex functions in reflexive Banach spaces. Bull. Austr. Math. Soc. **51**(1), 55–72 (1995)
[29] Páles, Z., Zeidan, V.: Co-Jacobian for Lipschitzian maps. Set-Valued Var. Anal. **18**(1), 57–78 (2010)
[30] Penot, J.-P.: Second-order derivatives: comparisons of two types of epi-derivatives. In: Oettli,W., Pallaschke, D. (eds.) Advances in Optimization, Proceedings 6th French-German Conference on Optimization, Lambrecht, FRG, 1991. Lecture Notes in Economics and Mathematical Systems, vol. 382, pp. 52–76. Springer, Berlin (1992)
[31] Penot, J.-P.: Optimality conditions in mathematical programming and composite optimization. Math. Program. **67**(2), 225–246 (1994)
[32] Penot, J.-P.: Sub-hessians, super-hessians and conjugation. Nonlinear Anal. Theor. Methods Appl. **23**(6), 689–702 (1994)
[33] Penot, J.-P.: Generalized convexity in the light of nonsmooth analysis. In: Duriez, R., Michelot, C. (eds.) 7th French -German Conference on Optimization, Dijon, July 1994. Lecture Notes on Economics and Mathematical Systems, vol. 429, pp. 269–290. Springer, Berlin (1995)
[34] Penot, J.-P.: Sequential derivatives and composite optimization. Revue Roumaine Math. Pures Appl. **40**(5–6), 501–519 (1995)

[35] Penot, J.-P.: Are generalized derivatives useful for generalized convex functions? In: Crouzeix, J.-P., Martinez-Legaz, J.-E., Volle, M. (eds.) Generalized Convexity and Generalized Monotonicity, Marseille, June 1996, pp. 3–59. Kluwer, Dordrecht (1997)
[36] Penot, J.-P.: On the mean value theorem. Optimization **19**(2), 147–156 (1988)
[37] Penot, J.-P.: Glimpses upon quasiconvex analysis. ESAIM Proc. **20**, 170–194 (2007)
[38] Penot, J.-P.: Calculus Without Derivatives. Graduate Texts in Mathematics, vol. 266. Springer, New York (2013)
[39] Penot, J.-P.: A geometric approach to second-order generalized derivatives (preprint)
[40] Penot, J.-P.: Directionally limiting subdifferentials and second-order optimality conditions. Appear in Optimization Letters. DOI 10.1007/s11590-013-0663-0
[41] Penot, J.-P.: Towards a new era in subdifferential analysis? in D.H. Bayley et al. (eds), Computational and Analytical Mathematics, Springer Proceedings in Mathematics & Statistics 50, (2013) DOI 10.1007/978-1-4614-7621-4_29
[42] Penot, J.-P., Quang, P.H.: On generalized convexity of functions and generalized monotonicity of set-valued maps. J. Optim. Theory Appl. **92**(2), 343–356 (1997)
[43] Rockafellar, R.T.: First- and second-order epi-differentiability in nonlinear programming. Trans. Amer. Math. Soc. **307**, 75–107 (1988)
[44] Rockafellar, R.T.: Second-order optimality conditions in nonlinear programming obtained by the way of pseudo-derivatives. Math. Oper. Res. **14**, 462–484 (1989)
[45] Rockafellar, R.T., Wets, R.J.-B.: Variational Analysis. Grundlehren der Mathematischen Wissenschaften, vol. 317. Springer, Berlin (1998)
[46] Warga, J.: Fat homeomorphisms and unbounded derivate containers. J. Math. Anal. Appl. **81**(2), 545–560 (1981)
[47] Yang, X.Q.: On relations and applications of generalized second-order directional derivatives. Nonlin. Anal. Theor. Methods Appl. **36**(5), 595–614 (1999)

# The Nonsmooth Path of Alex M. Rubinov

**S.S. Kutateladze**

**Abstract** This is a short tribute to Alex Rubinov (1940–2006) on the occasion of the 50 years of nonsmooth analysis.

**Keywords** Convex functional • Equilibrium • Minkowski duality • Generalized convexity

Rubinov was born on March 28, 1940, in St. Petersburg. His father was a college teacher and his mother worked in a court of justice. Rubinov graduated from high school in 1957 and was admitted to the Physics and Mathematics Department of the Leningrad State Pedagogical Institute. In 1958 he arranged his transfer to Leningrad State University. He graduated from the Department of Mechanics and Mathematics in 1962 and entered into his post graduate study in functional analysis. His supervisor was Gleb Akilov, a student and coauthor of Leonid Kantorovich. In these years Rubinov started his lifelong collaboration with Vladimir Demyanov. They were involved then in application of functional analysis to optimization and numerical methods.

Kantorovich joined the Siberian Division of the Academy of Sciences which was organized in 1957 and moved to Novosibirsk. Akilov followed his tutor and so did Rubinov. From June of 1964 he was on the staff of the Mathematical–Economic Department headed by Kantorovich, continuing his postgraduate study distantly in Leningrad State University. In May of 1965 he maintained his thesis "Minimization of Convex Functionals on Some Classes of Convex Sets in Banach Spaces" and received the Kandidat Degree.

S.S. Kutateladze (✉)
Sobolev Institute of Mathematics, 4 Koptuyg Avenue, Novosibirsk 630090, Russia,
e-mail: sskut@math.nsc.ru

V.F. Demyanov et al. (eds.), *Constructive Nonsmooth Analysis and Related Topics*, Springer Optimization and Its Applications 87, DOI 10.1007/978-1-4614-8615-2_15,

The "Siberian period" formed the main areas of Rubinov's research. He narrated this in one of his last interviews as follows:

> My first book: "Approximate Methods in Optimization Problems" (with V.F. Demyanov) was written in 1966–1967 and published in 1968.
>
> My second book "Mathematical Theory of Economic Dynamics and Equilibria" (with V.L. Makarov) was written in 1968–1970 and published in 1973.
>
> My third book "Minkowski Duality and Its Applications" (with S.S. Kutateladze) was written in 1970–1972 and published in 1976.
>
> My research interests are mainly concentrated around indicated fields: Optimization, Mathematical economics, Abstract convexity. (It seems that "Minkowski Duality" was the first book in the world dedicated to abstract convexity.) I also work in related topics. I tried to contribute to economics (one of my textbooks is called "Elements of Economic Theory (a textbook for students of mathematical departments)" (with A. Nagiev)), I had some papers and books in nonsmooth analysis (quasidifferential calculus, jointly with V.F. Demyanov), some papers in dynamical system theory etc. Last years I am involved in application of optimization to data analysis and telecommunication.

Rubinov chose his scientific itinerary in his green years and travelled along the route up to his terminal day.

In 1970 Rubinov was on his rapid creative uprise—he submitted and maintained the doctorate thesis "Point-to-Set Mappings Defined on Cones."

The defense took place in Novosibirsk and was quite a success. So the chances were very high that the degree would be awarded by the Higher Attestation Committee in Moscow. The Committee sent Rubinov's thesis to Nikita Moiseev for giving an official "blind" review, which was a standard procedure. Moiseev rang to Rubinov, which was in violation of the procedure, and told Rubinov that he would give a favorable review. Moiseev was a corresponding member of the Academy of Sciences of the USSR and was interested in good relations with Kantorovich who was an influential full member. After the call Rubinov decided to move from Novosibirsk back to European Russia right away, not awaiting the Moscow decision on the new degree which would open many favorable positions for Rubinov. This was hasty but it was not Rubinov's own decision. He was under the pressure of his family. But the plans to return to the Northern capital of Russia failed, and Rubinovs moved to Tver (then Kalinin) which is between Moscow and St. Petersburg. Rubinov found a position in Kalinin State University which was then an asylum for the capable mathematicians mostly of Jewish origin who could not find positions in Moscow and Leningrad. When Moiseev became aware of the departure of Rubinov from Novosibirsk, which was groundlessly interpreted as the end of Kantorovich's interest in Rubinov's fate, Moiseev lost his compassion to Rubinov and abstained from giving any review.

These years in the academic life of the USSR were poisoned with careerism whose most disgusting manifestation was Anti-Semitism. Quite a few "successful" figures of the epoch were so unscrupulous that traded over xenophobia and other similar techniques of emptying the career lane. Rubinov suffered from these deceases of the Soviet life for many years. The pressure on him was aggravated by this refusal of Moiseev. In result, Rubinov decided to vacate his thesis formally to ensure his family from losing everything had he continued asking for the doctorate. Of course, Rubinov's decision was a relief to his academic adversaries and native Anti-Semites.

A few years later Rubinov found a modest position in Leningrad. He was supported by his friends and began to work at the Institute for Social-Economic Problems in the laboratory headed by Nikolai Vorob′ëv. Rubinov wrote a new thesis for his doctorate in demography, but it turned out impossible to arrange the official defense in the rotten atmosphere of the academic life of the USSR.

Rubinov received his Doctor degree only in 1986 after maintaining his thesis in the Computer Center of the Academy of Sciences of the USSR. Moiseev had become a full member of the Academy in 1984, his career was crowned, and he supported Rubinov this time.

In 1988 Rubinov moved to Azerbaijan where he worked successfully in the Institute of Mathematics and Baku State University. After disintegration of the USSR Rubinov emigrated to Israel where he spent a few years in the Ben-Gurion University of the Negev. In 1996 Rubinov moved to Australia and took a position in Ballarat University where he worked up to his untimely death from cancer on September 9, 2006.

The creative talent of Rubinov had blossomed in his "Australian period." His efforts made Ballarat a noticeable world center of global optimization and nonsmooth analysis. In 2006 the European Working Group on Continuous Optimization chose Rubinov as a EUROPT Fellow 2006.

Rubinov was an outstanding example of the working mathematician. He contributed to the theory and numerical methods of optimization. He developed the technique of studying the sublinear functionals that are defined and monotone on a cone. He elaborated the duality technique for them which belongs to subdifferential calculus. Rubinov paid much attention to abstract local convex analysis to quasidifferentiable functions whose derivatives can be represented as differences of sublinear functionals. Rubinov was one of the leading figures in abstract convexity—he extended duality to the upper envelopes of the subsets of some given family $H$ of simple functions. The conception of $H$-convexity happened to be tied with the deep problems of Choquet theory and approximation by positive operators through the new concept of supremal generator. Rubinov invented monotonic analysis. He gave the nicest form for the theorem of characteristics of optimal trajectories of models of economic dynamics and other discrete dynamic problems. His research enriched the important section of mathematical economics whose progress was connected with the theory of convex processes by John von Neumann, David Gale, and Terry Rockafellar. Rubinov published about twenty books and more than 150 papers. But what stands above all Rubinov's scientific contributions is his path of academic service impeccable despite all obstacles and pits of his rough world line.

Rubinov remains in the memory of those who knew and understood him not only a prominent scholar but also a brilliant, loving, faithful, and charming personality.

# Two Giants from St. Petersburg

**S.S. Kutateladze**

**Abstract** This is a short tribute to Alexandr Alexandrov and Leonid Kantorovich for their contribution to the field of nonsmooth analysis, convexity and optimization.

**Keywords** Convex body • Curvature • Minkowski problem • Vector lattice • Linear programming • Boolean valued models • Optimal planning

## 1 Introduction

Nonsmooth analysis and nondifferentiable optimization are next of kin to convexity and linear programming. This year we celebrate not only the 50 years of nonsmooth analysis but also the centenary of Alexandr Alexandrov and Leonid Kantorovich, the mathematical giants whose contributions are indispensable for the fields of convexity and optimization.

Alexandrov was a fast-rising star in mathematics who had maintained his second DSc thesis by the age of 30 and became the youngest rector of Leningrad (now St. Petersburg) State University a decade later. The Mathematics Subject Classification, produced jointly by the editorial staffs of *Mathematical Reviews* and *Zentralblatt für Mathematik* in 2010, has Section 53C45 "Global surface theory (convex surfaces à la A.D. Aleksandrov)." None of the other Russian geometers, Lobachevsky inclusively, has this type of acknowledgement. Alexandrov became the first and foremost Russian geometer of the twentieth century.

S.S. Kutateladze (✉)
Sobolev Institute of Mathematics, 4 Koptuyg Avenue, Novosibirsk 630090, Russia
e-mail: sskut@math.nsc.ru

V.F. Demyanov et al. (eds.), *Constructive Nonsmooth Analysis and Related Topics*, Springer Optimization and Its Applications 87, DOI 10.1007/978-1-4614-8615-2_16,

Kantorovich was a prodigy who graduated from St. Petersburg University at the age of 18, became a professor at the age of 20, was elected a full member of the Department of Mathematics of the Academy of Sciences of the USSR, and was awarded with a Nobel Prize in economics.

These extraordinary events of the lives of Alexandrov and Kantorovich deserve some attention in their own right. But they hardly lead to any useful conclusions for the general audience in view of their extremely low probability. This is not so with their creative legacy, for what is done for the others remains unless it is forgotten, ruined, or libeled. Recollecting the paths and contributions of persons to culture, we preserve their spiritual worlds for the future.

## 2 Alexandr D. Alexandrov (1912–1999)

### *2.1 Life's Signposts*

Alexandr Danilovich Alexandrov was born in the Volyn village of the Ryazan province on August 4, 1912. His parents were high school teachers. He entered the Physics Faculty of Leningrad State University in 1929 and graduated in 1933. His supervisors were Boris Delauney, a prominent geometer and algebraist, and Vladimir Fok, one of the outstanding theoretical physicists of the last century. The first articles by Alexandrov dealt with some problems of theoretical physics and mathematics. But geometry soon became his main speciality.

Alexandrov defended his Ph.D. thesis in 1935 and his second doctorate thesis in 1937. He was elected to a vacancy of corresponding member of the Academy of Sciences of the USSR in 1946 and was promoted to full membership in 1964.

From 1952 to 1964, Alexandrov was rector of Leningrad State University. These years he actively and effectively supported the struggle of biologists with Lysenkoism. The name of Rector Alexandrov is connected with the uprise of the new areas of science such as sociology and mathematical economics which he backed up in the grim years. Alexandrov was greatly respected by established scholars as well as academic youth. "He led the University by moral authority rather than the force of direct order," so wrote Vladimir Smirnov in the letter of commendation on the occasion of Alexandrov's retirement from the position of rector.

In 1964 Mikhail Lavrentyev invited Alexandrov to join the Siberian Division of the Academy of Sciences of the USSR. Alexandrov moved with his family to Novosibirsk where he found many faithful friends and students. By 1986 he headed a department of the Institute of Mathematics (now, the Sobolev Institute), lectured in Novosibirsk State University, and wrote new versions of geometry textbooks at the secondary school level. Alexandrov opened his soul and heart to Siberia, but was infected with tick-borne encephalitis which undermined his health seriously. From

April of 1986 up to his death on July 27, 1999, Alexandrov was on the staff of the St. Petersburg Department of the Steklov Mathematical Institute.

## 2.2 *Contribution to Science*

Alexandrov's life business was geometry. The works of Alexandrov made tremendous progress in the theory of mixed volumes of convex figures. He proved some fundamental theorems on convex polyhedra that are celebrated alongside the theorems of Euler and Minkowski. While discovering a solution of the Weyl problem, Alexandrov suggested a new synthetic method for proving the theorems of existence. The results of this research ranked the name of Alexandrov alongside the names of Euclid and Cauchy.

Another outstanding contribution of Alexandrov to science is the creation of the intrinsic geometry of irregular surfaces. He suggested his amazingly visual and powerful method of cutting and gluing. This method enabled him to solve many extremal problems of the theory of manifolds of bounded curvature.

Alexandrov developed the theory of metric spaces with one-sided constraints on curvature. This gave rise to the class of metric spaces generalizing the Riemann spaces in the sense that these spaces are furnished with some curvature, the basic concept of Riemannian geometry. The research of Alexandrov into the theory of manifolds with bounded curvature prolongates and continues the traditions of Gauss, Lobachevsky, Poincaré, and Cartan.

## 2.3 *Retreat to Euclid*

Alexandrov accomplished the turn-round to the ancient synthetic geometry in a much deeper and subtler sense than it is generally acknowledged today. The matter is not simply in transition from smooth local geometry to geometry in the large without differentiability restrictions. In fact Alexandrov enriched the methods of differential geometry by the tools of functional analysis and measure theory, driving mathematics to its universal status of the epoch of Euclid. The mathematics of the ancients was geometry (there were no other instances of mathematics at all). Synthesizing geometry with the remaining areas of the today's mathematics, Alexandrov climbed to the antique ideal of the universal science incarnated in mathematics.

Alexandrov contributed to nonsmooth analysis, developing the theory of DC surfaces and inspecting the first- and second-order differentiability of convex functions in the classical and distribution senses. He also was a pioneer of using the functional-analytical technique for studying the spaces of compact convex bodies.

Alexandrov overcame many local obstacles and shortcomings of the differential geometry based on the infinitesimal methods and ideas by Newton, Leibniz, and

Gauss. Moreover, he enriched geometry with the technique of functional analysis, measure theory, and partial differential equations. Return to the synthetic methods of *mathesis universalis* was inevitable and unavoidable as illustrated in geometry with the beautiful results of the students and descendants of Alexandrov like Misha Gromov, Grisha Perelman, Alexei Pogorelov, Yuri Reshetnyak, and Victor Zalgaller.

### *2.4 Alexandrov and the Present Day*

Alexandrov emphasized the criticism of science and its never-failing loyalty to truth. Science explains "how the thingummy's actually going on" with greatness and modesty, using experience, facts, and logic. The universal humanism of the geometer Alexandrov, stemming from the heroes of antiquity, will always remain in the treasure-trove of the best memes of the humankind.

## 3 Leonid V. Kantorovich (1912–1986)

### *3.1 Life's Signposts*

Kantorovich was born in the family of a venereologist at St. Petersburg on January 19, 1912. The boy's talent was revealed very early. In 1926, just at the age of 14, he entered Leningrad State University. After graduation from the university in 1930, Kantorovich started teaching, combining it with intensive scientific research. Already in 1932 he became a full professor at the Leningrad Institute of Civil Engineering. From 1934 Kantorovich was a full professor at his *alma mater*.

The main achievements in mathematics belong to the "Leningrad" period of Kantorovich's life. In the 1930s he published more papers in pure mathematics, whereas his 1940s are devoted to computational mathematics in which he was soon appreciated as a leader in this country.

In 1935 Kantorovich made his major mathematical discovery—he defined *K-spaces*, i.e., vector lattices whose every nonempty order bounded subset had an infimum and supremum. The Kantorovich spaces have provided the natural framework for developing the theory of linear inequalities which was a practically uncharted area of research those days. The concept of inequality is obviously relevant to approximate calculations where we are always interested in various estimates of the accuracy of results. Another challenging source of interest in linear inequalities was the stock of problems of economics. The language of partial comparison is rather natural in dealing with what is reasonable and optimal in human behavior when means and opportunities are scarce. Finally, the concept of linear inequality is inseparable from the key idea of a convex set. Functional analysis implies the existence of nontrivial continuous linear functional over the space under consideration, while the presence of a functional of this type amounts to the existence of nonempty proper open convex subset of the ambient space. Moreover, each convex set is generically the solution set of an appropriate system of simultaneous linear inequalities.

From the end of the 1930s the research of Kantorovich acquired new traits in his audacious breakthrough to economics. The novelty of the extremal problems arising in social sciences is connected with the presence of multidimensional contradictory utility functions. This raises the major problem of agreeing conflicting aims.

Kantorovich's booklet *Mathematical Methods in the Organization and Planning of Production* which appeared in 1939 is a material evidence of the birth of linear programming. Linear programming is a technique of maximizing a linear functional over the positive solutions of a system of linear inequalities. It is no wonder that the discovery of linear programming was immediate after the foundation of the theory of Kantorovich spaces.

At the end of the 1940s Kantorovich formulated and explicated the thesis of interdependence between functional analysis and applied mathematics and suggested three new techniques: the Cauchy method of majorants, the method of finite-dimensional approximations, and the Lagrange method for general optimization problems in topological vector spaces. Kantorovich based his study of the Banach space versions of the Newton method on domination in general ordered vector spaces. His analysis had been so profound that the term "Newton–Kantorovich method" was often used in common parlance. Kantorovich had applied his general approach to the Monge problem, which led to the modern transport theory.

The economic works of Kantorovich were hardly visible at the surface of the scientific information flow in the 1940s. But the problems of economics prevailed in his creative studies. During the Second World War he completed the first version of his book *The Best Use of Economic Resources* which led to the Nobel Prize awarded to him and Tjalling C. Koopmans in 1975.

In 1957 Kantorovich accepted the invitation to join the newly founded Siberian Division of the Academy of Sciences of the USSR. He moved to Novosibirsk and soon became a corresponding member of the Department of Economics in the first elections to the Siberian Division. Since then his major publications were devoted to economics with the exception of the celebrated course of functional analysis, "Kantorovich and Akilov" in the students' jargon.

The 1960s became the decade of recognition. In 1964 Kantorovich was elected a full member of the Department of Mathematics of the Academy of Sciences of the USSR, and in 1965 he was awarded the Lenin Prize. In these years he vigorously propounded and maintained his views of interplay between mathematics and economics and exerted great efforts to instill the ideas and methods of modern science into the top economic management of the Soviet Union, which was almost in vain.

At the beginning of the 1970s Kantorovich left Novosibirsk for Moscow where he was deeply engaged in economic analysis, not ceasing his efforts to influence the everyday economic practice and decision making in the national economy. His activities were mainly waste of time and stamina in view of the misunderstanding and hindrance of the governing retrogradists of this country. Cancer terminated his path in science on April 7, 1986. He was buried at Novodevichy Cemetery in Moscow.

## *3.2 Linear Programming*

The principal discovery of Kantorovich at the junction of mathematics and economics is linear programming which is now studied by hundreds of thousands of people throughout the world. The term signifies the colossal area of science which is allotted to linear optimization models. In other words, linear programming is the science of the theoretical and numerical analysis of the problems in which we seek for an optimal (i.e., maximum or minimum) value of some system of indices of a process whose behavior is described by simultaneous linear inequalities. It was in 1939 that Kantorovich formulated the basic ideas of the new area of science.

The term "linear programming" was minted in 1951 by Tjalling C. Koopmans, an American economist with whom Kantorovich shared in 1975 the Nobel Prize for research of optimal use of resources. The most commendable contribution of Koopmans was the ardent promotion of the methods of linear programming and the strong defense of Kantorovich's priority in the invention of these methods.

In the USA the independent research into linear optimization models was started only in 1947 by George B. Dantzig who also noted the priority of Kantorovich.

It is worth observing that to an optimal plan of every linear program there correspond some optimal prices or "objectively determined estimators." Kantorovich invented this bulky term by tactical reasons in order to enhance the "criticism endurability" of the concept. The conception of optimal prices as well as the interdependence of optimal solutions and optimal prices is the crux of the economic discovery of Kantorovich.

## *3.3 Mathematics and Economics*

Mathematics studies the forms of reasoning. The subject of economics is the circumstances of human behavior. Mathematics is abstract and substantive, and the professional decision of mathematicians does not interfere with the life routine of individuals. Economics is concrete and declarative, and the practical exercises of economists change the life of individuals substantially. The aim of mathematics consists in impeccable truths and methods for acquiring them. The aim of economics is the well-being of an individual and the way of achieving it. Mathematics never intervenes into the private life of an individual. Economics touches his purse and bag. Immense is the list of striking differences between mathematics and economics.

Mathematical economics is an innovation of the twentieth century. It is then when the understanding appeared that the problems of economics need a completely new mathematical technique.

Homo sapiens has always been and will stay forever homo economics. Practical economics for everyone as well as their ancestors is the arena of common sense. Common sense is a specific ability of a human to instantaneous moral judgement. Understanding is higher than common sense and reveals itself as the adaptability of behavior. Understanding is not inherited and so it does not belong to the inborn traits of a person. The unique particularity of humans is the ability of sharing their understanding, transforming evaluations into material and ideal artifacts.

Culture is the treasure-trove of understanding. The inventory of culture is the essence of outlook. Common sense is subjective and affine to the divine revelation of faith that is the force surpassing the power of external proofs by fact and formal logic. The verification of statements with facts and by logic is a critical process liberating a human from the errors of subjectivity. Science is an unpaved road to objective understanding. The religious and scientific versions of outlook differ actually in the methods of codifying the artifacts of understanding.

The rise of science as an instrument of understanding is a long and complicated process. The birth of ordinal counting is fixed with the Paleolithic findings that are separated by hundreds of centuries from the appearance of a knowing and economic human. Economic practice precedes the prehistory of mathematics that became the science of provable calculations in ancient Greece about 2,500 years ago.

It was rather recently that the purposeful behavior of humans under the conditions of limited resources became the object of science. The generally accepted date of the birth of economics as a science is March 9, 1776—the day when there was published the famous book by Adam Smith *An Inquiry into the Nature and Causes of the Wealth of Nations*.

## 3.4 Consolidation of Mind

Ideas rule the world. John Maynard Keynes completed this banal statement with a touch of bitter irony. He finished his most acclaimed treatise *The General Theory of Employment, Interest, and Money* in a rather aphoristic manner: "Practical men, who believe themselves to be quite exempt from any intellectual influences, are usually the slaves of some defunct economist."

Political ideas aim at power, whereas economic ideas aim at freedom from any power. Political economy is inseparable from not only the economic practice but also the practical policy. The political content of economic teachings implies their special location within the world science. Changes in epochs, including their technological achievements and political utilities, lead to the universal proliferation of spread of the emotional attitude to economic theories, which drives economics in the position unbelievable for the other sciences. Alongside noble reasons for that, there is one rather cynical: although the achievements of exact sciences drastically change the life of the mankind, they never touch the common mentality of humans as vividly and sharply as any statement about their purses and limitations of freedom.

Georg Cantor, the creator of set theory, remarked as far back as in 1883 that "the essence of mathematics lies entirely in its freedom." The freedom of mathematics does not reduce to the absence of exogenic restriction on the objects and methods of research. The freedom of mathematics reveals itself mostly in the new intellectual tools for conquering the ambient universe which are provided by mathematics for liberation of humans by widening the frontiers of their independence. Mathematization of economics is the unavoidable stage of the journey of the mankind into the realm of freedom.

The nineteenth century is marked with the first attempts at applying mathematical methods to economics in the research by Antoine Augustin Cournot, Carl Marx, William Stanley Jevons, Leon Walras, and his successor in Lausanne University Vilfredo Pareto.

John von Neumann and Leonid Kantorovich, mathematicians of the first calibre, addressed the economic problems in the twentieth century. The former developed game theory, making it an apparatus for the study of economic behavior. The latter invented linear programming for decision making in the problems of best allocation of scarce resources. These contributions of von Neumann and Kantorovich occupy an exceptional place in science. They demonstrated that the modern mathematics opens up broad opportunities for economic analysis of practical problems. Economics has been drifted closer to mathematics. Still remaining a humanitarian science, it mathematizes rapidly, demonstrating high self-criticism and an extraordinary ability of objective thinking.

The turn in the mentality of the mankind that was effected by von Neumann and Kantorovich is not always comprehended to full extent. There are principal distinctions between the exact and humanitarian styles of thinking. Humans are prone to reasoning by analogy and using incomplete induction, which invokes the illusion of the universal value of the tricks we are accustomed to. The differences in scientific technologies are not distinguished overtly, which in turn contributes to self-isolation and deterioration of the vast sections of science.

## *3.5 Universal Heuristics*

The integrity of the outlook of Kantorovich was revealed in all instances of his versatile research. The ideas of linear programming were tightly interwoven with his methodological standpoints in the realm of mathematics. Kantorovich viewed as his main achievement in this area the distinguishing of Dedekind complete vector lattes, also called $K$-spaces or Kantorovich spaces in the literature of the Russian provenance, since Kantorovich wrote about "my spaces" in his personal memos.

The abstract theory of $K$-spaces, linear programming, and approximate methods of analysis were particular outputs of Kantorovich's universal heuristics.

More recent research has corroborated that the ideas of linear programming are immanent in the theory of $K$-spaces. It was demonstrated that the validity of one of the various statements of the duality principle of linear programming in an abstract mathematical structure implies with necessity that the structure under consideration is in fact a $K$-space.

The Kantorovich heuristics is connected with one of the most brilliant pages of the mathematics of the twentieth century the famous problem of the continuum. Recall that some set $A$ has the cardinality of the continuum whenever $A$ in equipollent with a segment of the real axis. The continuum hypothesis is that each subset of the segment is either countable or has the cardinality of the continuum. The continuum problem asks whether the continuum hypothesis is true or false.

The continuum hypothesis was first conjectured by Cantor in 1878. He was convinced that the hypothesis was a theorem and vainly attempted at proving it during his whole life. In 1900 the Second Congress of Mathematicians took place in Paris. In the opening session Hilbert delivered his epoch-making talk "Mathematical Problems." He raised 23 problems whose solution was the task of the nineteenth century bequeathed to the twentieth century. The first on the Hilbert list was open the continuum problem. Remaining unsolved for decades, it gave rise to deep foundational studies. The efforts of more than a half century yielded the solution: we know now that the continuum hypothesis can neither be proved nor refuted.

The two stages led to the understanding that the continuum hypothesis is an independent axiom. Gödel showed in 1939 that the continuum hypothesis is consistent with the axioms of set theory, and Cohen demonstrated in 1963 that the negation of the continuum hypothesis does not contradict the axioms of set theory either. Both results were established by exhibiting appropriate models, i.e., constructing a universe and interpreting set theory in the universe.

Cohen's method of forcing was simplified in 1965 on using the tools of Boolean algebra and the new technique of mathematical modeling which is based on the nonstandard models of set theory. The progress of the so-invoked Boolean valued analysis has demonstrated the fundamental importance of the so-called universally complete $K$-spaces. Each of these spaces turns out to present one of the possible noble models of the real axis and so such a space plays a similar key role in mathematics. The spaces of Kantorovich implement new models of the reals, this earning their eternal immortality.

Kantorovich heuristics has received brilliant corroboration, this proving the integrity of science and inevitability of interpenetration of mathematics and economics.

## 3.6 *Memes for the Future*

Kantorovich's message was received as witnessed by the curricula and syllabi of every economics or mathematics department in any major university throughout the world. The gadgets of mathematics and the idea of optimality belong to the tool-kit of any practicing economist. The new methods erected an unsurmountable firewall against the traditionalists that view economics as a testing polygon for the technologies like Machiavellianism, flattery, common sense, or foresight.

Economics as an eternal boon companion of mathematics will avoid merging into any esoteric part of the humanities, or politics, or belles-lettres. The new generations of mathematicians will treat the puzzling problems of economics as an inexhaustible source of inspiration and an attractive arena for applying and refining their impeccably rigorous methods. Calculation will supersede prophesy.

## 4 Conclusion

Alexandrov and Kantorovich were livelong friends and collaborators. They had mutual students and followers who made great contributions to pure and applied mathematics. This led to the extraordinary fusion of various ideas from geometry, functional analysis, and optimization which resides now within the realm of nonsmooth analysis and nondifferentiable optimization.

# Convex Analysis and Optimization in the Past 50 Years: Some Snapshots

**Jean-Baptiste Hiriart-Urruty**

**Abstract** We offer some personal historical snapshots on the evolution of convex analysis and optimization in the past 50 years. We deliberately focus on some specific periods or dates which resonate in our memories.

**Keywords** Convex analysis • Optimization • Nonsmooth analysis • History of optimization

## 1 Fifty Years Ago: 1962–1963

We all know that the development of convex analysis during the last 50 years owes much to W. Fenchel (1905–1988), J.-J. Moreau (1923-) and R.T. Rockafellar (1935-). Fenchel was very "geometrical" in his approach; Moreau used to say that he did applied mechanics: he "applied mechanics to mathematics", while the concept of "dual problem" was a constant leading thread for Rockafellar. The years 1962–1963 can be considered as the date of birth of *modern convex analysis with applications to optimization.* The now familiar appellations like subdifferential, proximal mappings, infimal convolution date back from this period, exactly 50 years ago. In two consecutive notes published by the French Academy of Sciences [16, 17], Moreau introduced the so-called proximal mappings and a way of regularizing a convex function defined on a Hilbert space by performing an inf-convolution with the square of the norm; these preliminary works culminated with the 1965 paper [19], which remains for me the archetype of elegant mathematical paper.

J.-B. Hiriart-Urruty (✉)
Institut de Mathématiques University Paul Sabatier 118, route de Narbonne
31062 Toulouse Cedex 9, France,
e-mail: jbhu@math.univ-toulouse.fr

V.F. Demyanov et al. (eds.), *Constructive Nonsmooth Analysis and Related Topics*, Springer Optimization and Its Applications 87, DOI 10.1007/978-1-4614-8615-2_17,

The short paper by Hörmander [14], on the support functions of sets in a general context of locally convex topological vector spaces, published (in French) some years earlier (1954), was influental in modern developments of convex analysis. These thoughts came to my mind these days since L. Hörmander just passed away (on November 2012); he was very young (less than 23 years old) when he wrote this paper, his Ph.D. thesis on PDE was not yet completed. I remained impressed by the maturity of this mathematician at this early age.

Various names appeared in 1963 to denote a vector $s$ satisfying

$$f(y) \geqslant f(x) + \langle s, y - x \rangle \text{ for all } y.$$

R.T. Rockafellar in his 1963 Ph.D. thesis [21] called $s$ "a differential of $f$ at $x$"; it is J.-J. Moreau who, in a note at the French Academy of Sciences (in 1963) [18], introduced for $s$ the word "sous-gradient" (which became "subgradient" in English). Even the wording "la sous-différentielle" (a feminine word in French, closer to the classical "la différentielle" for differentiable functions) was used in the early days, it became later "le sous-différentiel" (a masculine word in French). As it often happens in research in mathematics, when times are ripe, concepts bloomed in different places of the world at about the same time; in the former USSR, for example, institutes or departments in Moscow and Kiev were on the front; just to give a name, N.Z. Shor's thesis in Kiev is dated 1964. A little bit earlier, in 1962, N.Z. Shor published a first instance of use of a subgradient method for minimizing a nonsmooth convex function (a piecewise linear one actually).

One of the most specific constructions in convex or nonsmooth analysis is certainly taking the supremum of a (possibly infinite) collection of functions. In the years 1965–1970, various calculus rules concerning the subdifferential of sup-functions started to emerge; working in that direction and using various assumptions, several authors contributed to this calculus rule: B.N. Pshenichnyi, A.D. Ioffe, V.L. Levin, R.T. Rockafellar, A. Sotskov, etc.; however, the most elaborated results of that time were due to M. Valadier (1969); he made use of $\varepsilon$-active indices in taking the supremum of the collection of functions.

The transformation $f \longmapsto f^*$ has its origins in a publication of A. Legendre (1752–1833), dated from 1787. Since then, this transformation has received a number of names in the literature: conjugate, polar, maximum transformation, etc. However, it is now generally agreed that an appropriate terminology is *Legendre-Fenchel transform.* In preparing the books with Lemaréchal [12], I remember to have asked by letter L. Hörmander whether the appellation should be *Fenchel transform* or *Legendre-Fenchel transform*; he answered that the name of Legendre should be added to that of Fenchel, which we adopted subsequently. In a letter to C. Kiselman (a colleague from the University of Uppsala, Sweden), dated 1977, W. Fenchel wrote: *"I do not want to add a new name, but if I had to propose one now, I would let myself be guided by analogy and the relation with polarity between convex sets (in dual spaces) and I would call it for example* parabolic polarity". Fenchel was influenced by his geometric (projective) approach and also by the fact that the "parabolic" function $f(x) = 1/2 \|.\|^2$ is the only one satisfying the relation $f = f^*$.

We have intended to mark this 50th birthday of modern convex analysis by editing a special issue in the Mathematical Programming series B [5].

## 2 Forty Years Ago: 1971–1973

My first contact with the name of J.-J. Moreau was via his mimeographed lecture notes [20] in the academic year 1971–1972. I was beginning my doctoral studies at the University of Bordeaux, and J.-L. Joly presented his course to us by saying: "These are the notes corresponding to my lectures", and he gave to each of us a copy. I still have this copy, typed on an old typewriter, by some secretary at Collège de France in Paris (I suppose), comprising my own handwritten annotations. I remember that, with another student next to me, we were impressed by the long list of references authored by J.-J. Moreau and R.T. Rockafellar and posted at the end of lecture notes ([18, 22] references respectively). As beginners in research, we did not know that one of the objectives of researchers in mathematics (the only one?) is to publish as many papers as possible. However, I do not think that the way of publishing at that time was (what is sometimes called) "salami publishing" like it is nowadays. These lecture notes were widely spread in France and elsewhere but never published by an editing house; only in 2003 they were published by a group in Italy (University "Tor Vergata" in Roma). J.-L. Joly was a young professor, just settled in Bordeaux, coming from the University of Grenoble (like others, B. Martinet, A. Auslender, C. Carasso, P.-J. Laurent, C.F. Ducateau, etc.). After some time devoted to convex analysis, he moved to the PDE area. He did some works on convex analysis with P.-J. Laurent; they were presented (some of them exclusively there) in the book entitled "Approximation et Optimisation", authored by P.-J. Laurent and published in 1972 [15]. I remember exactly when and where (in a bookstore in Bordeaux) I bought this book (students at that time used to buy books, not just photocopy them…). I still have this personal copy; the chapters VI (on convex functionals) and VII (on stability and duality in convex optimization) are the most worn ones. This book has been translated into Russian, never into English, I believe. The exam session of June 1972 (a 4-h long written examination) on the lectures in Joly's course consisted into two parts: the first one was devoted to the construction of some geometrical mean of two convex functions; the second one had for objective to explore the link between "local uniform convexity" of a convex function and the Fréchet-differentiability of its Legendre-Fenchel conjugate…A tough exam indeed. I discovered some time later that the matter of the exam was directly taken from a paper by E. Asplund…My Master's thesis was presented in 1972–1973, my first readings of works of mathematical research were those of R.J. Aumann ("Integration of set-valued mappings") and Z. Artstein ("Set-valued measures", 1972). I was to cross paths with Z. Artstein several times in my career.

The long papers by A. Ioffe-V. Tikhomirov (Russian Math. Surveys, 1968) and A. Ioffe-V. Levin (Trans. Moscow Math. Soc., 1972), the classical ones by V. F. Demyanov and A. M. Rubinov (1967–1968), were also at our disposal.

R.T. Rockafellar's book, entitled "Convex Analysis", was published in 1970. It was quickly spread among interested mathematicians in France. Interestingly enough, this book remains one of the most sold ones in mathematics.

A bit later, in 1974, convex analysis and duality in variational problems were presented in the book by Ekeland and Temam [6], two influential mathematicians from J.-L. Lions' research group in Paris. The book has been translated into English and Russian.

In those years, techniques and results from convex analysis illuminated several other areas of mathematics: that of monotone operators and PDE (with students and collaborators of H. Brezis), stochastic control theory (Bismut [4] for example), etc.

## 3 Thirty Years Ago: 1981–1983

I always have been a fan of Russian mathematics. At the end of the 1970s years, I began exchanging letters with B.M. Mordukhovich (Belarus State University in Minsk), colleagues in Kiev (B.N. Pshenichnyi, Yu. Ermoliev, E. Nurminski), and elsewhere. In February 1980, a meeting entitled "Convex Analysis and Optimization" was organized in London in honour of A. Ioffe (Moscow) (see [3]); I presented there (and published in [3]) a survey paper on $\varepsilon$-subdifferential calculus. I like to write survey papers from time to time. Only some years later I had the opportunity to meet (for the first time) B.N. Pshenichnyi, V.F. Demyanov and A. Ioffe; it was at the occasion of these charming meetings organized from time to time in Erice (Sicily).

After doctoral studies under the supervision of A. Auslender and some additional years in Clermont-Ferrand (1973–1981), I moved to Toulouse in September 1981. I left the city of B. Pascal (Clermont-Ferrand) for that of P. Fermat (Toulouse); after all, both lived in the same century, the seventeenth century, the one where the physical notion of motion (velocity, acceleration) was "made mathematics" (birth of differential calculus, tackling extremum problems). In the meantime, between 1973 and 1980, Clarke's approach of generalized subdifferentials of nonsmooth nonconvex functions had been introduced and solidified. I delivered my first lectures on that subject at the Master level in Toulouse between 1981 and 1983. I also began supervising Ph.D. theses, as it is the role of university professors. The first one, by R. Ellaia (period 1981–1984), was devoted to the analysis and optimization of differences of convex functions [7], a topic I tried to develop and follow for years.

Some years later, in June 1987 precisely, a large meeting on "Applied nonlinear analysis" was organized in Perpignan (extreme south of France), at the occasion of the retirement of J.-J. Moreau[1]. With my Ph.D. student Ph. Plazanet, we presented

[1] A meeting celebrating the 80th birthday of J.-J. Moreau has been organized later (in 2003), by colleagues in mechanics this time, in Montpellier (where Moreau had spent the major part of his university life). See [1] for more on that.

there and published in [2] a converse to Moreau's theorem, a *factorization theorem* in a way. Since I like this theorem, I reproduce it here.

**Theorem (Hiriart-Urruty and Plazanet).** *Let $g$ and $h$ be two convex functions defined on a Hilbert space $H$, satisfying*

$$g(x) + h(x) = \frac{1}{2} \|x\|^2 \text{ for all } x \in H.$$

There then exists a lower-semicontinuous convex function $F$ such that:

$$g = F \Diamond \frac{1}{2} \|.\|^2 \text{ and } h = F^* \Diamond \frac{1}{2} \|.\|^2.$$

Here, $\Diamond$ stands for the infimal convolution operation, and $F^*$ designates the Legendre-Fenchel conjugate of $F$.

Moreover, an expression of $F$ can be obtained, via $g$ (or $h$) and $\frac{1}{2}\|.\|^2$, by performing a "deconvolution" of a function by another.

## 4 Twenty Years Ago: 1993

In 1993 was published the two-volume book co-authored with Lemaréchal [12] (CL in short), final point of 7 years of wrestling with convex analysis, optimization, computers and editing difficulties. We used to call "the HULL" this book (from the initials of our names). So, here is the occasion of some reminiscences of relationships with CL during years. I already told these stories and anecdotes at the occasion of the "CL festchrisft" which took place in Les Houches (Alps region in France) in January 2010 (see [8] for a follow-up as a special issue in the Mathematical Programming series B).

I met CL for the first time in a meeting in the Alps region, during the "Convex analysis days" which took place in January 1974. J.-P. Aubin and P.-J. Laurent were the organizers of this meeting[2]. This was my first international meeting... I remember well that it took place in a charming village called St Pierre-de-Chartreuse and the talks were delivered in a movie theatre or village hall. For me, it was the first time that I saw mathematicians I knew the names (or mathematical results) of: among the 70 participants [confirmed by Laurent (Personal communication, 2010)] were R.T. Rockafellar (who was on sabbatical leave in Grenoble); students or collaborators of H. Brezis (H. Attouch, Ph. Bénilan, A. Damlamian, etc.), E. Zarantonello, J.-P. Penot, J.-J. Moreau, J.-Ch. Pomerol, M. Valadier, J. Cea, L. Tartar, etc. I remember that I. Ekeland had a pertinent question at each delivered talk.

[2]Another meeting, with the same title, took place in Murat-Le-Quaire (centre of France) in March 1976; it was organized by A. Auslender. Soon after, in May 1977, a meeting of the same kind, but devoted to nonconvex analysis, was organized in Pau (southwest of France) by J.-P. Penot.

At breakfast, J.-P. Aubin was drinking all the left coffees. I ventured into discussing a bit with M. Valadier (and his inevitable anorak jacket) on the "continuous infimal convolution". CL was there, a young researcher (just 29 years old) at the research institute called IRIA close to Versailles. The talk by him (in French) was on some "steepest ascent method on the dual function", the matter of which was written in a research report of IRIA (with a red and white cover). I remember the following anecdote. At the end of the talk, a colleague, the kind of mathematician "who-has-understood-everything-better-and-before-everyone" (you see what I mean), asked CL the following question: "Why do you call that a "steepest ascent"?…I understand that wording only for "steepest descent" "…CL answered straight out: "Well…take for example "a deep sky"…" (in French, it is even more striking "méthode de plus profonde montée", "un ciel profond"…The one who posed the question (I won't reveal the name) remained speechless…During the lunch, I heard a colleague pursing his lips: "Yeah, we know that some people look for "descent directions"…". That anecdote leads me to a first theorem.

**Theorem 4.** *Beware, in meetings, young students or colleagues may be listening to what you are saying… They might remember what you said.*

As a corollary, aimed at beginners.

**Corollary 1.** *Do not believe that all your colleagues (mathematicians) are fond of mathematics you are doing or theorems you are proving.*

Some of these colleagues just could say: "What you are doing is just routine work, boring…" or "a trivial matter, I can prove it easily".

About 10 years later, in May 1985, I organized a 1-week long congress in Toulouse, entitled "Mathematics for Optimization"; the main topics of the meeting were variational problems and optimization. Many colleagues came, among the best known ones: P. Ciarlet, J.-P. Aubin, J. Borwein, A. Bensoussan, I. Ekeland, A.B. Kurzhanski, L.C. Young, J. Warga, F. Clarke, B. Dacorogna, J.-P. Penot, R. Temam, H. Tuy, etc. Some participants were there for one of their first meetings abroad their countries: H. Frankowska (Poland and University of Paris IX), M. Lopez and M. Goberna (Valencia, Spain), J.E. Martinez-Legaz (Barcelona), etc. CL was also there, as well as some of our collaborators and colleagues from Chile (R. Correa, A. Jofre). I here would like just to recall the atmosphere during this period, concerning the relationships with other countries, especially with Soviet Union (including Ukraine at that time). Some colleagues from Soviet Union were officially invited: V. Tikhomirov, B.N. Pshenichnyi, V.F. Demyanov…None of them could come, the access to visas was denied. It was typical of what used to happen during those years: you invite (officially) colleagues A, B or C, you get an acknowledgement and answer letter from D, and finally E offers to come…This happened to me several times, especially with Kiev, despite the fact that Kiev and Toulouse are twin cities. I also remember that, during this meeting in Toulouse, a telephone call was organized from J.-P. Aubin, J. Warga, F. Clarke to A. Ioffe (Moscow). All these stories or details are hard to believe nowadays, and yet they took place less than 30 years ago.

CL continued also relationships with colleagues from Soviet Union via IIASA, a research institute close to Vienna in Austria; several meetings on nonsmooth optimization were organized there, with R.T. Rockafellar, R. Wets, B.T. Polyak, Yu. Ermoliev, B.N. Pshenichnyi, V.F. Demyanov, etc.

Another snapshot I would like to offer concerns the two-volume book we wrote together with CL. The initial project was just a 150–200-page book presenting the basics of nonsmooth convex optimization (fundamentals and algorithms); it finally ended up with two volumes, more than 800 pages altogether.

**Theorem 5.** *If you have a project of writing a book, do not believe it will be finished on time and that its length will be the one you had in mind.*

For the project, I used to go to INRIA (close to Versailles), about one week per month for some years. The INRIA barracks (formerly the NATO headquarters in France) were located in Rocquencourt. The Rocquencourt appellation was known to me because, every morning on the radio news, were evocated the traffic jams at "the triangle of Rocquencourt". The whole country of France was supposed to be informed of the traffic around this "triangle of Rocquencourt". So, for me, this triangle was as familiar as the "Bermuda triangle" or the "equilateral triangle". In CL's office at INRIA, a large sheet of handwritten paper was posted on the wall, with the list of chapters we had to write for the book project. In front of this office, the one of C. Sagastizabal, doing mathematics on the screen of her computer but also permanently listening at music with her ear flaps. The manuscript of the projected book was written (*he* typed everything) on an Apple Mac$^+$ (the screen was just like a stamp!) using Microsoft Word3 and CricketDraw for the pictures. It was then converted to TeX with the help of some home-made code. This took place only about 20 years ago! Here is an excerpt from a letter we exchanged, as the project proceeded: "Like a horse, I feel the smell of the stable, even though the rate of efficiency decreases as and when the tiredness increases". There is a possible advantage when you write a paper or a book with a co-author (I feel it is difficult to write a book with more than two co-authors, complexity increases a lot, at least that's my experience), this is what I call the "max rule": when you are inactive on the project, you may think that your co-author is active... so the max of the activities is continuously non-zero.

During that period (around the 1990s), faxes arrived at CL's office: A. Nemirovski and Yu. Nesterov were organizing their first trips to France (and the West).

A revised printing of the book was published in 1996, but that was not a new edition. Actually, my experience is that a new edition of a book is always... augmented, never reduced; C. Byrne from Springer certainly could confirm this statement. J. Dennis commented this statement at the Les Houches meeting in January 2010: "A new edition of a book is always augmented... and sometimes worse!". Later, Springer asked us to write an abridged version of our book, a student version. The project was finalized during a skiing holiday in 2000 in a family house of CL in the Alps. The booklet that we used to call "the soft HULL" was published in 2001. It contained exercises... a couple of them are wrong [13].

Despite our numerous exchanges on optimization during years, CL and I never wrote a specific research paper together, except [11]... which remained unpublished.

Here is a further statement that I excerpt from one letter from CL: "I'm sorry but that's my way of doing research and one has to get used to that... I give punches in all directions to find the hole; many may go on the wall but may also are for my unlucky partner". A final point: CL liked to stud his letters with metaphors or Latin sentences, here is one, a French pun actually, written after some extensive search on properties of the epsilon subdifferential: "Caecum saxa fini (= At this point, that won't get me anywhere)".

## 5 As an Epilogue

I cannot report on all the books (research books or textbooks) written on convex and/or nonsmooth analysis and optimization. The theory and practice are now well established, even if the fields are relatively young if you compare with other fields in applied and/or fundamental mathematics. By experience, I can say that (advanced) students like the geometry and elegance of topics such as Moreau's decomposition in Hilbert spaces (a typical illustration of techniques in convex analysis). Tools from nonsmooth analysis are now used to handle *nonconvex* variational problems; they can be considered as "basics" when beginning to study variational analysis and optimization. This was precisely the aim of my latest (published) lecture notes on the subject [10].

## References

[1] Alart, P., Maisonneuve, O., Rockafellar, R.T. (eds.): Nonsmooth Mechanics and Analysis: Theoretical and Numerical Advances. Springer, New York (2006)

[2] Attouch, H., Aubin, J.-P., Clarke, F., Ekeland, I. (eds.): Analyse Non linéaire. CRM université de Montréal et Gauthier-Villars, Paris (1989)

[3] Aubin, J.-P., Vinter, R. (eds.): Convex Analysis and Optimization. Pitman publishers, London (1982)

[4] Bismut, J.-M.: Conjugate functions in optimal stochastic control. J. Math. Anal. Appl. **44**, 384–404 (1973)

[5] Combettes, P.L., Hiriart-Urruty, J.-B., Théra, M. (eds.): Modern convex analysis and its applications. Special issue Math. Program. ser. B (2014, to appear)

[6] Ekeland, I., Temam, R.: Analyse convexe et Problèmes variationnels. Dunod et Gauthier-Villars, Paris (1974)

[7] Ellaia, R.: Contributions à l'analyse et l'optimisation de différences de fonctions convexes. Ph.D. thesis, Paul Sabatier University Toulouse (1984)

[8] Frangioni, A., Overton, M.L., Sagastizabal, C. (eds.): Convex analysis, optimization and applications. Special issue Math. Program. ser. B **140**(1), (2013)

[9] Hiriart-Urruty, J.-B.: Reminiscences... and a little more. J. Set-Valued Var. Anal. **18**(3–4), 237–249 (2010)

[10] Hiriart-Urruty, J.-B.: Bases, Outils et Principes pour l'analyse Variationnelle. SMAI Series. Springer, New York (2012)
[11] Hiriart-Urruty, J.-B., Lemaréchal, C.: Testing necessary and sufficient conditions for global optimality in the problem of maximizing a convex quadratic function over a convex polyhedron. Seminar of Numerical Analysis. University Paul Sabatier, Toulouse (1990)
[12] Hiriart-Urruty, J.-B., Lemaréchal, C.: Convex Analysis and Minimization Algorithms, vol. I and II. Springer, Berlin (1993). New printing in 1996
[13] Hiriart-Urruty, J.-B., Lemaréchal, C.: Fundamentals of Convex Analysis, Text editions. Springer, New York (2001)
[14] Hörmander, L.: Sur la fonction d'appui des ensembles convexes dans un espace localement convexe. Arkiv för matematik **3**, 181–186 (1954)
[15] Laurent, P.-J.: Approximation et Optimisation. Hermann, Paris (1972)
[16] Moreau, J.-J.: Fonctions convexes duales et points proximaux dans un espace hilbertien. Note aux C.R. Acad. Sci. Paris **255**, 2897–2899 (1962)
[17] Moreau, J.-J.: Propriétés des applications "Prox". Note aux C.R. Acad. Sci. Paris **256**, 1069–1071 (1963)
[18] Moreau, J.-J.: Fonctions sous-différentiables. Note aux C.R. Acad. Sci. Paris **257**, 4117–4119 (1963)
[19] Moreau, J.-J.: Proximité et dualité dans un espace hilbertien. Bull. Soc. Math. France **93**, 273–299 (1965)
[20] Moreau,J.-J.: Fonctionnelles convexes. Séminaire sur les équations aux dérivées Partielles, Collége de France, 1966–1967. Published in 2003 by the university "Tor Vergata" in Roma
[21] Rockafellar, R.T.: Convex functions and dual extremum problems. Ph.D. thesis, University Paul Sabatier, Harvard (1963)